Studien zur theoretischen und empirischen Forschung in der Mathematikdidaktik

Series Editors

Gilbert Greefrath, Münster, Germany

Stanislaw Schukajlow, Münster, Germany

Hans-Stefan Siller, Würzburg, Germany

In der Reihe werden theoretische und empirische Arbeiten zu aktuellen didaktischen Ansätzen zum Lehren und Lernen von Mathematik – von der vorschulischen Bildung bis zur Hochschule – publiziert. Dabei kann eine Vernetzung innerhalb der Mathematikdidaktik sowie mit den Bezugsdisziplinen einschließlich der Bildungsforschung durch eine integrative Forschungsmethodik zum Ausdruck gebracht werden. Die Reihe leistet so einen Beitrag zur theoretischen, strukturellen und empirischen Fundierung der Mathematikdidaktik im Zusammenhang mit der Qualifizierung von wissenschaftlichem Nachwuchs.

More information about this series at http://www.springer.com/series/15969

Cathleen Heil

The Impact of Scale on Children's Spatial Thought

A Quantitative Study for Two Settings in Geometry Education

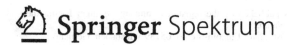

Cathleen Heil
Lüneburg, Germany

Dissertation, Faculty of Education, Leuphana Universität Lüneburg, 2019

ISSN 2523-8604   ISSN 2523-8612   (electronic)
Studien zur theoretischen und empirischen Forschung in der Mathematikdidaktik
ISBN 978-3-658-32647-0   ISBN 978-3-658-32648-7   (eBook)
https://doi.org/10.1007/978-3-658-32648-7

Responsible Editor: Marija Kojic
This Springer Spektrum imprint is published by the registered company Springer Fachmedien
Wiesbaden GmbH part of Springer Nature.
The registered company address is: Abraham-Lincoln-Str. 46, 65189 Wiesbaden, Germany

... my methods of navigation have their advantage. I may not have gone where I intended to go, but I think I have ended up where I needed to be.

—*Douglas Adams*

Danksagung

Mein besonderer Dank gilt meiner Doktormutter Prof. Dr. Silke Ruwisch für ihre kontinuierliche fachliche wie auch persönliche Unterstützung während meiner Promotionszeit. Sie half mir mit sehr viel Geduld und Vertrauen, den richtigen Weg für mein Vorhaben und mich zu finden, ohne das Steuer an sich zu reißen. Das ist nicht selbstverständlich und gab mir die Freiheit, der es für meine Navigation durch den (wissenschaftlichen) Ideendschungel bedurfte.

Darüber hinaus danke ich meinem Zweitprüfer Prof. Dr. Andreas Büchter, der mir nicht nur einen kritischen Zugang zur Thematik ermöglichte, sondern mich stets konstruktiv unterstützte und den ich auch für den persönlichen Austausch auf Tagungen sehr zu schätzen lernte. Meine Drittprüferin Prof. Dr. Eva Neidhardt begleitete diese Studie ebenso fast von Anfang an. Sie brachte nicht nur wertvolle Ideen aus psychologischer Perspektive ein, sondern ermutigte mich stets, nach vorne zu schauen und weiter zu gehen.

Meine Eltern haben mich mit sehr viel Liebe auf meiner gesamten akademischen Laufbahn begleitet. Ich danke Ihnen für ihre uneingeschränkte und vielseitige Unterstützung, die mir stets Kraft gab, meine Ideen zu verfolgen.

Mein Dank geht ebenso an all die Kinder und helfenden Studierenden, die diese Studie nicht nur ermöglichten, sondern die Wochen der Datenerhebung zu einem ganz besonderen Moment für mich machten. Und an all diejenigen Menschen, die Ideen gaben, kritisch nachfragten, Freiräume schufen, Konferenzmomente mit mir teilten, mich mit Kaffee versorgten, mit Flurgesprächen erheiterten und den Feierabend einläuteten; die sich mit mir verliefen, gemeinsam neue Perspektiven einnahmen, Ideen unterstützten und sich weitere mit steter Neugierde anhörten; die aus der Heimat anriefen, Daumen drückten, Gästebetten zur Erholung bereithielten und Stunden am Telefon ausharrten; die Feierabende zu beschwingten Momenten und Heimkommen zu einer Herzensangelegenheit machten; die Mut zusprachen, mein Gejammer ertrugen, Seiten dieser Arbeit korrigierten und mir beim Schreiben in Englisch halfen.

Das waren viele. Wie dankbar ich bin, mit euch an meiner Seite an genau dieser Stelle in meinem Leben zu stehen!

Abstract

Children's thought about space is influenced by their abilities to perceive, encode, and mentally manipulate spatial relations they experience and explore in everyday life. Geometry education in primary school aims to support children as they organize those experiences at an abstract level and develop cognitive abilities to consciously manipulate spatial information in different spatial settings, that is, their spatial abilities. Many studies have investigated children's abilities to mentally manipulate spatial relations in tabletop settings but not those required when the self is located or moving in real space. Addressing this gap in the literature, this study proposes map-based spatial tasks in real space and examines the relations of individual differences in the corresponding underlying cognitive abilities used to solve spatial tasks at both scales of space, small-scale and large-scale spatial abilities, in greater detail.

Using a correlational study design, this study investigates the relation between performances of 240 fourth graders on a mid-sized German university completing paper-based tasks in a classroom setting and map-based orientation tasks in a real space setting. The former test consisted of a subset of tasks that required the children to mentally manipulate object-based transformations and another asking the children to transform the imagined self. The latter test mimicked the practical use of maps such as indicating the direction toward unseen locations, finding one's position and viewing direction on the map, or navigating toward a predefined goal. The test also included tasks without a map that required the children to make inferences on directions to landmarks from survey knowledge acquired during movement in space.

Descriptive results revealed that paper-and-pencil tasks requiring multistep mental transformations of abstract and complex spatial information were appropriate means to measure individual differences in children's performances reflecting small-scale spatial abilities. Moreover, maps were found to be potentially powerful cognitive tools for teachers and researchers to stimulate and measure children's spatial thought in real space. By comparing different models in confirmatory factor analyses, the study showed that at both scales of space, spatial abilities should not be treated as an undifferentiated construct but rather be understood as multidimensional.

The results suggested that a two-factor model distinguishing between object manipulation and perspective transformation abilities might be an option to model small-scale spatial abilities. They also confirmed a three-factor model distinguishing between the abilities to make inferences on directions from survey knowledge and two subclasses of map use, namely the abilities to transform information from the map to the referent space, comprehension abilities, and the ones to use information from the referent space to reason about spatial locations on the map, production abilities.

The results of multivariate statistical analyses at the manifest and latent level indicated that children's spatial abilities at both scales of space are partially but not fully related. These results specify the degree of overlap between subclasses of small-scale and subclasses of large-scale spatial abilities, clarify the role of visuospatial working memory as a mediator when it comes to relations with abilities to use a map in real space, and emphasize the predictive role of particular spatial tasks. The results provide new insights regarding the similarities and differences between both classes of spatial abilities. The findings of this study contribute to the literature in the study of spatial thought in mathematics education and provide empirical evidence for the development of pedagogical interventions both in geometry education and beyond.

Contents

List of Figures

List of Tables

Introduction

> *"Tracks," said Piglet. "Paw-marks." He gave a little squeak of excitement. "Oh, Pooh! Do you think it's a – a – a Woozle?" "It may be," said Pooh. "Sometimes it is, and sometimes it isn't. You never can tell with paw-marks." With these few words he went on tracking, and Piglet, after watching him for a minute or two, ran after him. Winnie-the-Pooh had come to a sudden stop, and was bending over the tracks in a puzzled sort of way. "What's the matter?" asked Piglet. "It's a very funny thing," said Bear, "but there seem to be two animals now. This – whatever-it-was – has been joined by another – whatever-it-is – and the two of them are now proceeding in company. Would you mind coming with me, Piglet, in case they turn out to be Hostile Animals?"*
>
> *(Winnie-the-Pooh by Milne & Rojahn-Deyk 1988, pp. 46–48)*

Children and Space

Space plays an important role in children's everyday life and behavior since it is subject to their experiences, observations, and activities. Starting in toddlerhood, children explore and interact with the space that surrounds them, first by crawling, then by touching objects of different sizes and shapes and observing their positions and locations in space. Later on, children begin to understand that these objects exist beyond their direct experience by reasoning about them and their visual-spatial characteristics. In other words, they develop a conception of space.

© The Author(s), under exclusive license to Springer Fachmedien Wiesbaden GmbH, part of Springer Nature 2020
C. Heil, *The Impact of Scale on Children's Spatial Thought*, Studien zur theoretischen und empirischen Forschung in der Mathematikdidaktik, https://doi.org/10.1007/978-3-658-32648-7_1

Seeking to understand the space they interact with, children commonly face two types of cognitive challenges: first, the perception of forms, colors, and structures and second, the handling of spatial relations. The first challenge typically refers to characteristics of objects that are mostly time-invariant. Successfully dealing with the second challenge, however, allows children to gain a deeper understanding of the space they explore, since spatial relations may change over time whenever an object is moved or the child moves. In the latter case, merely perceiving spatial relations is not sufficient, because children also have to cognitively engage with the spatial situation by mentally encoding and transforming it to keep track of or to anticipate changing spatial relations (D. H. CLEMENTS 1999). Hereby, children have to understand that relations between various objects and the self are organized in different spatial frames of reference, which constantly change over time (HERSHKOWITZ, PARZYSZ, & VAN DORMOLEN 1996).

Children's cognitive abilities of knowing where they are and where objects are located with respect to their surrounding environment and of drawing purposeful spatial inferences from those configurations, *spatial abilities*, are indispensable practical life abilities and are developed from toddlerhood to early adulthood (e. g., NEWCOMBE 1982). When children solve *spatial tasks*, spatial abilities enable *spatial thinking* during solving these, which can be described as reasoning about space by mentally manipulating a broad class of task-specific spatial information, that is, reasoning about relations between locations and configurations of objects and the self and how they can vary over time (NEWCOMBE & SHIPLEY 2015). Spatial tasks may differ in terms of the spatial information available, but reasoning about these pieces of information usually involves the construction of mental representations and subsequent inference processes in transformations of those (HEGARTY & WALLER 2005).

Whenever spatial abilities are developed only to a little extent, situations that require the safe handling of varying spatial relations may, as in the case of Winnie-the-Pooh and Piglet, become challenging. Indeed, spatial challenges located in real space such as the one experienced by Pooh and Piglet, seem to be particularly difficult since they require the spatio-temporal integration of information that is only partly perceived from different viewpoints while moving through space. This might not be the case for settings involving very limited spatial information such as written material, where all spatial relations can be observed from a single vantage point. Solving spatial tasks in both settings can be assumed to be enabled by spatial abilities, but it is not clear whether there are indeed different classes of spatial abilities required to master them or not. Gaining a deeper understanding of this relation is the primary goal of the present study.

Spatial Abilities in Conceptions of Geometry Education

> What is geometry? [...] There can be no doubt what I should then answer—geometry is *grasping space*. And since it is about the education of children, it is grasping that space in which the child lives, breathes and moves. The space that the child must learn to know, explore, conquer, in order to live, breathe and move better in it.
> (FREUDENTHAL 1973, pp. 402–403, emphasis added)

Having had a range of spatial challenges in everyday experiences from infancy to toddlerhood, children have acquired an early spatial knowledge that they bring to school. The fact that spatial thinking is pervasive in the children's everyday life does, however, neither imply that it is mastered without efforts nor that there is no need for spatial education at school. Geometry education in primary school typically focuses on organizing and structuring this implicit experience-based knowledge by fostering cognitive abilities that allow children to perform mental inferences on spatial information in various contexts of their everyday and school life (P. BRYANT 2009). School geometry can therefore be a means to help children using spatial objects and sucessfully navigate spatial situations encountered in an informal way, to help them to develop the abilities to analyze observed relationships and transformations in a formalized, mathematical manner, and ultimately, to develop a clear conception of space.

Undoubtedly, these ideas of geometry education at primary school are closely linked to the ideas of FREUDENTHAL (1971), who called for a geometry curriculum in which "geometry should be related to physical space [...] nobody can deny that we live in space, that we move in space, that we analyze space, to be better adapted to it" (p. 418). Freudenthal's vision has been at the core of modern geometry classes in primary school and cited within the conceptual framework of international comparative studies (e. g., OECD 2004).

Spatial abilities have been generally acknowledged as important cognitive abilities contributing to geometry learning (P. BRYANT 2009; D. H. CLEMENTS & BATTISTA 1992; D. H. CLEMENTS 1999; HATTERMANN, KADUNZ, & REZAT 2015; HERSHKOWITZ ET AL. 1996; K. JONES & TZEKAKI 2016; PINKERNELL 2003; SOURY-LAVERGNE & MASCHIETTO 2015). As a matter of fact, they have been systematically incorporated into conceptions of geometry education and development of geometry curricula since the early 80s (SINCLAIR & BRUCE 2015). TAHTA (1980), for example, emphasized the role of spatial abilities by proposing that visual images are fundamental in mathematics education and that geometry in particular is all about becoming aware of imagery that "arises from a dynamic process of the mind" (p. 6). In the German mathematics education literature, the role of spatial abilities was emphasized by BESUDEN (1999a) who stated that "understanding geometry is primarily about understanding spatial relations and their interdependencies which is impossible without involving spatial abilities" (BESUDEN 1999a, p. 3, translation by the author)[1]. This idea was further supported in the study by WITTMANN (1999) and MAIER (1999b).

The idea that geometry education should engage spatial abilities has been embedded in the curricula of different countries worldwide. For example, The National Council of Teachers of Mathematics (NCTM 2000) has considered spatial abilities by emphasizing practices related to two core domains: first, using "visualization, spatial reasoning, and geometric modeling to solve problems" in real space while "building and manipulating mental representations of two- and three-dimensional objects and perceiving an object from different perspectives" (p. 41) and second specifying locations and describing relationships of spatial configurations, that is, being aware of changing relative positions in space (p. 42). Similar ideas, which partly specify the use of maps and construction plans, can be found in mathematics curricula for primary school in Australia (NEW SOUTH WALES BOARD OF STUDIES 2012), England (DEPARTMENT FOR EDUCATION 2014), Canada (MINISTÈRE DE L'ÉDUCATION ET DU DÉVELOPPEMENT DE LA PETITE ENFANCE 2016), or Germany (KMK 2004).

Despite this general consensus regarding the importance of spatial abilities in geometry education, the relationship between these abilities and geometry learning is less well understood (MULLIGAN 2015). Addressing this epistemological question about the cognitive nature of spatial and geometrical abilities and their relation in the context of geometry education, PERRIN- GLORIAN, MATHÉ, AND LECLERCQ (2013) proposed a basic model of spatial and geometrical thinking in the context of geometry education (Figure 1.1). They enlarged the theoretical framework of BERTHELOT AND SALIN (1993) who distinguished between spatial and geometrical knowledge and conceptualized teaching and learning of geometry as a problem-solving and modeling process that takes place in three types of spaces: the physical, geometrical and graphical one (PERRIN- GLORIAN ET AL. 2013). According to their model, spatial and geometrical abilities are closely related but not interchangeable and activities in geometry education involve the three spaces mentioned above.

According to PERRIN- GLORIAN ET AL. (2013), the physical space is the three-dimensional space of concrete action with objects and the space in which real problems and activities are defined. The geometrical space is the space of Euclidean axioms and formal deductions that serve as abstract-cognitive tools to solve a

---

[1]German original: „Geometrieverständnis ist vor allem die Einsicht in Beziehungen und Zusammenhänge, was ohne Raumvorstellung nicht denkbar ist."

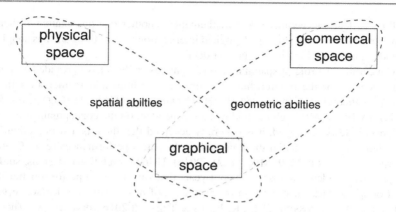

Figure 1.1 Spatial and geometric abilities in geometry education. (Redrawn from SOURY-LAVERGNE & MASCHIETTO 2015)

given problem. Finally, the graphical space contains all kinds of external visual representations such as diagrams, drawings, maps, and schemata. They are used in problem-solving processes and can therefore be interpreted as a bridge between the two other spaces and even seen as a space of creative experimentation (SOURY-LAVERGNE & MASCHIETTO 2015). Spatial tasks require children to use their spatial abilities to successfully master the interaction between physical space and graphical space, whereas in the case of geometrical tasks they have to think about the relation between graphical and geometrical space. The concept of spatial abilities is therefore not only bound to physical space but involves also graphical space. The latter, however, is not limited to two-dimensional representations, but might be any possible and useful way of representing real space (BERTHELOT & SALIN 1993).

To sum up, the development of children's spatial abilities has been identified to be a fundamental aspect of geometry education that is closely related to acquiring geometric knowledge. Understanding spatial abilities in different settings in greater detail, the primary concern of this study, might offer new insights regarding the conceptual framework of 'grasping space' in geometry education.

Perspectives of Mathematics Education Research on Spatial Abilities

Spatial abilities have been a consistent topic of interest in mathematics education for many decades (e. g., BRUCE ET AL. 2017; DAVIS & SPATIAL REASONING STUDY GROUP 2015). They have been investigated using three fundamental different perspectives; whereas one strand of the literature has investigated the role of spatial abilities in learning geometry and mathematics, another one has examined their development with the help of pedagogical interventions, and a third one has sought to understand spatial abilities in greater detail.

Concerning the role of spatial abilities, many researchers have provided empirical evidence for the relationship between these abilities and mathematics that had been theoretically suggested (D. H. CLEMENTS 1999; D. H. CLEMENTS & BATTISTA 1992). Although the findings are dependent on the conceptualization of the specific tests employed, it is generally accepted that there is a strong empirical relation between spatial thinking and mathematics performances (e. g., CARR ET AL. 2018; FRICK 2019; MIX ET AL. 2016). Using correlational designs, studies provided evidence across all school years that students who perform better in a set of spatial tasks also perform better in a set of mathematical tasks (see, e. g., BÜCHTER 2011; GRÜSSING 2012; K. JONES & TZEKAKI 2016, for a review). These performance-based findings have been shown for tasks that involve geometry (e. g., PITTALIS & CHRISTOU 2010), but also for non-geometric tasks that require number sense and arithmetics (FRICK 2019; LOGAN 2015; LOWRIE, LOGAN, & RAMFUL 2016) or problem solving (BISHOP 1980). Even though recent studies have raised the question whether those relations might be mediated by variables such as basic numeric abilities (GRASS & KRAMMER 2018), the mode of test delivery (LOWRIE & LOGAN 2017), cognitive style (see HEGARTY & WALLER 2005), or sex differences (BÜCHTER 2011), the importance of spatial abilities—or specific subcomponents of those—has been sufficiently emphasized. Spatial visualization, for example, has been shown to be an important predictor particularly for mathematical tasks that are difficult for students (MANGER & EIKELAND 1998). Overall, the performance-based empirical findings outlined above have been substantiated by results of intervention studies showing that training spatial abilities improved performances in mathematics (e. g., CHENG & MIX 2014; LOWRIE, LOGAN, & RAMFUL 2017). All of these findings have been completed by theoretical arguments regarding the spatial nature of mathematics itself (K. JONES 2003).

Concerning the development of spatial abilities with the help of pedagogical interventions, research has been grounded on arguments from the developmental literature which has shown that spatial abilities emerge in early childhood and improve rapidly during primary school (e. g., LIBEN 2006). They have been found to be malleable in various ways (UTTAL ET AL. 2013). Although this literature has suggested that spatial abilities may be improved through training, it is not clear how to implement intervention programs at school (LIBEN 2006). The mathematics education research literature has suggested some spatial training interventions (e. g., HELLMICH & HARTMANN 2002; K. JONES & TZEKAKI 2016; MEISSNER 2006; ROST 1977) exclusively concerning spatial abilities enabling solving spatial tasks in tabletop settings. Improvement has often been evaluated for certain subcomponents such as mental rotation or spatial visualization (e. g., BRUCE & HAWES 2015). This has, however, not been the case for perspective taking (LOWRIE ET AL. 2017), which is important for mathematical problem solving (TARTRE 1990). All of these pedagogical approaches share the idea of being direct spatial training programs, that is, interventions requiring spatial thinking in tabletop environments and exclusively intended to improve these cognitive abilities. Although relations to other mathematical and non-mathematical disciplines have been acknowledged, there is an abundance of indirect pedagogical approaches, that is, spatial training programs that infuse spatial abilities into interdisciplinary instructions that are linked to educational content that is already in the national mathematics education curricula (LIBEN 2006). A first step to design those would be to examine how spatial abilities are used in spatial activities in settings other than tabletop environments.

Concerning the understanding of spatial abilities, D. H. CLEMENTS (1999) proposed that there are two conceptually distinct classes of spatial abilities that need to be investigated:

(1) Abilities that allow children to think about the spatially invariant relations between manipulable two- and three-dimensional objects and themselves.
(2) Abilities that allow children to navigate the world by reflecting on different positions in space with respect to the self and other important features of the surrounding environment.

The first class of abilities typically involves tabletop settings with small two- or three-dimensional objects (D. H. CLEMENTS 1999). Abilities that allow children to think about the spatially invariant relations between manipulable two- and three-dimensional objects and themselves have been subject to intense research in mathematics education and have been studied from a wide range of different perspectives (e. g., MULLIGAN 2015). For example, this involves studies that have examined strategies while solving spatial tasks or spatially constructing (e. g., GRÜSSING 2002; PLATH 2014; REINHOLD 2007), sex differences (e. g., BÜCHTER 2011), spatial language (MIZZI 2017), spatial thought in the classroom (PINKERNELL 2003), or thought about dimension (e. g., PANORKOU & PRATT 2008).

In contrast, D. H. CLEMENTS (1999) second class of abilities, that is, navigation through real world, has received less attention by researchers, and this conception has not been properly addressed in the literature. A large number of studies either dramatically limited the initial meaning of those abilities and, as a consequence, studied them as perspective taking abilities, that is, they focused on one concrete aspect of reflection of different positions in space while again using written settings or tabletop tasks with manageable material (e. g., NIEDERMEYER 2015; VAN DEN HEUVEL-PANHUIZEN, ELIA, & ROBITZSCH 2015). Some also proposed definitions that include practical spatial challenges in the real world but suggested written spatial problems that did not involve a concrete relation to spatial orientation in real space (e. g., BESUDEN 1999b; MAIER 1999b). In both cases, the conceptualization of spatial abilities can be interpreted to be more related to D. H. CLEMENTS'S (1999) first class of spatial abilities since spatial tasks are posed in written format or with the help of small, easy-to-handle material that largely ignores the actual physical movement of the self in real space. The corresponding conceptualizations of the notion of spatial abilities have often been related to the one's in the psychological literature that have dealt originally with paper-and-pencil tasks.

To sum up, although researchers agree that spatial abilities are crucial not only for everyday life but also in geometrical and mathematical learning, and although there is a consensus agreement that mathematics education should be related to realistic situations of the student's life, most approaches to investigating and developing spatial abilities in mathematics education have relied on spatial tasks in settings that merely involved geometrical abstractions of real space. These included figural representations of objects or physical environments or manipulable objects in graphical space that, for the most part, were perceivable from one single vantage point. In other words, the approaches researchers used to address student's spatial abilities were rather limited given the complex spatial situations that students experience every day.

Findings from the cognitive and environmental psychology literature have suggested that the conceptions outlined above refer to a special class of spatial abilities that are only related to some extent to spatial abilities required in everyday life (e. g., HEGARTY, MONTELLO, RICHARDSON, ISHIKAWA, & LOVELACE 2006). There is a gap in the mathematics education literature considering this extended conceptualization of spatial abilities that allow students to complete spatial tasks in real space in which the available spatial information has to be perceived and processed from multiple vantage points. Although it has been acknowledged in the mathematics education literature that spatial tasks may vary depending on the perspective children have to take to solve them and that this might address different classes of spatial abilities (LOHAUS 1999; SOUVIGNIER 2000), recent conceptualizations of spatial abilities and corresponding spatial tasks have failed to involve spatial information in real space.

In line with FREUDENTHAL'S (1973) argument that geometry education should enable children to 'grasp space' and to help them to structure their everyday spatial experiences, it is astonishing that research on spatial abilities in real-world settings has been neglected so far. One reason for this gap might be that experimental settings to examine these would require a framework that differs from classical experimental settings such as written tests, behavioral observations, and oral transcripts. How can we study the way in which children navigate the real world, how they reason on different positions in space? How can we know what they experience in a space that is much larger than themselves and requires the integration of a considerable amount of visual-spatial information? And how can we relate 'movement in real space' to graphical space that has been emphasized to be an important element in the conceptualization of geometry learning? One way to bridge this gap, this study argues, is the use of maps.

Map Use as One Example of Spatial Activity in Real Space

Maps are a unique form of symbolic representation of a referent space that show in a one-to-one relationship a certain subset of features of that space (DAVIES & UTTAL 2007). From a practical point of view, given that they represent information that can be difficult to see or unknown to users but necessary for orientation, they represent a powerful means of finding one's way (BLADES & SPENCER 1986). Map-based orientation in real space, that is, reasoning about one's position and heading in the referent space with respect to important features, combines direct action in the environment with abstract spatial thought that relies on spatial and geometrical transformations of information from the map to reality and vice versa (LIBEN, KASTENS, & STEVENSON 2002). Mastering to read a map requires individuals not only to understand concepts that are directly related to fundamental ideas of mathematics such as abstraction, generalization, symbolization, distance, and measurement, but also requires them to apply spatial concepts such as perspective, direction, and location (D. H. CLEMENTS 1999; MUIR & CHEEK 1986). Those latter become important in a practical context, for example during the establishment of map alignment, which is a prerequisite for successful wayfinding and identification of one's location in space even, perhaps even more so, in the age of digital devices (see CHRISTENSEN 2011, p. 5–6).

Apart from their practical use, maps "influence how children come to think about space beyond their immediate experience" (DAVIES & UTTAL 2007, p. 220). Since they visualize what cannot be seen and since they depict multiple locations and their interrelations in real space simultaneously and independently from actually navigating through space, maps are integral to the way children come to think spatially in primarily two ways (BLADES 1989; UTTAL 2000a). First, since they provide spatial information irrespective of directly interacting with space, children might develop the idea that space exists as an abstract concept independently of their experience. This is an important aspect of cognitive development of a clear conception of space (BRUNER 1964; PIAGET & INHELDER 1948/1956).

Second, the use of spatial abstractions such as maps may influence how children think about the depicted space. By using them as cognitive tools, they "extend their reasoning about space in a new way. Over time, children can internalize the tool and think about space in map-like ways, even if they are not looking at a map at the time" (UTTAL 2000b, p. 248). Because they allow children to simultaneously perceive a large number of spatial relations from a single, rather static point of view, maps can help to direct their focus on what to perceive during movement in space. This, in turn, might help them to pre-structure direct experience. Conversely, although children have been shown to be able to navigate real space even in toddlerhood, the literature emphasized that children's spatial thought is informal and experience-based rather than abstract and formal. When children move around in space, they experience it in terms of concrete locations. Maps might help children not only to integrate spatial information that they perceive at multiple points of view but also to structure it in an abstract way (UTTAL 2000a). Therefore, maps allow children to go beyond their initial direct experience of real space during movement and to reflect on different locations in space with respect to the self and other important features in a way that involves more and more mental rather than physical actions.

That thought about reality might be mediated and shaped by iconic and symbolic representations has already been stated by BRUNER (1964). He argued that children's learning undergoes three different modes of representation, that is, three different ways in which information is processed and knowledge is encoded in memory (BRUNER 1964; BRUNER, OLVER, & GREENFIELD 1966). He suggested that children systematically elaborate an internal conception of the outer world through action (enactive representation), imagery (iconic representation), and the use of symbols (symbolic representation). Although BRUNER (1964) mainly referred to language as a means of symbolic representation, he acknowledged those might also include arbitrary symbolic systems, including maps. Because maps follow a dual existence and spatialization principle[2], they have, according to BRUNER (1964), a double role in shaping children's thought about reality, being both iconic and symbolic representations.

Because maps depict spatial information on physical space in an abstract way and because geometry education aims to foster the development of abstract conceptions of space, the conscious use of maps is a reasonable starting point for this study to conceptualize spatial thinking in real space. This conceptualization would be in line with the model proposed in Figure 1.1 (p. 5) with maps as elements of graphical space

---

[2]Maps do not only represent a referent space by standing for it using symbols. In addition, maps are something, in their most simplest way a flat sheet of paper. Moreover, they follow a double spatial essence in a way that they not only represent spatial relations of the referent space, but are also spatial in a sense of being in relation to space (LIBEN 2006).

that mediate concrete challenges of children in real space, for example, identifying their position or navigating toward a pre-defined goal. To solve those tasks, children have to mentally manipulate information from both the map and the corresponding real space to reason about how relations between objects and the self may vary over time and while moving in that space.

So far, the mathematics education literature has theoretically acknowledged the use of maps as one opportunity to engage students in realistic activities that are close to students' experience in everyday life but are also related to mathematical and geometrical concepts and ideas (LOWRIE & LOGAN 2007). The use of maps in geometry education has, however, been addressed only in a limited manner. Map items have, for example, been emphasized in the literature since they involve basic spatial abilities such as spatial orientation (LOGAN, LOWRIE, & RAMFUL 2017). Some authors conceptualize D. H. CLEMENTS'S (1999) second class of spatial abilities as map-reading abilities, thus focusing on the concrete aspect of abstract perspectives when reading maps. These studies, however, do not involve real spaces, and conclusions typically focus on map-reading abilities rather than on the underlying spatial abilities (DIEZMANN & LOWRIE 2008; LOWRIE, DIEZMANN CARMEL M., & LOGAN 2011).

Maps, then, can be understood as cognitive tools for examining how children reason spatially in real space, but is has yet to be determined which experimental setting should be chosen to delineate spatial abilities that enable students to solve spatial tasks not only in tabletop settings but also in map-based settings in real space. Among other possible perspectives,[3] studying individual differences in performances when solving spatial tasks, as addressed in the following section, is a reasonable starting point.

Studying Individual Differences in Children's Spatial Abilities
Considering diversity in classroom, that is, to consider each child and his or her respective individual cognitive and social pre-requisites, which affect individual performances when solving tasks, has become a major challenge in modern

---

[3] According to LINN AND PETERSEN (1985), those are the (a) differential, (b) psychometric, (c) cognitive, and (d) strategic perspectives. Differential studies aim to explain differences between groups of individuals, for example the effects of gender, by using a large portfolio of research methods (e. g., COLUCCIA & LOUSE 2004). Psychometric studies assume that space can be measured and quantified by means of spatial tests ELIOT (1987). They aim to detect different traits of spatial abilities that show homogenity within measures in pre-defined spatial tests. Cognitive studies of spatial cognition describe cognitive processes that are typical to particular spatial tasks, although they might be solved with different performances by individuals (e. g., CARPENTER & JUST 1982). Finally, strategic studies focus on detecting qualitatively different ways of solving spatial tasks by individuals (e. g., BARRATT 1953).

mathematics education, and, by implication, in geometry education. Studying individual differences, that is, "the natural variation in performance among individuals" (HEGARTY & WALLER 2005, p. 122) in pre-defined tasks that were chosen to reflect the underlying cognitive abilities, make sense from a practical point of view; understanding individual differences might help researchers and teachers to determine whether these differences matter for educational goals. This, in turn, might lead to different pedagogical interventions that would help children to master a variety of spatial tasks.

Investigating the phenomenon that individuals differ in terms of their performances when solving spatial tasks requires a two-step modeling process (EDWARDS & BAGOZZI 2000). In a first step, the underlying cognitive abilities that enable solving these tasks, spatial abilities, must be modeled based on findings from the literature. Since those are latent and not observable, modeling them allows researchers to establish, in a second step, a theoretical relationship between constructs and possible measures reflecting them. Individual differences can then be interpreted to be different quantitative expressions of these measures.

Although children's spatial abilities have been examined in the mathematics education literature, only a few models have been proposed to describe them by means of subcomponents.[4] MAIER (1999b) proposed a model for the German mathematics education community that is similar to NEWCOMBE'S (2018) 2×2-typology of spatial abilities proposed in the international literature (see also DAVIS & SPATIAL REASONING STUDY GROUP 2015; NEWCOMBE & SHIPLEY 2015; UTTAL ET AL. 2013). Within this typology, the authors distinguish between different solution processes in which the child has to engage to solve a given spatial problem in two dimensions: a static versus dynamic thinking dimension and a dimension focusing on whether intrinsic or general configurational properties of an object might be transformed. Although both models draw on empirical findings of separable subcomponents of spatial abilities, they have not been confirmed in empirical studies (see also BÜCHTER 2011). Studies that measured pre-defined models of children's spatial abilities in tabletop settings have faced problems during empirical validation of the hypothesized structure (e. g., GRÜSSING 2012). These problems might be explained by the fact that there are only a few age-specific and reliable measures for children at the end of primary school (NEWCOMBE & SHIPLEY 2015). No models of map-based orientation tasks in real space have been proposed in the literature.

---

[4]PINKERNELL (2003) proposed another important model of spatial abilities in the German mathematics education literature but rather addressed their relations to geometric and mathematical thinking from a conceptual point of view.

To model spatial abilities, combining theoretical insights from cognitive psychology and methods from psychometry has been proven to be a convincing approach in the psychology literature (e. g., HEGARTY & WALLER 2004). These models assume that spatial abilities are governed by a conglomerate of underlying cognitive processes that allow individuals to solve different pre-defined psychometric tasks. Examining the resulting data, researchers can identify homogenous patterns of individual performances using multivariate statistical analyses. This approach does not only provide theoretical models of latent constructs based on groups of different tasks but also allows researchers to determine the concrete cognitive processes that underlie them. Studying individual differences becomes therefore closely tied to research on spatial cognition.

Contributions of This Study
In this thesis, I aim to understand in greater detail how spatial abilities, allowing children to perform spatial tasks in tabletop environments, are related to spatial abilities allowing them to navigate in real space. In particular, I seek to investigate whether children's abilities to solve spatial tasks in written settings are dissociable from their abilities to solve spatial tasks in map-based orientation settings in real space. In both settings, children need spatial abilities to complete the corresponding tasks, it is not clear whether they use the same class of spatial abilities, or whether different settings require different classes of abilities. Although it is known from the literature that children develop spatial abilities during childhood, but it has yet to be determined whether these abilities evolve into two different classes or not.

There are two primary motivations to examine the relationship between different classes of spatial abilities. The first one, which is a basic research one, is to deepen the understanding of how the construct of spatial abilities may be extended to involve spatial thinking in real space. As I argued before, this would correspond much more to one of the goals in geometry education, namely that children develop the abilities to 'grasp space'. Second, by gaining a better understanding of what characterizes the corresponding underlying spatial abilities and by quantifying the extent of the possible relation between both classes of spatial abilities, researchers and educators may leverage them more effectively. The educational objective of this study is to establish an empirical basis for an indirect spatial intervention involving spatial abilities in a broader context instead of increasing pedagogical attention toward solving spatial problems in tabletop settings only.

By addressing these unresolved issues, this study contributes to the mathematics education literature in three important ways. The first one is a theoretical contribution to the field of spatial cognition in mathematics education. Besides providing definitions for the two classes of spatial abilities outlined above and integrating them into a general theoretical framework on spatial cognition, I identify cognitive processes that govern these two classes and integrate them into two models of them. I further outline similarities and differences between both classes of spatial abilities and suggest a hypothesis concerning their relation based on arguments derived from cognitive psychology.

The second contribution of this study is a set of age-specific measures that reflect the theoretical models of both classes of spatial abilities. Having developed and analyzed a paper-and-pencil test and a map-based orientation test for fourth graders, I provide statistical evidence showing that those tests can be used to measure individual performances in a differentiated manner. Analyzing patterns of correlations of performances in the tasks of the paper-and-pencil and map-based orientation tests using confirmatory factor analyses, I demonstrate that small-scale spatial abilities can be modeled in a two-factor model and large-scale spatial abilities in a three-factor model.

The third contribution of this study is to offer empirical insights concerning the question whether and to which extent both classes of spatial abilities are separable, which factors influence possible relations, and which subclasses of each class of spatial abilities are particularly influential. Analyzing patterns of correlations in performances in the paper-and-pencil and map-based orientation tests both at the manifest (using zero-order correlations and multiple regression analysis) and latent (using structural equation models) levels, I demonstrate that both classes of spatial abilities are partially, but not completely related and quantify the extent of this overlap. I further provide empirical evidence for the important role of visuospatial working memory and identify a set of spatial tasks that might be considered for future design of pedagogical interventions.

Organization of This Study

This study is organized in three main parts representing the theoretical background, the methodological framework as well as the results and subsequent discussion. Starting with the basic assumption that individual differences in performance when solving a particular set of spatial tasks reflect individual differences in a conglomerate of cognitive processes that act upon spatial representations in memory, I contribute to the literature by establishing a theoretical framework that describes how the relative size of spatial information involved in different spatial problems, scale, indirectly affects spatial behavior. In the first chapter of the theoretical part, I argue that spatial problems in settings of different relative size can be attributed to two conceptually different classes of spatial abilities: small-scale and large-scale spatial abilities. Those are defined based on the general psychological framework of this study, which links fundamental concepts of spatial cognition, such as spatial representations, spatial behavior, frames of reference, and spatial memory with the construct of spatial abilities (Ch. 2). In the following chapter, I show that scale, when considered as a psychological construct, does not only differentiate space into different classes with respect to the type of interaction that is afforded by the corresponding spatial information available, but might lead to differences in the kind of information-processing (Ch. 3). To do so, I review the cognitive literature concerning small-scale and large-scale spatial abilities. I first approach both classes of spatial abilities by comparing spatial behavior. Then, I describe the processing of small-scale spatial information, thereby analyzing sources of individual differences concerning the encoding of pieces of spatial information in a depictive representation, and their subsequent mental transformation and maintenance in working memory. In particular, I explain possible mental transformations involved when solving small-scale spatial tasks and develop the assumption that there seem to be two qualitatively distinct types of them: transformations of objects and transformations of the imagined self. Finally, I summarize empirical findings concerning the influence of visuospatial working memory. I subsequently delve into the understanding of large-scale spatial abilities. I argue that they might be approached by analyzing cognitive processes with a visuospatial component underlying map-based orientation during navigation in real space, thereby distinguishing between the question of how large-scale spatial information is encoded in a cognitive map and how information is processed during map-reading. Concerning the latter question, I identify important cognitive processes in the literature and integrate them into a process model of map-reading. This model does not only describe two possible distinct subclasses of transformations involved in map-reading, but proposes that mental transformations of objects and the imagined self may also play an important role during map use in real space. Finally, I summarize empirical findings concerning the role of visuospatial working memory. I conclude this chapter by showing that both classes of spatial abilities rely to a considerable extent, but not entirely, on similar cognitive processes. Besides proposing models to describe small-scale and large-scale spatial abilities, I argue for a possible partial dissociation concerning the relation of both classes of spatial abilities. In Chapter 4, I review the psychometric, individual differences, and developmental literature that investigated whether distinct cognitive processes used for describing possible subclasses of spatial abilities at both scales of space do, indeed, tap different types of abilities involved. To do so, I discuss the literature concerning the models for both classes of spatial abilities that I proposed before. Finally, I review studies that investigated a possible relationship between small-scale and large-scale spatial abilities in children and identify a gap in the literature. This study addresses this important gap.

In the second part, I present the methodological framework of this study. In Chapter 5, I specify the research questions and discuss important theoretical components of the quantitative research design, which was based on psychometric testing, multivariate statistical analyses, and multiple imputation treatment of missing data. Addressing several issues regarding the methodological framework, I then describe the empirical study design. In Chapter 6, I document how the paper-and-pencil test was developed. Moreover, I provide a statistical analysis of the data set of the pilot testing. In Chapter 7, I describe the test conceptualization of the map-based orientation test and provide insights concerning the pilot testing. In Chapter 8, I provide information on the additional testing material used for measuring visuospatial working memory capacity and perspective taking with manipulable material. I close the second part of this thesis in Chapter 9 by providing details on the sample and data collection.

In the third part, I present the results of this study and discuss them vis-a-vis the literature. I report empirical findings on the psychometric characteristics of the items in the paper-and-pencil and map-based orientation test, thereby providing information on the quality of the instruments I developed. I also present the results of confirmatory factor analyses, which justify the differentiation of small-scale and large-scale spatial abilities into two and three components (Ch. 10 and 11). In Chapter 12, I report the empirical findings on the relation between small-scale and large-scale spatial abilities at both the manifest and latent level, showing how both constructs might be related. Moreover, I report the findings on the influence on single predictors and clarify the role of spatial working memory. I also summarize the results of the analyses of specific components of both classes of spatial abilities that seem to be important predictors in both directions.

In the final chapter of the third part, I summarize the empirical findings and relate them to the research questions, describe the limitations of this study, derive practical implications for geometry classes in primary school, and outline implications for future research in mathematics education and psychology (Ch. 13). Chapter 14 concludes this study.

Part I
Theoretical Background

Conceptualizing Spatial Cognition

2

> Because space literally surrounds us, and because we so often use spatial metaphors and spatial representations automatically, we often fail to notice or to analyze these concepts, or to define terminology explicitly. (LIBEN 2006, p. 199)

Children have to engage in spatial cognition when they interact with space and especially when they face spatial challenges and problems. Spatial problems "are those that have a significant amount of spatial information in the original presentation of the problem or in the way a person represents it" (CARPENTER & JUST 1982, p. 221). As outlined in the introduction, primary children face spatial problems not only in the mathematics classroom but also in their day-to-day experience. The problems that they encounter might not only differ in the way potential solutions might be applied in real life, but also in the way they are presented. Spatial problems in the geometry classroom in particular typically involve spatial information that can be perceived from one single vantage point since the visual representations or objects used tend to be small in relation to the child. Solving spatial problems in real space require children to perceive and process spatial information often while moving because the surrounding space is much larger than they are and requires them to take various vantage points. This study seeks to answer the question whether the underlying spatial abilities that allow children to solve spatial problems in those different settings might vary and aims to understand structural similarities and differences.

This chapter establishes a general theoretical framework of spatial cognition for this study, that is, in line with BAGOZZI AND PHILLIPS (1982), "a system of concepts, hypotheses and observations all related among themselves in a meaningful way" (p. 461). Providing a more nuanced perspective on some of the issues addressed in the introduction, this chapter has three objectives. The first objective is

C. Heil, *The Impact of Scale on Children's Spatial Thought*, Studien zur theoretischen und empirischen Forschung in der Mathematikdidaktik, https://doi.org/10.1007/978-3-658-32648-7_2

to theoretically substantiate one of the key assumptions in the introduction, namely that there seem to be different classes of spatial abilities. It remains to be clarified why the scale of spatial information is an important determinant of spatial cognition. More specifically, the framework establishes why the scale of spatial problems might affect spatial behavior indirectly by explaining how spatial behavior is related to the cognitive processing of spatial information. The second objective is to provide theoretical insights concerning the key concepts of spatial cognition. By clarifying their relation in the context of solving spatial problems, it is possible to develop an information-processing-based definition of spatial abilities at different scales of space. The third objective is to introduce the theoretical constructs that can be used to explain individual differences in solving spatial problems.

This chapter is organized as follows. Section 2.1 discusses how the space that children explore and interact with is related to spatial cognition and how scale affects this relation. Section 2.2 introduces the concept of spatial representations as the basic underlying cognitive structure of spatial cognition and summarizes theoretical points of view concerning its conceptual pluralism, thereby also limiting the scope of this study. This section also explains how the notions of spatial behavior and spatial tasks can be integrated into a theoretical framework of spatial cognition. Section 2.3 describes how spatial knowledge is encoded and stored in memory, thereby introducing the important concept of frames of reference. It further summarizes the key theoretical model of representations and use of knowledge in human memory. Based on all of these insights, the notion of spatial abilities and spatial thinking at different scales of space is defined and conceptually distinguished from other similar and related notions and concepts in Section 2.4.

2.1   Space in Cognition

The definition of the notion of space itself has a long philosophical history (see e. g., ELIOT 1987; LIBEN 2006; O'KEEFE & NADEL 1978, for further elaboration). In line with previous studies on spatial cognition (e. g., LIBEN 2006), this study defines space as a relative concept, that is, space is conceptualized through relations among objects and the self contained in space. Space is not assumed to exist if there are no objects or subjects. As a consequence, space changes whenever objects or the self change or move due to a shift of the corresponding relations (e. g., O'KEEFE & NADEL 1978). Since the self and objects are related, it is possible to speak of the notions of perspective and vantage points.[1]

---

[1] In contrast, absolute space is conceptualized as a container that might have objects included but is assumed to exist independently of them. Whenever objects are included, they are

In a first step, this section explains how space can be conceptualized from a psychological point of view. In a second step, it shows how this idea is related to the notion of spatial cognition. In a third step, the notion of scale is introduced as an important determinant of spatial cognition.

2.1.1 Physical and Psychological Space

To understand how space becomes a matter of subjective experience, that is, to conceptualize the psychological nature of space, the distinction between physical and psychological space deserves particular consideration. According to O'KEEFE AND NADEL (1978), psychological space is "any space which is attributed to the mind [...] and which would not exist if minds did not exist" (O'KEEFE & NADEL 1978, p. 6), and physical space is "any space attributed to the external world independent of the existence of minds" (p. 7).[2] Physical space can, according to the definition, be understood as something external which should be conceptually differentiated from a subject's[3] thought about it or how the subject responds or behaves in it (ELIOT 1987).

Psychological space has also been denoted as "space-as-experienced" (WELWOOD 1977, p. 98) or as "space of the mind–'space' as it is perceived, imagined, and stored in memory" (MONTELLO 1992, p. 137). Similarly, ELIOT (1987) described psychological space as an "amalgam of physical capacities, mental processes, learned skills, forms of representation, and dimensions of thought at different levels of awareness for different tasks in changing surroundings" (p. 6). The latter two definitions indicate that the notion of psychological space refers to a whole set of cognitive processes. This idea has been further underlined by O'KEEFE AND NADEL's (1978) description of what characterizes psychological space:

---

localized with respect to the container or to other objects. In this case, the notion of a vantage point or viewpoint is irrelevant as the movement of the self or objects does not alter the container or the objects included (e. g., O'KEEFE & NADEL 1978).

[2] Although both notions use the expression space, the specification that space is relative in its nature applies ostensively to real space. In contrast, it is not mandatorily clear whether psychological space is relative or not. This question demands for further theoretical elaboration (LIBEN 1981). Some models will be addressed in this study.

[3] This study uses the notion of 'the individual' and 'the subject' when addressing phenomena that are valid for all ages. The notions of examinee and participant are used to report about psychological studies and concepts. The expression 'the child' is used to refer to age-specific phenomena.

Included are concepts which the mind constructs on the basis of reflections on experience, abstractions from sensations, organizing principles which impose unified perceptions upon otherwise diverse sensory inputs, organized sensory arrays which derive
their structure from the nature of peripheral receptors, or a particular set of sensations
transduced by a specialized (spatial) sense organ. (p. 7)

Although the concept of psychological space captures the idea of subjective experience of an individual within its physical surrounding, it is rather a theoretical
concept, even more, a psychophilosophical[4] expression for a whole set of cognitive
processes which are related in a complex way. The definitions given above point
toward a multitude of concepts that have to be considered, involving perception, sen-
sation, imagination and memory. This study does not seek to further elaborate the
concept of psychological space from a philosophical point of view (see WELWOOD
1977, for further discussion) itself but uses the theoretical concept to derive a general
idea of the notion of spatial cognition. To do so, this study addresses psychological
space in its intertwined relation to physical space.

The Ontogeny of Psychological Space

O'KEEFE AND NADEL (1978) introduced the notion of psychological space using
a strict dichotomous distinction between physical space as studied by physicists,
and psychological space as studied by psychologists. In contrast, LIBEN (1981)
emphasized that physical space was important for all those psychologists who pos-
tulated that psychological space is generated by direct interaction with physical
space instead of being innate. Any sharp definition between physical and psychological space would become senseless in this context and "since physical space
can never be measured independently of a mind, the distinction becomes blurred"
(LIBEN 1981, p. 5). According to her, the hypothetical relation between psychological and physical space relates to the epistomological position that the researcher
takes when studying the origins of psychological space. There are three different
epistomological positions to distinguish: the empirist, the nativist, and the constructivist.

The first one, the empirist position, is the one taken by LIBEN (1981). Empirists
have presumed that psychological space was an isomorphic copy of the subjective
experiences with physical space. Physical space has therefore been understood to
be necessary for setting up psychological space since the latter one is generated by
subject exploring physical space.

---

[4] 'Psychophilosophical' is used here to stress that the concept of psychological space combines
psychological and philosophical aspects. It is not readily measurable by any way of data, yet
is is clearly experiencable by the subject (WELWOOD 1977).

In contrast, when taking a nativist position, psychological space is hypothesized to be innate. Nativists believe that "the structure and function of those parts of the brain involved in generating the space are specified by information in the organism's genes" (O'KEEFE & NADEL 1978, p. 8). Researchers taking the nativist position therefore make no assumptions on the nature of the psychological space but study the evolution and development of psychological space in the brain in detail. Different assumptions on what is innate have been pronounced. Among those, one important example is the assumption that there are broad cellular structures that organize spatial experience (O'KEEFE & NADEL 1978).

Researchers taking the constructivist position asssume that psychological space is actively constructed by the individual in interaction with its environment. Constructivists combine the position from empirist and nativist theory and hypothesize therefore inherited structures that are matured by processes of active interaction with the physical space (e. g., BRUNER 1973; HART & MOORE 1973; PIAGET & INHELDER 1948/1956; SIEGEL & WHITE 1975).

2.1.2 Spatial Cognition

From an etymological point of view, the notion of cognition can be derived from the Latin origin *cognitio*, which means any kind of knowledge, concept and insight. According to NEISSER (2014), cognition is defined by the cognitive dynamics that allow a subject to individually experience a given sensory input:

> Whatever we know about reality has been *mediated*, not only by the organs of sense but by complex systems which interpret and reinterpret sensory information. The activity of the cognitive systems results in and is integrated with the activity of muscles and glands that we call "behaviour". [...] As used here, the term "cognition" refers to all the processes by which the sensory input is transformed, reduced, elaborated, stored, recovered, and used. It is concerned with these processes even when they operate in the absence of relevant stimulation, as in images and hallucinations. (NEISSER 2014, pp. 3–4, emphasis in original)

Consequently, cognition is an umbrella term covering a broad range of mental processes, such as perception, attention, thinking, language, planning, problem solving, recognition of places and objects, decision making, and some memory processes, such as encoding and representing stimuli in mind. Cognition is therefore closely related to the ongoing and everlasting transformation of information that the subject experiences, which, in turn, stimulates a certain behavior of the subject engaged in spatial cognition, that is, the individual's "sensorimotor activity in the environment, such as the manipulation of objects in space, or the locomotion of the self through the environment" (LIBEN 1981, p. 8). Using the two concepts introduced above, cognition refers therefore to the complex intertwined connection between the physical and psychological space by describing how (sensory) information from physical space is cognitively interpreted by the subject interacting with it. In summary, studying psychological space as the individual's 'world in the head' does not make physical space irrelevant. Indeed, it is the connection between both spaces, mediated by thought, that is the central issue of cognition.

Consequently, this study conceptualizes spatial cognition as referring to the complex interrelationship between the physical spatial world and the corresponding psychological space. Investigating spatial cognition refers therefore to studying the connection between pieces of spatial information of the physical world, how the individual subjectively experiences them, and how this stimulates a certain behavior in space.

Information or stimuli can be, in their most general sense, conceptualized as structured data (e. g., R. E. ANDERSON 2008). Spatial information, then, refers to locational data of objects or configuration of objects[5]. It is conceptually different from object information, such as sizes of objects, their shapes and appearances. Given the fact that space is understood as a relative concept, this study focuses on spatial information because it is time-variant whenever objects or the subject move in space and does not consider the time-invariant object-information.[6] Spatial information refers to much more sensory modalities than visual information:

> [S]patial information from the various sources of sensory information provided in the learning experience. Vision is probably the main sensory modality in humans for sensing the spatial layout of an environment. However [...] one also senses one's own movement through non-visual senses. Movement is sensed by the vestibular system, which provides information about linear and angular accelerations, kinesthesis, which senses movement of the limbs, and efference copy, which is based on signals from the central nervous system to the muscles. (HEGARTY ET AL. 2006, p. 155)

This study focuses on spatial information that can be visually perceived, that is, information about locations and distances of objects and configurations of objects. In

---

[5]The expression spatial configuration refers to a set of objects that are organized in a coherent entity.

[6]This distinction is further in line with the two-stream hypothesis. According to this hypothesis, those two types of information are neurally differently processed, with object information being processed in the ventral what-stream and spatial information relative to the subject being processed in the dorsal where-stream (e. g., FREUD, PLAUT, & BEHRMANN 2016; KOZHEVNIKOV, KOSSLYN, & SHEPHARD 2005; UNGERLEIDER & HAXBY 1994)

addition, other sensory inputs will be considered whenever it becomes meaningful. When an individual cognitively engages with a given piece of spatial information, the individual is said to solve a spatial task or spatial problem. This also indirectly involves the overall spatial environment in which a problem is posed. To directly emphasize different surrounding environments in which a spatial task is posed, this study also uses the expression of spatial setting.

To sum up, space can be interpreted as an abstract conceptual notion that govern the relationship between a physical environment, denoted as physical space, and a category of subjective experience of it, denoted as psychological space. Space relates to the notion of spatial cognition by means of the spatial properties of physical space that are individually experienced by the subject in their 'world in the head'. The major theoretical contribution of this relation is the fact that physical space with its properties is related to the internal, cognitive world of the subject who experiences it to become cognitively useful. Their characteristics are indeed important when it comes to studying settings in which the individual engages with different spatial information.

Indeed, whenever physical space is linked to human psychology, that is, when considering it "as a problem for human perception, thought, and behavior (i.e., when studied as a *psychological* problem)" (MONTELLO 1993, p. 313, emphasis in original), its attributes do become important. The subject's psychology of space, in particular the way the subject experiences, perceives, thinks about and memorizes space, becomes interlinked to the attributes of the physical space itself. The following section discusses different conceptualizations of this idea.

2.1.3 Scale as a Psychological Construct

> "Space is not space is not space" when it comes to human psychology.
> (MONTELLO 1993, p. 313)

One might imagine a setting in which a child searches for a hidden object such as a small colored ball. If the ball is hidden within the child's playpen or on a table, the child will probably move around objects to find the toy, but the child is unlikely to move themself. Imagine now that the ball is hidden in a large, probably unknown environment, such as the neighborhood of the child's residence. The child will need to understand that objects and buildings obscure other objects and buildings and, consequently, understand the need to move around to find the toy. Although the goal is the same for both cases (find the small ball), qualitatively different spatial behaviors result from the different natures of both spatial settings.

Since space has been conceptualized as a category of subjective experience of the individual, a differentiation of a spatial setting based on the concept of size, which is an absolute rather than a relative one, might be misleading (MONTELLO 1993). Instead, a different concept constitutes a stronger base, namely the concept of scale of space. Scale of space relates the projected dimension of the given pieces of spatial information in a setting relative to a subject dealing with it. In other words, scale is the relative size of spatial information with regard to a subject and may therefore impact the way the subject interacts with it (see also WEATHERFORD 1982). In line with FREUNDSCHUH AND EGENHOFER's (1997) classification of the literature, the following sections present different conceptualizations of how scale might affect psychological space.

Differentiation into Small-Scale and Large-Scale Spaces

ITTELSON (1973) was one of the first authors who underlined the importance to consider scale in the discussion of spatial cognition. He emphasized the relationship between scale, space, and the subject's multimodal experience in his own conceptualization of space. As he stated, "the distinction between *object* and *environment* is crucial" because in the study of objects, the subject may perceive them as a lifeless, passive things while "one cannot be a subject of an environment, one can only be a participant" (ITTELSON 1973, pp. 12–13, emphasis added). He argued that the space of objects engulfs objects that were smaller than the subject's body (e. g. tabletop spaces or room-size spaces) and that the large-scale environment surrounds the body. Besides this defining criterion for large-scale spaces, ITTELSON (1973) emphasized that movement is necessary to explore the whole space with all its parts which "forces the observer to become a participant" (p. 12).

Similarly, DOWNS AND STEA (1977) distinguished dichotomously small-scale and large-scale spaces to discuss contributions of Piagetian theory to their theory of cognitive mapping. The authors emphasized the perceptual character of small-scale spaces and the environmental or 'transperceptual' character of large-scale spaces (p. 197).

Differentiation into Small-Scale, Medium-Scale, and Large-Scale Spaces

MANDLER (1983) differentiated the former concepts by subsuming the existing literature into a conceptualization in three scales of space, namely small-scale, medium-scale, and large-scale spaces. According to her, small-scale spaces are apprehended from one single vantage point which is outside of the considered space itself (e. g., tabletop models, small models of larger spaces). Medium-scale spaces are apprehended by the subject while moving through the space and taking a number of vantage points. However, spatial relations in such a space can directly be observed from one single viewpoint (e. g., rooms, a gym) which is not longer true for large-scale spaces which must be 'learned' over time from locomotion in space (e. g., house, campus, town). Although it was not yet conceptualized in her classification, MANDLER (1983) stressed the fact that very large large-scale spaces such as countries probably account for further classification as they require the subject to apprehend them via maps or aeriel views.

Similar to MANDLER (1983), GÄRLING AND GOLLEDGE (1989) further differentiated ITTELSON's (1973) concepts of environments to review environmental perception and cognition research. Although they also distinguished between small-scale (single rooms), medium-scale (inside of a building) and large-scale (neighborhoods) environments, the given examples demonstrate that the argumentative specification, though not explicitly stated, must be different from MANDLER's (1983) point of view and more related to a synonymous use of scale with absolute size. However, they concluded that large-scale environments demand for a piecewise integration of spatial configurations perceived and require therefore a higher level of mental activity because more spatial information has to be integrated over time.

Models of Multiple Scales
MONTELLO (1993) further differentiated former frameworks to propose a taxonomy that consists of four classes of scale of psychological space, namely figural space, vista, environmental and geographic space. According to him, figural space is the space that is projectively smaller than the body of the subject and therefore perceivable from one single vantage point. Figural space can be further subdivided into pictorial (two-dimensional figures) and object (three-dimensional things) space. Figural space is therefore the space of figures, graphic representations, maps, small objects, school books but also distant landmarks (MONTELLO 1993). Vista space is projectively as large as the body of the subject but can still be perceived from a single vantage point without locomotion in space. It is the space of single rooms, a gym, town squares and street views (MONTELLO 1993). Environmental space surrounds the body of the subject as it is projectively larger than it. It is therefore too large to be perceived from a single vantage point and requires the subject to move around and to integrate information of time to apprehend spatial properties. Direct experience with the space becomes necessary. Environmental space is the space of buildings, neighborhoods, school grounds, campuses and towns (MONTELLO 1993). Geographic space is projectively huge in comparison to the subject's body

and cannot be apprehended only by direct experience.[7] It must be learned by graphic representations such as maps and models that reduce the given geographic space to figural space. Geographic space is the space of countries, continents and solar systems (MONTELLO 1993).

FREUNDSCHUH AND EGENHOFER (1997) derived their own framework out of the study of MONTELLO (1993) and embedded each class of scale into the existing literature. They introduced six classes of scales of space, namely manipulable object space (similar to figural space), non-manipulable object space (similar to pictorial space), environmental space (similar to former concepts), geographic space (similar to former concepts), and panoramic space (similar to vista space). In a sixth class of scale of space, they emphasized the particular role of maps and models as representations of large-scale spaces. Whereas maps were classified as small-scale spaces in the studies of MANDLER (1983) and MONTELLO (1993), FREUNDSCHUH AND EGENHOFER (1997) defined a separate type, denoted as map spaces, within their typology. According to the authors, map spaces represent non-manipulable, small-scale and large-scale spaces at the same time that do not account for locomotion. Map spaces stand for the unique role of maps as symbolic representations of larger space in a manageable form.

Conceptualization of Space In This Study
To sum up, this section demonstrated that scale is not only a spatial attribute, but that it becomes an important construct[8] when considering spatial settings, or space in general, as a psychological issue. The models that have been outlined above indicated that scale is a broad continuous spectrum that becomes a determinant of psychological space when studying subjective experiences of the spatially interacting individual. Therefore, there seems to be a continuum of different classes of psychological space and hence, since spatial cognition involves the relation between physical and psychological space, spatial cognition itself.

---

[7]Strictly speaking, it is never possible to study geographic space at it becomes figural space when mapping it and it becomes vista space when seeing its surface from an airplane, for example.

[8]In line with EDWARDS AND BAGOZZI (2000), a construct is defined as "a conceptual term used to describe a phenomenon of theoretical interest" (p. 157). Constructs refer to real existing phenomena that of scientific interest, but rather then being real in an objective, manipulable sense, "they are elements of scientific discourse that serve as verbal surrogates for phenomenon named by the construct" (p. 157). In the context of sciences, the notions of construct and concepts are accepted to be used as synonyms (P. KLINE 2013, p. 25).

In the literature, the criteria for large-scale space has been twofold; first, it has been described as a space that engulfs a subject moving within that space. Second, it has been emphasized that a subject needs to take multiple vantage points to apprehend the space (see also ACREDOLO 1981; WEATHERFORD 1982). ACREDOLO (1981) emphasized that a clear definition of large-scale spaces must encompass both criteria. Whereas an engulfing space usually implies that the subject needs to move, a definition based only on the second criterion does not imply the 'surroundingness'. In HUTTENLOCHER AND PRESSON's (1979) 'Room and Cube' study, for example, third graders were presented either a large cube on a table, that had to be viewed from outside, or a room viewed from inside. Both apparatus require multiple vantage points to be completely apprehended, but only the room surrounded the subject.

Even when defining large-scale space using a combination of both criteria, there remains the problem that large-scale space becomes small-scale when viewed from far away (e. g., from an airplane). According to WEATHERFORD (1982), a stipulation must be added to former definitions, the requirement that the subject "is standing on the same plane as the given spatial layout itself" (WEATHERFORD 1982, p. 6).

The conceptual distinctions that were presented theoretically substantiate the differentiation into the two settings in which individuals engage in spatial cognition, as outlined in the introduction, spatial challenges in the classroom and in real space. This study addresses the question how and to which extent scale affects how a subject engages in spatial cognition in both settings. To do so, this study differentiates dichotomously between two classes of psychological space, that is, small-scale and large-scale space.

The following two conceptualizations are chosen:

1. **Small-scale space**, that is defined in the sense of MONTELLO's (1993) figural space as the space of small figures and objects that can be perceived from one single viewpoint. Small-scale space corresponds therefore to classroom settings involving spatial tasks with figures and manipulable material.

2. **Large-scale space**, that is defined in the sense of WEATHERFORD (1982) as the space that surrounds the body of the subject standing on the same plane as the spatial layout and that requires the individual to apprehend it from multiple vantage points while moving. Large-scale space is the psychological definition of settings involving spatial tasks in real space.

Conclusion

Space is never merely subject to passive recipience but is constantly explored when individuals interact with it. From a psychological perspective, space is therefore far more than just a theoretical-philosophical three-dimensional construct, but a category of subjective experience. Spatial attributes of the experienced space, that is, pieces of spatial information consciously or unconsciously perceived in a spatial setting cannot be understood as disembodied. When engaging in spatial cognition, subjects process this type of information by assigning a psychological meaning to it and representing it in the spatial world in their mind. Since spatial cognition is concerned with linking attributes of physical space to human psychology, the construct of scale of space, that is the relative size of available pieces of information in relation to the subject, becomes an important determinant. This determinant can be investigated by looking at spatial cognition at different scales of space, for example small-scale and large-scale spaces.

2.2 Spatial Representations and Spatial Behavior

By now, scale of space has been identified as a determinant of spatial cognition. However, the theoretical relationship between spatial information of different scales of space and spatial behavior remain unclear. The concept of psychological space described space as a category of subjective experience but does not explain this relationship. The following section presents this theoretical junction, the concept of representation.

Cognitive science is closely bound to the concept of representation (e. g., J. R. ANDERSON 1987; ENGELKAMP & PECHMANN 1993; KNAUFF 1997; MANDLER 1983), and as a consequence, spatial cognition is tied to the concept of spatial representation. Psychologists (e. g., JANZEN 2013; LIBEN 1981, 1997; MALLOT 1999; MAY & KLUWE 2000) and researchers in geography and cartography (e. g., MACEACHREN 1995) have emphasized the central theoretical character of the concept of representation. Because the notion is omnipresent, there exists a conceptual pluralism for it (e. g., J. R. ANDERSON 1978; ENGELKAMP & PECHMANN 1993; JANZEN 2013) which has led to difficulties in interpreting results from different studies in the literature (HERRMANN 1993; LIBEN 1981; PRESSON & SOMERVILLE 1985).

In this section, the notion of representation is introduced and related to the concept of psychological space in a first step. In a second step, a possible relationship between physical space, spatial representations, and models of these is outlined. In a third step, another model addressing the conceptual pluralism of the notion based on conscious and unconscious use of spatial information is presented. Finally, this section explains how the notion of spatial behavior can be related to the concept of spatial representation.

2.2.1   The Concept of Representation

The concept of mental representation is anchored in the core assumptions of cognition (KNAUFF 1997, p. 18).[9] The two core assumptions in the study of cognition are:

(1) adaptive behavior[10] relies on a set of cognitive processes that
(2) operate on mental representations.[11]

In brief, behavior can be interpreted to be an output of an information-processing cognitive activity. The first central assumption of cognitive science implies that cognitive processes are an appropriate theoretical description of adaptive behavior (see also KNAUFF 1997; PALMER 1978). This assumption does not imply that the mind is actually considered to be a computer by cognitive psychologists (see KNAUFF 1997, for further elaboration).

---

[9]This viewpoint is different from the one of behaviorists who have conceptualized the mind as a black-box and have studied cognition as a stimulus-response-process (e. g., MALLOT 1997, 1999).

[10]'Adaptive' is not understood as the optimal output toward information-processing demands given a certain situation, but emphasizes that action of the individual is the output of a conscious information-processing procedure.

[11]J. R. ANDERSON (1987) suggested that concerning the structure of human cognition, the pair of mental procedures and representations allow individuals to behave adaptively. Similar to what is outlined above, he indicated that mental representations are the basis for adaptive behavior. In contrast, he did not conceptually distinguish mental procedures (transformations) from representations, but considered them to be a fixed entity. This intertwined nature of both core assumptions has been emphasized by other authors by defining a representation as a conglomerate of both assumptions. MARR (1982), for example, stated that "a representation is a formal system for making explicit certain entities or types of information, together with a specification of how the system does this" (p. 20).

The second assumption emphasizes the central character of the notion of representation in cognitive science (see also HERRMANN 1993)[12], that is, in its broadest sense, "something that stands for something else" (PALMER 1978, p. 262). This general definition provided by PALMER (1978) stresses that there are external 'states', that is information that is somehow internalized (see also ENGELKAMP & PECHMANN 1993). Internalization of information, however, must be, in some form, materialized in the brain. This is realized by neurophysiological structures in the brain during an encoding procedure (ZIMMER 1993).

In Section 2.1.1, the theoretical concept of psychological space was introduced as a category of subjective experience and as the space attributed to the mind and can be interpreted as being a psycho-philosophical concept. Mental representations also reflect subjective experiences, that is, the individual interpretation of information. The notion can therefore be interpreted as being a concept of cognitive science. As this study aims to address cognitive abilities, the notion of (mental) representation[13] instead of psychological space is used.

2.2.2 Palmer's Model of (Spatial) Representations

Following the definitions of the concept representations outlined above, a spatial representation is an internal representation of external spatial information. Moreover, it has been emphasized in the literature that spatial representations are the facilitator of spatial behavior (e. g., ALLEN 1999b; SIEGEL & WHITE 1975).

Spatial representation feature a spatial dualism (e. g., McCLOSKEY 2015). On the one hand, the attribute spatial emphasizes that the corresponding mental representation is based on the internalization of spatial information. On the other hand, one might wonder whether and in which sense a spatial representation is spatial itself. One might therefore ask whether and to which extent spatial information is represented, thus addressing the first meaning. Or, one might wonder how spatial information is presented, thus addressing the second meaning under a modeling perspective (see also MANDLER 1983). Since a spatial representation is not directly observable, investigating this question comes to proposing models of spatial representations. As an example, different models have been proposed for hypothesizing

---

[12]The authors emphasize a similarity to the notion of images (*German 'Vorstellungen'*), that has often been used in mathematics education (e. g., PLATH 2014). However, the notion is implicitely bound to an assumption of a certain format in which knowledge is stored in memory (see imagery-debate as elaborated in TYE 2000).

[13]In the following sections, 'mental' is omitted whenever it is evident within the context, that mental representations are meant.

how spatial information is stored in memory (e. g., MEILINGER 2007; O'KEEFE & NADEL 1978; TYE 2000). Unfortunately, models of spatial representations have often been denoted as (spatial) representation themselves.

Coming back to PALMER's (1978) general definition of mental representation as 'stand-for'-relation, it is important to emphasize that he did not define this relation as an isomorphism, but as being a model that is ideally specified in the research process.[14] According to him, the notion of representation is then a complex amalgam specifying the following five characteristics:

> (1) what the represented world is; (2) what the representing world is; (3) what aspects of the represented world are being modeled; (4) what aspects of the representing worlds are doing the modeling; and (5) what are the correspondences between the two worlds. (PALMER 1978, p. 262)

PALMER (1978) further addressed the conceptual pluralism of the notion of spatial representation by distinguishing between the mental world, which is "some kind of representation of the 'real world'" (p. 276) and the cognitive model that is "in turn, a representation of that mental world. Thus, the mental model is a representation of a representation of the real world. Almost by accident, the mental model is a representation of the real world in its own right" (p. 276). Both representations are conceptualized within a cognitive theory that "should simultaneously be the proper definition of both the mental world and the mental model" (p. 276).

Speaking of spatial information and re-addressing the question in which sense spatial representations are spatial themselves, the answer is that the attribute spatial is also tied to the chosen model of the spatial representation. This study uses the notions of 'spatial representation' instead of 'mental model', denotes 'real world' as 'physical space', 'cognitive model' as 'model of the spatial representation' and 'cognitive theory' as 'representational theory'. Figure 2.1 presents an analysis of the conceptual pluralism of spatial representations according to PALMER's (1978) model by distinguishing between the spatial representation and its model.

Concerning the spatial dualism of the concept of spatial representation, one might therefore use it in the context of studying what kind of spatial information is represented and to which extent, thus addressing the concept of spatial representation as internalized in the brain itself. Or one might ask how a given piece of information

---

[14]See also MANDLER (1983) who emphasized the complex nature of mental representations and pointed out that the definition includes not only what is known but also how it is intrinsically structured.

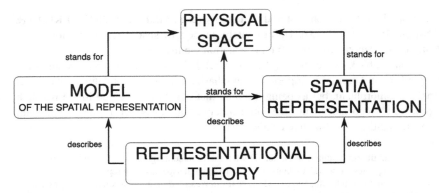

Figure 2.1 Relationship between physical space, spatial representation, its model and the corresponding representational theory. (Redrawn from PALMER 1978, p. 275)

is represented, thus studying the models of spatial representations.[15] As emphasized by PALMER (1978), studying what information is represented, which is the primary concern of this study, is ideally based on a theoretical assumptions of an underlying theoretical model.

2.2.3 Liben's Model of Spatial Representations

The notion of representation has not only been used in the context of mental representation of information in a 'stand-for'-relation, but is also used for external representations using symbol systems or symbolic depictions (e. g., MANDLER 1983).

LIBEN (1981) points towards the need to understand various interpretations of spatial representations in spatial cognition to reflect research outcomes in the spatial cognition literature in an appropriate way. She distinguishes among three types of spatial representations, namely spatial products, spatial thought, and spatial storage,

---

[15]MCNAMARA (1986) proposed to structure models of spatial representations within a certain representational theory with respect to four features: form, function, structure and content. Form refers to the "type of mental code used to represent knowledge in memory" (MCNAMARA 1986, p. 88) and referes therefore to the imagery-debate of whether mental codes are analogue or abstract propositional (see TYE 2000). The discussion is closely bound to the function of a representation. Structure and content play a crucial role when describing different theoretical models of spatial representations. Structure investigates the "kinds of relations allowed among codes" (MCNAMARA 1986, p. 88). Questions such as hierarchical vs. non-hierarchical representation of knowledge refer to the structure of a representation. Content refers to the elements that are stored in a representation and is bound to the question whether or not extra computational processes are important for the subject to show a certain behavior.

and between two contents of spatial representations, denoted as concrete and abstract content (Figure 2.2). According to LIBEN (1981), these clear theoretical boundaries might be permeable and fuzzy in reality. Boundaries between all distinctions have to be interpreted in a permeable way and can be crossed.

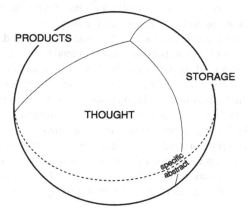

Figure 2.2 Conceptual differentiation of spatial representations (in LIBEN 1981, p. 11). ©1981 L. S. Liben, Elsevier. Reprinted with permission

The first type of spatial representations, spatial products, refers to "external products that represent space in some way [...] regardless of the medium; it includes [...] sketch maps, miniature models, and verbal descriptions" (LIBEN 1981, p. 11). Research that addresses spatial products is usually either concerned with the relation between the product and the space or how ready-made spatial products such as maps are understood by the individual. The spatial product itself is in the focus of all research and is studied in its own right.

The second type, spatial thought, refers to "thinking that concerns or makes use of space in some way [...] knowledge that individuals have access to, can reflect upon, or can manipulate, as in spatial problem solving or spatial imagery" (LIBEN 1981, p. 12). Whereas spatial thought is conscious and differs from the third type, spatial storage, which refers to any unconscious[16] information about space. Spatial storage is implicit knowledge that is not consciously accessible but assumed to exist.

---

[16]This distinction has been proposed in many concepts in psychology. SMITH AND QUELLER (2000), for example, emphasized that "the explicit/implicit distinction refers to *uses* of memory - the consciously recollective use of memory versus its use in performing some other task without conscious awareness of memory *per se*." (p. 112, emphasis in the original).

LIBEN (1981) further distinguishes between the content of those representations, that is, what is being represented. She proposed a theoretical distinction
between specific content, as addressed in environmental cognition, and abstract
content addressed by cognitive or developmental psychologists, which is "a distinction between locations or places in particular, and spatial concepts or abstractions in
general" (p. 5). According to her, even though the two contents are conceptualized
to be distinct, it is not possible to conclude that both are unrelated.

Besides PALMER's (1978) important ideas on mental representations, LIBEN's
(1981) conceptualization of spatial representation is a core model of the theoretical
framework of this study. Two major implications can be derived from the model.
First, the model emphasizes that spatial representation, and thus spatial cognition,
involves conscious and unconscious components. Second, the model conceptualizes
spatial cognition in different contexts that can be assumed to be related to different
types of spatial representations indirectly elicited by the scale of spatial information.
Consequently, spatial representation of large-scale spaces is of specific in content,
whereas spatial representation of small-scale spaces is abstract in content. According to the model, both contents represent a conceptually different semi-sphere in
Figure 2.2.

2.2.4 Spatial Behavior

Spatial behavior has two primary senses. First, in its broadest sense, it can be seen as
the outcome of the processing of spatial information by a set of cognitive processes
acting on a spatial representation (e. g., KNAUFF 1997). Hereby, spatial behavior
might occur spontaneously, even unconsciously. Second, since spatial representations can neither be observed directly by the individual nor the experimenter, the
question of externalizing a representation becomes important. Among different pos-
sibilities, physiological and behavioral approaches have the advantage that they are
non-verbal and therefore not correlated with the individual's abilities to commu-
nicate on cognitive processes (HERRMANN 1993).[17] Spatial behavior, in its second
sense, can therefore be interpreted as a means to study the subject's representation.

---

[17]Physiological approaches study the neural basis of representations. Neural activity is interpreted as being a materialized representation. Interpreting the relationship between the mental
representation and neural activities can be challenging. Although it can be acknowledged that
neurons enable cognitive processes, their activity should not be interpreted as being a sufficient condition for the existence of a representation. Physiological approaches to human
cognition need to be grounded on explicit theoretical assumptions that are derived from an
analysis of possible cognitive processes (see ZIMMER 1993, for further elaboration).

In this sense, for example, spatial behavior might be the result of conscious thought about a given task (Figure 2.3).

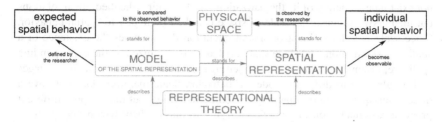

Figure 2.3 Assumed relationship between observed behavior and the researchers intended behavior under the assumption of a representational cognitive theory

Behavioral approaches assume that a subject's spatial representation is external-ized by the researcher's observation of the subject's spatial behavior. The researcher then concludes from the observed spatial behavior of the individual on the underlying spatial representation and cognitive processes (MONTELLO, WALLER, HEGARTY, & RICHARDSON 2004; OTTOSSON 1987). This approach typically involves pre-defined spatial tasks (ENGELKAMP & PECHMANN 1993). Every interpretation of the sub-ject's spatial behavior is closely bound to the model of the spatial representation that the researcher assumes within a certain representational theory. By assuming a certain model of the spatial representation, the researcher hypothesizes a 'model behavior' that the subject should show in the experimental setting. The expected spatial behavior is then compared to the real, observed behavior of the subject and becomes interpretable (see also ENGELKAMP & PECHMANN 1993).

Three implications can be derived from this explanation. First, the interpretation of an observed spatial behavior is only meaningful within a certain representational framework and "[r]esearchers must therefore be cautious in assuming that a specific outcome measure necessarily reflects a particular type of internal representation" (MONTELLO ET AL. 2004, p. 262). Second, the observed and interpreted spatial behavior of the individual is the outcome of the subject's individual conception of a spatial setting encountered, for example a pre-defined task. It is a result of how an individual relates to a certain (pre-defined) spatial situation and, consequently, a subjective expression of the individual's psychological space (see OTTOSSON 1987, p. 61, for further elaboration). In line with OTTOSSON (1987), this study emphasizes that "the conception [of a task] is an ascribed aspect of the individual's behavior. The conception does not exist beyond this ascribed aspect of the individuals behavior. The conception does not exist beyond this ascription by an observer, that is, it does not exist independently of its thematization" (p. 45). Third, the theoretical relations outlined above suggest that scale of space is linked indirectly to spatial behavior. Since scale is a determinant of spatial cognition that assumes the existence of a spatial representation, and since spatial behavior is organized by the corresponding spatial representation, scale of a given setting in which a subject cognitively engages with might lead to different modes of behavior. Scale therefore does not have a direct consequence on the spatially reasoning individual but may impose different processing demands on the individual trying to master them which may, in turn, result in different spatial behaviors (see also HEGARTY ET AL. 2006; LOCKMAN & PICK 1984).

Conclusion

Spatial cognition defines the subjective experiences of and in space as a psychological information-processing phenomenon. The core assumption of spatial cognition is that there are spatial representations on which cognitive processes operate. Spatial representations describe what and how spatial information becomes embodied, or how this information enters the individual's subjective 'world in the head.' Moreover, spatial representation organizes a subject's behavior in space. Spatial cognition assumes that behavior in physical space is defined as the sensorimotor outcome of cognitive processes that operate upon the spatial representations. Spatial behavior is therefore linked indirectly to the spatial information in physical space by a corresponding spatial representation. Consequently, when scale is not only considered as a spatial dimension of physical space, but as psychological construct, it has an indirect effect on spatial behavior.

Because it is a powerful theoretical construct, spatial representation has been subject to a conceptual pluralism and has been defined as both conscious and unconscious embodied spatial information. It has also been understood as theoretical models or external representations of spatial information.

2.3   Encoding and Maintainance of Spatial Knowledge

Since spatial representations refer to subjective internalization of spatial information, they have also been interpreted as a form of spatial knowledge stored and
organized in memory (e. g., MEILINGER 2007; THAGARD 2005) that is activated
whenever required to solve a particular spatial problem (TERGAN 1993). The notion
of (spatial) knowledge has further been defined as being related to the individual's
training, thus to some form of individual acquisition of it (R. E. ANDERSON 2008).
Finally, some authors have emphasized that this act of acquiring knowledge corresponds to "the construction of a stable mental representation of a possible world"
(DENHIÈRE, LEGROS, & TAPIERO 1993, p. 312). There are two things that can be
emphasized with regard to these definitions. First, since knowledge can be acquired,
the source of information provided seems to be important. Second, since some definitions stressed that knowledge is stable, that is, probably long-term representation,
whereas other definitions do not, one might hypothesize that the way in which
knowledge is stored in memory can be conceptualized from different viewpoints.

Encoding refers to the process of internalizing external spatial information in
a spatial representation stored in memory. Encoding, however, is far more than
a 'storage problem' but involves cognitive challenges such as employing suitable
frames of reference. The latter ones are reference systems that are assumed to be the
definitional basis for describing and specifying spatial relations when it comes to
their perception, encoding and mental manipulation (ALLEN 1999a). Consequently,
the concept of frames of reference is meaningful for many aspects of spatial cognition and related disciplines such as object recognition (e. g., MARR 1982), navigation
(e. g., O'KEEFE & NADEL 1978), spatial language (e. g., LEVINSON 1996), and imagined manipulation of objects (e. g., ZACKS, MIRES, TVERSKY, & HAZELTINE 2000).

As outlined in Section 2.1.2, there are two types of spatial information, locations
and attributes of a certain information. Whereas locational information describes
where a piece of information is located in space, attributive information refers to the
kind of information that is given. Locational information is typically specified in
terms of distance and direction information. A combination of both helps to localize
a certain information and it is completed by attributive information that specifies
why the particular information might be meaningful (ITTELSON 1973, pp. 16–17). To
solve spatial problems that involve the movement of objects and the self, being the
primary concern of this study, individuals have to cognitively engage with locational
information of objects or configurations of those and the self.

In a first step, this section presents the concepts of frames of reference and
describes different ways of encoding the locational information given in a spatial
stimuli. Three different ways to encode the resulting distance and direction infor-

mation are presented. In a second step, this section presents a conceptual distinction concerning the acquisition of spatial knowledge based on two different sources of spatial information. In a third step, this section introduces three models of how spatial knowledge may be stored in human memory and hypotheses how these might be related.

2.3.1 Frames of Reference

Frames of reference[18] are "a means of representing the locations of entities in space" (KLATZKY 1998, p. 1). EASTON AND SHOLL (1995) emphasized that they help individuals "extracting and storing knowledge of stable interobject relations from continually changing patterns of self-to-object relations" (p. 483).

Frames of reference can be understood as relational systems that specify spatial relations between objects in space using coordinate systems (e. g., SHELTON & MCNAMARA 2001). Their basic entities, objects, can be single objects, particular features of the environment, so-called landmarks. Those might refer to particularly interesting locations in real space, but also to topological features such as corners of a building or crossing streets (e. g., BELL 2002; DAVIES & UTTAL 2007; MEILINGER 2007). According to MAY AND KLUWE (2000), such a coordinate system consists of (1) a specification of the origin $\mathcal{O}$ of the system; (2) a specification of at least one ray that passes the origin and constitutes a reference direction; and (3) a scale along the ray (or the rays) that allows for a definition of topological (often ordinal or metrical) relations. Typical examples of coordinate systems are Cartesian and polar coordinate systems. Although some researchers narrow the concept of reference frames to a Cartesian understanding of them, this study understands frames of reference as a basic concept with the case of orthogonal axes being a special one.

One way to classify different frames of reference is to distinguish them according to their origin and consequently, whether they change whenever the origin moves in space (Figure 2.4). By doing so, frames of reference can be categorized into absolute and relative frames of reference (e. g., BELL 2002; HUNT & WALLER 1999). In an absolute frame of reference, objects and subjects are located with respect to a frame that is external to the situation, thus being unmovable. In a relative reference frame, the direction and location of an object is encoded relative to another object or a

---

[18]This section refers to spatial frames of reference, that rely on perceptual processes. LEVINSON (1996) showed that there are logically similar reference systems in language. However, it is not clear whether representations from one modality (e. g., manipulable) can be 'translated' into another (e. g., visual). To approach the term itself, the scientific literature from many disciplines is cited.

subject's body. Relative frames of reference change therefore whenever a subject or object moves in space. Depending on the origin of the relational system, relative frames of reference are subdivided into egocentric and intrinsic reference frames. A reasonable, but rare subdivision of absolute frames of reference has been proposed by BELL (2002) who distinguishes between local absolute and global absolute reference frames. In the former, locations are described with respect to an invariant frame of reference in the environment, such as the borders of the campus, whereas in the latter, locations are described with the help of the cardinal directions (N, E, S, and W).[19]

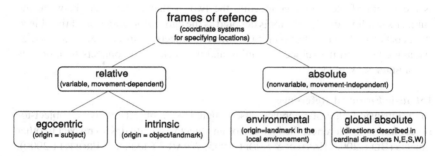

Figure 2.4 A classification of frames of reference

Egocentric Frames of Reference

In an egocentric frame of reference, also denoted as 'self-to-object' reference frame, the individual themselves constitutes the origin (e. g., SHELTON & MCNAMARA 2001). All spatial stimuli are defined with respect to the reference direction of the subject's heading or movement direction (KLATZKY 1998).[20] Their location is therefore represented from the individual's perspective, is subject-centered and represented with respect to the up–down, front–back, and left–right axis of the observer's body (e. g., D. J. BRYANT & TVERSKY 1999; EASTON & SHOLL 1995). Objects represented in an egocentric frame of reference might therefore be 'in front of me', 'to my left', or 'just behind me.'

As the orientation of the egocentric frame of reference is determined by the subject's heading, the human brain sets up multiple frames of reference that are

---

[19]The children's conception of cardinal directions is out of the scope of this study. Therefore, no further information is provided concerning this frame of reference.

[20]More specifically, a heading in space is "the angle between the object's axis of orientation and some reference direction external to the object" (KLATZKY 1998, p. 3).

centered around different parts of the body, for example an eye-centered one, a head-centered one, or a core-centered one (LEVINSON 1996; MCCLOSKEY 2015; ZACKS & MICHELON 2005). The particular orientation of each egocentric frame of reference is therefore body-part specific. For the sake of simplicity, this study assumes that the subjects heading direction is representative of all different egocentric reference systems; thus implying that all body parts are oriented in the same way (see also WOLBERS & WIENER 2014).

Egocentric frames of reference can be represented in polar coordinate systems (Figure 2.5). The location of an external object can be then expressed as following: the distance between the subject and the object, denoted as egocentric distance, is the length of the vector connecting the two of them. The angle between the subject's heading and that vector is denoted as the egocentric direction of the object. An egocentric direction between two different objects in space is then defined as the angle between the subject's heading and the vector that connects both objects (WOLBERS & WIENER 2014).

Intrinsic Frame of Reference
In an intrinsic frame of reference, also denoted as an object-relative or object-to-object frame of reference, the location of an object is specified with respect to other objects (BELL 2002; EASTON & SHOLL 1995; WRAGA, CREEM, & PROFFITT 2000). The intrinsic frame of reference is independent of the subject's body and of a global landmark in the environment. It relates objects locally with respect to each others heading (Figure 2.5).[21] A configuration of objects, represented in an intrinsic frame of reference, can therefore be interpreted as a network of interrelated intrinsic frames of reference (EASTON & SHOLL 1995). Since intrinsic frames of reference locate objects or subjects with respect to the intrinsic structure of an object, locations of objects might become 'on the top of the cube' or 'between the yellow and red cup.'

Environmental frame of reference
In an environmental frame of reference, also denoted as allocentric (e. g., MAY & KLUWE 2000; WOLBERS & WIENER 2014), exocentric, or geocentric (e. g., KLATZKY 1998), all given visual stimuli are encoded with respect to a particular environmental feature (for example a certain building on the campus or the corner of the room), which constitutes the origin of the coordinate system.[22] Environmental frames are therefore independent of the subject's heading. Whenever the

---

[21] More specifically, the heading of an arbitrary object in space "is the angle between its axis of orientation and some reference direction" (KLATZKY 1998, p. 7).

[22] Although the formal definition of an environmental frame of reference is plausible, it is not yet clear how it actually looks like. Whereas egocentric frames of reference can be interpreted

subject rotates or locomotes via translations in space, static objects and relations of objects remain unchanged within this frame of reference. Since objects and subjects are located with respect to the surrounding environment, their location might be described as 'at the end of the floor', or 'next to the library' (WOLBERS & WIENER 2014).

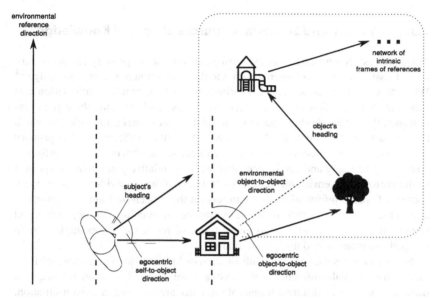

Figure 2.5 Environmental and egocentric frames of reference with network of intrinsic frames of reference (adopted from WOLBERS & WIENER 2014, p. 2). ©2014 Wolbers and Wiener. Adopted with permission

Environmental frames of reference can also be represented in polar coordinates (Figure 2.5). The environmental reference direction refers then to one arbitrarily set coordinate axis. The environmental direction of an object is defined by the angle between the environmental reference direction and the object's axis of orientation. The environmental distance is the length of the vector that connects the origin and the object. However, those definitions are only theoretically meaningful since most environments do not provide a reference point that serves as an unambiguous origin

---

as vista spaces in a particular spatial situation, it is still not clear what constitutes the origin of an allocentric reference frame in the same situation (WOLBERS & WIENER 2014, p. 2).

(see WOLBERS & WIENER 2014, p. 2). For this reason, the direction and distance of an object is often specified with respect to another object.

In summary, spatial locations can be encoded in three different frames of reference. They constitute the building blocks whenever describing the encoding of spatial information and when comparing spatial representations.[23]

2.3.2   Primary and Secondary Sources of Spatial Knowledge

PRESSON AND SOMERVILLE (1985) distinguished between primary and secondary uses of spatial information when it comes to the acquisition of spatial knowledge.[24] Whereas primary uses of spatial information "include practical orientation and action in direct relation of space", secondary uses "include symbolic representation and allow older children and adults to think about spatial information to which they are not directly oriented" (PRESSON & SOMERVILLE 1985, p. 15). The primary use of spatial information therefore refers to direct actions that do not require further cognitive processing and can be described as being relatively automatic responses to the surrounding environment (PRESSON 1987). When individuals use primary sources of spatial information, they can rely on the egocentric frames of reference as they are naturally established in their immediate surrounding. One object encoded to the left with respect to the egocentric frame of reference, for example, can be grasped and manipulated using the left arm.

Since secondary uses of spatial information include information from which individuals have to make inferences for guessing about other information, this requires them to be aware that different frames of reference are involved in a given situation. Furthermore, they have to understand that the meaning of a symbolic representation might depend on the frame of reference chosen. In particular, this requires the individual to be aware that the direct relation to the perceived spatial information, encoded in a respective frame of reference, might conflict with the frame of reference of the spatial information (PRESSON 1987).

The distinction proposed by PRESSON AND SOMERVILLE (1985) can be related to LIBEN's (1981) model of spatial representations. Primary uses of spatial representation correspond to LIBEN (1981) spatial storage, that is, unconscious thought about

---

[23]The definition of those frames of reference does not imply in which spatial situation a certain frame of reference is applied. A thorough discussion of spontaneous uses of frames of reference is beyond the scope of this study.

[24]They therefore acknowledged the theoretical distinction between practical/sensorimotor and conceptual/representational space that had been proposed by PIAGET AND INHELDER (1948/1956).

space that might occur in navigational situations based on perceptual processes such as walking along a gangway without bumping into the walls. Consequently, the focus on (unconscious) attention lies on perceivable aspects of the surrounding worlds with respect to direct actions that are related to those without involving further spatial processing (PRESSON 1987).

Secondary uses of spatial information refer to LIBEN's (1981) spatial thought mediated by a spatial product. Thus, secondary uses involve the conscious encoding and manipulation of spatial information, thus referring to rather abstract and formal use of spatial information. PRESSON AND SOMERVILLE (1985) did not conceptualize both ways to use spatial information to be distinct, but emphasized that applying abilities allowing secondary uses of spatial information requires the individual to integrate information gained from primary use of this information. Thereby, applying abilities to use secondary information depend on and must coordinate with uses of primary spatial information (PRESSON 1987).

The distinction proposed by PRESSON AND Somerville (1985) theoretically grasps the use of visual representations such as depictions of spatial situations or maps that have been argued to be meaningful in the study of spatial thought as mentioned in the introduction. Consequently, this study focuses on the subject's acquisition of knowledge through secondary uses of spatial information.

2.3.3 The Architecture of Spatial Memory

The previous statements left no doubt that spatial cognition implies that spatial information is stored in memory in forms of representations. These become useful when mental inferences are drawn from them. Based on this idea, it remains an open question where knowledge is stored in human memory.

Different models on human memory have been proposed in the literature (see NORMAN 2013, for further elaboration). This study addresses three of them to give an idea of which subcomponents of human memory are interacting when it comes to solving spatial problems.

Multi-Store Models of Memory
Multi-store models explain how memory processes might work by assuming a similarity between those processes and those doing computer-based computations. The three-part multi-store model of ATKINSON AND SHIFFRIN (1968) is among the most prominent models (Figure 2.6). They distinguished between knowledge structures located in long-term memory (LTM) and short-term memory (STM), and assumed the existence of a sensory register in which sensory information enters memory (as cited in KNAUFF 1997, p. 92).

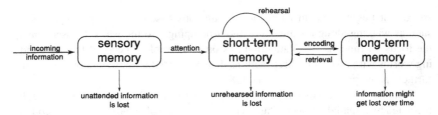

Figure 2.6 Multi-Store model of memory according to ATKINSON AND SHIFFRIN (1968)

When information is perceived by the subject, it enters into the sensory register where it either decays or is attended to. The attended pieces of information are then processed to the STM which holds them present for further reasoning processes by means of rehearsal, a control process that allows an individual to keep information in STM. Moreover, STM combines the information received from the sensory memory with knowledge stored in LTM. There is a general agreement that capacity of STM is limited to about seven chunks, that are, independent pieces of information that can be held simultaneously (MILLER 1956). Knowledge that is transferred from STM to LTM and finally stored is considered as permanent, stable knowledge that can be re-copied to STM where it might be manipulated and then re-stored in LTM. Over time, however, information might get lost from the LTM.

It has been doubted whether spatial knowledge should be conceptualized as a structure of STM or LTM (e. g., N. JONES, ROSS, LYNAM, PEREZ, & LEITCH 2011, for further discussion). This study assumes, in line with the hypothesis proposed by NERSESSIAN (2002), that (spatial) representations are knowledge structures that exist in either STM or LTM. Knowledge structures from LTM are assumed to be called upon to support inference processes from representations held in STM whenever the individual is reasoning about pieces of information.

Working Memory
Working memory (WM) can be described as

> a cognitive component that mainly serves two functions. It *temporarily maintains* information that was either perceived but is no longer present in the environment, or that was internally generated, and it supplies a *work space* for transforming and manipulating elements of perception and thinking. (ZIMMER, MÜNZER, & UMLA-RUNGE 2010, p. 13, emphasis in original)

BADDELEY AND HITCH (1974) emphasized that STM is an active part of human memory and should further be differentiated into different subcomponents, and

introduced the notion of working memory as a substitute for ATKINSON AND
SHIFFRIN's short-term memory. Similar to the role of STM, they described working
memory as a cognitive 'port' between perception, long-term memory, and action
(as cited in ZIMMER ET AL. 2010, p. 14).

Although several distinct models have been proposed to model working memory
(see MIYAKE & SHAH 1999, for a review), BADDELEY AND HITCH's (1974) model
has been the most considered model until today (Figure 2.7). According to them,
WM can be modeled to consist of three components: the central executive (CE)
and two domain-specific subsystems, the phonological loop (also denoted as verbal
working memory, VWM), and the visuospatial scratchpad or sketchpad (VSSP, also
denoted as visuospatial working memory, VSWM). Those slave systems are considered as mental workspaces for verbal and spatial information that keep pieces
of information mentally available for further interpretation and processing such as
mental updating. According to the model, both slave systems are controlled by the
CE that modulates attention of the individual (see also ZIMMER ET AL. 2010).
BADDELEY AND HITCH's (1974) model of working memory therefore assumes a
combined storage and processing, with the CE being tightly linked to both subsystems.

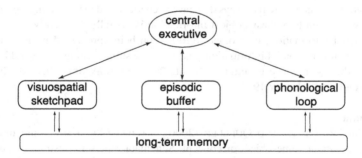

Figure 2.7 Model of working memory according to BADDELEY AND HITCH (1974), and
BADDELEY (2000)

WM has been emphasized to be "a limited resource that constrains cognitive
performance" (ZIMMER ET AL. 2010, p. 13). The fact that WM is limited has been
expressed by introducing the notion of working memory capacity that is individually restricted. Not only WM but each subsystem has been modeled to have its
own specific capacity, leading to suggestions that dual-task interferences may occur

whenever subjects face a task that uses the same subsystem, but single task effects should be found whenever the task involved different subsystems (see ZIMMER ET AL. 2010, for further elaboration). Due to these effects, BADDELEY AND HITCH's (1974) model has also been denoted as tripartite model (ZIMMER ET AL. 2010). In a more recent model, BADDELEY (2000, as cited in ZIMMER ET AL. 2010, p. 14) extended the tripartite model by a fourth component, the episodic buffer, which enables the integration of multi-model information from the different slave systems (ZIMMER ET AL. 2010).

Conceptualizations of VSWM have questioned whether the slave systems are best represented as an unitary construct or should be further differentiated. There is a converging agreement that VSWM can be further divided into a visual and spatial component. Maintaining a visual, surface property-based appearance information in VSWM is considered to be passive whereas spatial information has been modeled to be actively held in the visual buffer (see ZIMMER ET AL. 2010).[25] The visual buffer was hereby modeled to function as the visual working space for pro-active maintenance of spatial information, that is the processing and transformation of information. It is therefore related to the visuospatial sketchpad, the CE and conscious awareness processing (ZIMMER ET AL. 2010). One way to interpret this conceptualization of two subsystems of VSWM is close to the distinction between what pathways and where pathways that have been identified in the brain. According to the two-stream hypothesis, object information is neurally differently processed from spatial information, with object information being processed in the ventral 'what'-stream and spatial information relative to the subject being processed in the dorsal 'where'-stream (e. g., FREUD ET AL. 2016; KOZHEVNIKOV ET AL. 2005; UNGERLEIDER & HAXBY 1994).

Schemata

According to SMITH AND QUELLER (2000), a schema "is a structured unit of knowledge about some object or concept. Schemata represent abstract or generalized knowledge" (p. 114) and have been further described as higher-order cognitive structures representing "bundles of knowledge in memory" (MACDONALD & MACINTYRE 1999, p. 16). They can be activated explicitly by thinking about something or implicitly by encountering information that is related to the schema. Whenever a schema is activated, all structured knowledge that relates to the schema becomes accessible. The accessibility of a schema, that is the extend to which the schema is activated, can be increased by frequent use of it. Schemata are indepen-

---

[25]The notion of the visual buffer was particularly put forward in the discussion of the imagery debate, see TYE (2000).

dent, that is, whenever one schema becomes activated this does not necessarily imply that another, related one is also activated. Moreover, the perceptual interpretation of visual stimuli is affected by schemata; that is, they influence the attention which individuals select during the perception of information by triggering a comparison process between what is actually perceived and what is expected to be perceived. Whenever there is a mismatch, the individual might make inferences about missing or schema-inconsistent information. When a stimulus is schema-consistent, in other words, when it matches a person's schema, that schema is activated and used to perceive important features of the stimulus with greater ease by referencing to the encoded memory structures. Taken together, schemata direct the attention to individually relevant pieces of information and thus away from irrelevant ones, influence the way perceived stimuli are encoded and how missing stimuli are inferred from the available ones (see also SMITH & QUELLER 2000, p. 114).

Schemata have been conceived as particular long-term knowledge structures that exist besides long-term knowledge representing spatial information (N. JONES ET AL. 2011). Although the relation between both concepts remains an open question, it has been assumed that they differ with regard to their representational flexibility. Under this perspective, schemata have been described as being inflexible knowledge structures that are used in routine situations that an individual has encountered several times, whereas spatial representations in LTM have been described as flexible knowledge structures that might combine several schemata (see HOLLAND ET AL., 1986, as cited by N. JONES ET AL. 2011, Tab. 1, p. 4)

Conclusion
Since they represent internalized spatial information, spatial representations can also be understood as spatial memory. One possible model is that spatial information in a given spatial setting is encoded in a spatial representation with respect to different frames of reference. A spatial representation can then be interpreted as a form of spatial knowledge, stored in visuospatial working memory. Knowledge structuring and attentional control by the central executive allow individuals both to hold spatial information present in a mental working space and to retrieve knowledge and schemata that might support subsequent inference processes from long-term memory. Since all of these structures of human memory are bound to individual capacities, spatial information might get lost during encoding, maintenance, updating, and retrieval processes.

2.4   Spatial Abilities

> Spatial ability has been defined in such a variety of different ways that it is often
> difficult to be precise about the meanings which we ascribe to the term. (ELIOT &
> SMITH 1983, p. 1)

Since they have been conceptualized as a primary cognitive ability in the context of human intelligence research, there has been general agreement that spatial abilities represent a factor of an individual's intelligence, but the definition of the notion of spatial abilities has been subject to debate in different disciplines (e. g., CARROLL 1997; ELIOT 1987; HEGARTY & WALLER 2005). In a first step, this final section defines spatial abilities in the context of the concepts and the theoretical relations outlined in the previous sections. In a second step, the definitions provided are distinguished from related concepts dealing with spatial information, for example visual perception, spatial aptitude, imagery, and spatial intelligence.

2.4.1   Definition of Spatial Abilities at Different Scales of Space

In Section 2.1, spatial cognition was defined as the complex relationship between the physical spatial world and the corresponding psychological space. The previous sections showed that spatial cognition is closely tied to the assumption that spatial information is mentally encoded and stored in a spatial representation and subsequently further transformed by cognitive processes. This kind of processing allows an individual to causally draw inferences when engaging in spatial behavior, especially when solving spatial tasks.

Based on the theoretical assumptions outlined so far, this study uses the following definition of spatial cognition:

> **Spatial cognition** addresses the complex relationship between the physical
> spatial world and the corresponding psychological space by a set of cognitive processes on a spatial representation that organize an individual's spatial
> behavior.

As suggested by the etymology of the term ability (lat. *habilitās*), spatial abilities are cognitive abilities that enable an individual to engage in spatial cognition. They allow an individual to encode spatial information in a spatial representation and to hold this representation in working memory for further mental transformations, which, in turn, allows the individual to make inferences based on the subjective conceptualization of a spatial situation or a given spatial task.

This basic definition of spatial abilities is narrowed down in the following section. Recent conceptions of spatial abilities in the context of mathematics education have defined spatial abilities as cognitive abilities enabling spatial thinking (e. g., NATIONAL RESEARCH COUNCIL 2006). They are understood as a conglomerate of three components: (1) knowledge about space; (2) representations; and (3) reasoning (e. g., LIBEN 2006). This study adopts this definition of spatial abilities with the following theoretical specifications for each of these components:[26]

First, the component knowledge about space refers to spatial knowledge and schemata that an individual holds about space in long-term memory which can be retrieved when solving a spatial task. Knowledge about space also involves, among others, knowledge about different dimensions, coordinate systems, or distance (measures) in space.

Second, spatial abilities rely on spatial representations. These representations contain all information necessary to stimulate higher-order cognitive processes such as mental transformations, deduction or inference, that is, processes that contribute to individual behavior in space or the individual conception of a given spatial task. Given the conceptual pluralism of spatial representations outlined in Section 2.2 and 2.3, this study is based on the following specifications:

- In line with PALMER's (1978) model, this study examines what kind of spatial information is represented (denoted as spatial representation)
- This study addresses, in line with LIBEN (1981) model, the conscious manipulation of spatial information (denoted as spatial thought in Figure 2.2, p. 35).
- According to PRESSON AND SOMERVILLE (1985), this study understands spatial knowledge as being acquired from secondary uses of spatial information, thus involving symbolic representations which allow the subject to think about spatial information beyond the direct experience of them while reasoning about the relation of different frames of reference involved. This spatial knowledge is assumed to be stored in the visualspatial working memory and supported by long-term knowledge structures.

---

[26]Since the definition proposed above involves the notion of reasoning, this study uses the phrase 'spatial reasoning' as a synonym for 'spatial thinking.' Although this study addresses solving spatial problems at various scales of space, the notion of 'spatial problem-solving' is avoided to distinguish the focus of this study (spatial thought as a result of processing a significant amount of spatial information at different scales of space) from studies that conceptualized spatial thought as a problem-solving process (e. g., PERRIN- GLORIAN ET AL. 2013; SOURY-LAVERGNE & MASCHIETTO 2015).

Third, reasoning about space refers to the manipulation of spatial stimuli by means of mental transformations and inferences on a representation (TVERSKY 2005). Reasoning can be understood in the sense of BRUNER (1973, as cited in TVERSKY 2005, p. 209), that is, "going beyond the information given" (TVERSKY 2005, p. 209).

Those assumptions emphasize the conscious manipulation of spatial information that is maintained in working memory rather than unconsciously retrieving and using a spatial representation from long-term memory. However, the chosen conceptualization emphasizes that there are also unconscious components involved (knowledge about space as defined by LIBEN's (1981) spatial storage or by PRESSON AND SOMERVILLE's (1985) knowledge acquired from primary sources of spatial information). Spatial abilities are therefore understood as being closely linked to the unconscious use of spatial information. This idea has been expressed in definitions that were previously used in the literature. Spatial abilities have, for example, been defined as "the ability to formulate mental images and to manipulate these images in mind" (LEAN & CLEMENTS 1981, p. 267) and spatial reasoning as a "set of cognitive processes by which mental representations for spatial objects, relationships, and transformations are constructed and manipulated" (D. H. CLEMENTS & BATTISTA 1992, p. 420). ELIOT AND HAUPTMAN (1981) proposed "to consider spatial ability *both* as an intrinsic aspect of thinking *and* as a set of operations for solving spatial problems" (p. 46, emphasis in original). All definitions emphasize the importance of two aspects. On the one hand, the construction and storage of the spatial representation of the corresponding physical space and, on the other hand, the manipulation of the representation. Spatial abilities therefore involve far more than just storing spatial information.

Different scales of space provide different sources of information to the individual, and the degree of interaction between individual and physical space is scale-dependent. Small-scale spaces typically provide visual stimuli or stimuli of manipulable material and do not require movement by the individual, whereas large-scale spaces primarily provide visual but also multisensory information that must be integrated while moving. Distinguishing between small-scale and large-scale spatial abilities seems therefore not only expedient but in line with LIBEN's (1981) conceptual distinction of spatial representation into specific and abstract content.

Using these ideas and assumptions, this study is based on the following two definitions:

Small-scale spatial abilities are a class of cognitive abilities that enable an individual to represent and make inferences from spatial information that can be perceived from one single viewpoint.

Large-scale spatial abilities are a class of cognitive abilities that enable an individual to represent and make inferences from spatial information perceived and processed over time and from more than one viewpoint due to changes resulting from movement.

2.4.2   Conceptual Distinctions

Since spatial abilities have been a consistent topic of interest in the literature, several other conceptualizations have been used. Spatial abilities have also been addressed under the notion of spatial skills (e. g., UTTAL ET AL. 2013), spatial aptitude (e. g., PELLEGRINO & KAIL 1982), and spatial factors (e. g., DOGU & ERKIP 2000). Moreover, they have been addressed as being closely related to spatial perception, mental imagery, and spatial intelligence. Although there is legitimacy in all definitions, there is a need to explain the different conceptual ideas behind the other notions used. This study distinguishes the chosen definitions from other ones not only for the sake of clarity, but also to state three further assumptions that underlie this study.

First, spatial abilities involve a process of perceiving spatial information and are therefore closely related to spatial perception. According to HART AND MOORE (1973), spatial perception refers both to a form of processing visuospatial information and figurative knowing. HOFFER (1977, as cited in GAL & LINCHEVSKI 2010, p. 165) described visual perception as a conglomerate of abilities that contain figure-ground perception, perception of spatial relationships, visual discrimination, and visual memory that allow an individual to spot and interpret a given visual stimulus (see DOMNICK 2005, pp. 15–57, for further elaboration).

Spatial abilities as conceptualized and defined above involve higher cognitive processes than the physical-neural ones during the process of perception since they are grounded on inference processes based on mental representations of spatial stimuli (GAL & LINCHEVSKI 2010). ROST (1977) argued[27]

> [t]he idea of spatial abilities should not be equated with the perception of spatial stimuli, since it is not perception that is the main issue of those abilities, but imagining and moving beyond perception, that is, acting and handling with spatial objects, concepts and relations in a mental way. (ROST 1977, p. 21, translation by the author)

The distinction between spatial thinking and spatial perception is, of course, not dichotomous, since spatial perception can be seen as prerequisite, or as a constructing element of spatial thinking. "Perception is thus both a subsystem of cognition and a function of cognition. [...] Spatial perception and spatial cognition, therefore, are two separate but reciprocating processes" (HART & MOORE 1973, p. 250).[28]

Concerning the discrimination between the notions abilities, skills, factors, and aptitudes, three different perspectives are fruitful. First, from a psychometric intelligence research perspective, the notion of spatial factors represent a subject's cognitive abilities concerning the processing of spatial information (SEERY, BUCKLEY, & DELAHUNTY 2015).[29] Inspired by new statistical techniques, psychometric researchers tried to specify the factorial structure of spatial abilities by mapping groups of measures, provided by spatial tests, into homogenous dimensions, so-called spatial factors. Spatial abilities, skills, and aptitude can then be interpreted as the subject's capacity to complete pre-defined spatial tests. Second, from an ontogenetic and spatial training perspective, the notion of spatial aptitude has been differentiated from the notions of abilities and skills because it describes an "innate level of ability, where no education or training specifically designed to develop spatial ability has taken place" (SEERY ET AL. 2015, p. 2). Finally, from a perspective of universal usability, CARPENTER AND JUST (1982) stressed the need

---

[27] German original: „Raumvorstellung ist nicht mit der Perzeption räumlicher Gegebenheiten gleichzusetzen, da nicht die Wahrnehmung, sondern das über die Wahrnehmung hinausgehende Vorstellen und Bewegen, also das gedankliche Handeln und Hantieren mit räumlichen Objekten, Begriffen und Relationen im Vordergrund stehen."

[28] Although it is still a matter of debate, there is a consensus that mental activation during spatial thinking reflects brain areas that are also involved in spatial perception and attention (J. R. ANDERSON & FUNKE 2013; ZACKS & MICHELON 2005), leading to the conclusion that perceptual processes are involved in spatial thinking processes (see BARTOLOMEO 2002, for a full review). BEHRMANN (2000) proposed to interpret perception and spatial abilities as two distinct processes that overlap partially.

[29] More specifically, human cognitive abilities have been defined as "any ability that concerns some class of cognitive tasks", that is, tasks "in which correct or appropriate processing of mental information is critical to successful performance." (CARROLL 1993, p. 10).

to distinguish between the notion of spatial abilities and skills since "[a]bilities are conceived of as broad traits that enter into a *variety of performance measures*. [...] In contrast, skills are thought to reflect performance in *specific tasks*" (CARPENTER & JUST 1982, p. 226, emphasis added). For the reasons outlined, this study will use the notion of abilities rather than skills but acknowledges the legitimacy of this notion.[30]

Mental imagery is a concept that has been emphasized to be closely related to spatial abilities. LOHMAN (1979) definition of spatial abilities that "may be defined as the ability to generate, retain and manipulate abstract visual images" (LOHMAN 1979, p. 188) is only one example that makes use of the notion of images. Indeed, mental imagery that is the "occurrence of mental activity corresponding to the perception of an object, but when the object is not presented to the sense organ" (LEAN & CLEMENTS 1981, p. 267) can be seen as one research tradition that contributed to the process analysis of small-scale spatial abilities with important empirical findings. Studies typically focused on mental rotation experiments or investigations on transformations of small objects. However, small-scale spatial abilities have been shown to be richer in their inner differentiation and encompass also other abilities than mental rotation. Furthermore, mental imagery did not contribute to the understanding of large-scale spatial abilities. As a matter of fact, this study emphasizes the importance of mental imagery for precising mental transformations in the analyses of cognitive processes.

Finally, a conceptual distinction between spatial thinking and spatial intelligence was proposed by HEGARTY (2010). She defines spatial intelligence as "adaptive spatial thinking" (p. 1) and links it to the concept of metacognition by stressing that flexible strategy choice is fundamental to spatial intelligence. A closer investigation of spatial intelligence would result in taking the strategic perspective outlined by LINN AND PETERSEN (1985).

To sum up, the following conclusion can be drawn from the conceptual distinctions outlined above. First, spatial abilities are more than mere visual perception, since they involve higher-order cognitive processes that act on mental representations. Second, they can be assumed to be innate but being constantly developed during the lifespan by the individual in interaction with their environment. Addressing spatial abilities therefore means to deduce to an individual's current level of abilities that have been acquired during a constructivistic learning process. Third, spatial abilities, in particular when considered as one component of intelligence do not only involve cognitive abilities but also metacognitive abilities that are closely

---

[30]This is in line with international volumes on spatial thinking in which both notions are used as synonyms (e. g., DAVIS & SPATIAL REASONING STUDY GROUP 2015).

related. This study investigates spatial abilities under the strong assumption of strategy homogenity.

Conclusion

Spatial abilities enable the subject to engage in spatial cognition, that is, to perceive, represent, and mentally manipulate spatial information. When solving spatial tasks or facing spatial challenges, individuals use spatial abilities to consciously reason about and consciously interact with different types of spatial information provided in graphical space. Spatial abilities allow individuals to use them to think about space beyond the direct experience of it. Spatial abilities enable the subject to make inferences from spatial information at different scales of space, a process that involves cognitive engagement with information that can either be perceived from one single viewpoint, or one that requires the subsequent integration of spatial information from multiple vantage points during movement.

Understanding Spatial Abilities at Different Scales of Space

The previous chapter claimed that scale might indirectly influence spatial behavior by affecting the way how spatial information at different scales of space is processed. In other words, when treating scale not only as a spatial but also as a psychological construct, scale might have an impact on spatial thought, on the conscious reflection on the manipulation of spatial information. So far, the analysis of basic concepts of spatial cognition has led to this assumption, but it is not yet clear whether or not spatial information of different scales really imposes different demands on the information processing. This chapter explains how small-scale and large-scale spatial information is processed by the individual, and it identifies structural similarities and differences.

TVERSKY (2005) proposed two different approaches when it comes to studying spatial abilities from a cognitive perspective. One way to do so is to choose a bottom-up approach, that is, "studying the elementary representations and processes that presumably form the building blocks for more complex reasoning" (p. 211). The top-down approach aims to examine spatial abilities by disentangling the underlying cognitive processes and then addressing those ones that have a visuospatial basis. Both approaches have their advantages and disadvantages, and the decision to use one or the other depends on whether the spatial behavior induced by those abilities is a conglomerate of many different behavioral patterns, or whether it is limited to the extent that it can be observed and clearly described.

The primary goal of this chapter is to develop a cognitive model of small-scale and large-scale spatial abilities. It also has three other important objectives. First, it aims to identify possible sources of individual differences at both scales of space based on the models proposed here. By identifying the sources of individual differences, it might be possible to complement findings demonstrating that such differences exist (Chapter 4). Second, this chapter aims to provide theoretical insights that can be used to analyze and develop measures of spatial abilities. The cognitive analysis

© The Author(s), under exclusive license to Springer Fachmedien Wiesbaden GmbH, part of Springer Nature 2020
C. Heil, *The Impact of Scale on Children's Spatial Thought*, Studien zur theoretischen und empirischen Forschung in der Mathematikdidaktik, https://doi.org/10.1007/978-3-658-32648-7_3

suggests intended cognitive processes that have to be completed when solving a particular spatial task. Third, this chapter seeks to clarify whether spatial abilities at each scale of space should be understood as a unitary or multidimensional construct. In the latter case, it would be possible to propose models that describe spatial abilities at both scales of space.

This chapter is organized as follows: Section 3.1 provides more insights concerning spatial behavior at both scales of space. By describing the challenges that individuals face when engaging in small-scale and large-scale cognition, this section emphasizes that small-scale spatial behavior typically does not require the individual to make references to any kind of real space, which is unavoidable in the case of large-scale spatial behavior. Section 3.2 presents theoretical insights concerning the cognitive processes underlying small-scale spatial abilities using a bottom-up approach. There are three main sources of individual differences: the representation of small-scale spatial information, their mental transformation, and their maintenance in working memory. This section also identifies two distinct types of mental transformations when it comes to inference processes, that is, transformations of objects and the imagined self. Section 3.3 analyzes cognitive processes underlying large-scale spatial abilities using a top-down approach. The section describes how large-scale spatial information is represented and maintained in working memory. Describing processes involved in large-scale spatial cognition is, however, less straightforward. The section disentangles the underlying and most likely intertwined cognitive processes and shows that transformations of objects and the imagined self may also be involved when it comes to navigating real space with a map. Section 3.4 compares spatial abilities at both scales of space.

3.1   Spatial Behavior at Different Scales of Space

This section provides a more nuanced perspective on what might be spatial behavior at different scales of space. By now, this term was considered from a theoretical point of view (Section 2.2.4) and introduced as theoretical construct. It was assumed that spatial behavior is enabled by a set of cognitive abilities operating on a spatial representation. In line with both the definition of spatial abilities introduced in Section 2.4.1 as well as PERRIN-GLORIAN ET AL.'s (2013) model of spatial and geometrical thinking in the context of geometry education outlined in the introduction (Figure 1.1, p. 5), spatial abilities require the subject to master the interaction between physical and graphical space. This section uses examples of spatial behavior at both scales of space to identify the elements of graphical space and describes the relations that individuals have to establish to physical space. This section aims to show that because solving spatial tasks at different scales of space involves different types of spatial information, cognitive engagement is premised on an individual's ability to relate to the depicted spaces in a different way.

3.1.1   Object Manipulation in Small-Scale Spaces

Manipulating objects and figural stimuli is an important activity in the everyday life of children and adults (Figure 3.1). Manipulating information that can be perceived from one single vantage point is important during free play, when playing board games, during organization of work and private spaces (e. g. when grasping a cup or a toothbrush). In short, object manipulation in small-scale spaces is important whenever individuals manipulate and use objects as tools for everyday life. In other words, manipulating objects typically involves the active sensorimotor manipulation of objects in the surrounding environment, that is, primary uses of spatial information.

Figure 3.1 Free play in the sandbox, an everyday situation of manipulation of objects. *Source*: Pixabay

A special case of object manipulation in small-scale spaces are assembly instructions such as for LEGO buildings or IKEA furniture. Assembly instructions depict an object in a way that all spatial information required for construction can be perceived from one single viewpoint. A LEGO building, for example, is typically depicted from one perspective and the required LEGO bricks are depicted. The individual cannot directly perceive the position of some of the bricks but has to make inferences regarding that of others by cognitively engaging with the depicted situation. In a subsequent step, based on the information gained during reasoning, individuals begin to physically manipulate objects in the environment. Using assembly instructions is therefore an example for the secondary use of spatial information, involving manipulation of manipulable objects, that is, primary uses of spatial information.

Mental object manipulation in small-scale space might occur when individuals consult brochures, catalogues, or floor plans and start to reason about those, probably to wonder whether certain pieces of furniture can fit into a room or a certain spatial setting. In those cases, objects or object configurations in physical space are depicted, but the individual is not directly oriented toward them, and the outcome of the cognitive process is a mental one.

When individuals engage in pre-defined small-scale spatial tasks as, for example, those presented in text books or in riddles, they reason about spatial stimuli that are abstractions of spatial objects, object configurations or situations (understood as complete spatial scenes), which may be encountered in physical space (Figure 3.2).

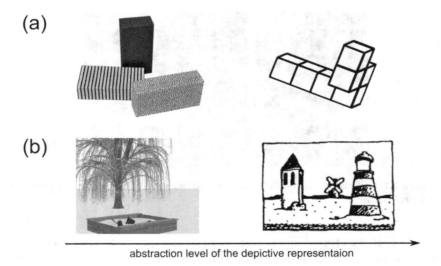

abstraction level of the depictive representaion

Figure 3.2 Two-dimensional depictions of spatial objects or configurations of those (a) and spatial situations (b) that may be encountered in physical space. Those depictions feature lower or higher levels of abstraction

To solve pre-defined tasks using small-scale spatial abilities, the individual can perceive and process all the necessary information on spatial relations from one single point of view (e. g., while sitting in front of a figural stimulus). This does not mean that individuals perceive all the spatial relations that are inherent in the depicted situation, but they can make inferences on the missing information based on the spatial relations provided in the task. While doing so, individuals have to imagine the corresponding situation for a specific purpose, spatial thought, but they actually do interact with the depicted space to assign a spatial meaning to what is depicted in the task.

As outlined in Section 2.2.4, individual's spatial behavior is always a reflection of their subjective conception of a situation or a pre-defined task. This might depend on previous knowledge, probably stored in long-term memory or in schemata (see Section 2.3.3), but also on the their level of cognitive ability. Since solving small-scale spatial tasks does not require the individual to make references to the depicted space, this conception of the situation can be understood to be rather static while completing the task. After all, consulting the available spatial information that is depicted does not considerably alter an individual's conception of the task as such.

Relating these insights to PERRIN-GLORIAN ET AL.'s (2013) model, one could argue that the particular characteristic of small-scale spatial behavior and the underlying class of abilities is that both the information available and the possible response behavior are situated only in graphical space. Although the elements contained in that space (the available pieces of information) refer to physical space, direct interacting is not necessary when engaging in small-scale spatial cognition.

3.1.2   Map-Based Navigation in Real Space

Getting from A to B is one of the fundamental spatial problems solved by human beings and mobile animals. Moving in space is essential for getting food, meeting friends and family, for reaching houses or shelter and other resources that are important bases of everyday life. Since 'getting from here to there' is so essential, navigation in real space is "the perhaps most prominent real-world application of spatial cognition" (WIENER, BÜCHNER, & HÖLSCHER 2009, p. 1). The term navigation can thereby be defined as a "coordinated and goal-directed movement of one's self (one's body) through the environment" (MONTELLO 2005, p. 258). To do so, people frequently use maps, particularly in unfamiliar environments (WOLBERS & HEGARTY 2010).

Navigation with maps, also denoted as map-based navigation[1], is a "combined task of map-reading and navigation" (LOBBEN 2004, p. 270). Consequently, all sub-facets of navigation without maps, denoted as unaided navigation, become also meaningful in the context of map-based navigation. This section elaborates sub-facets of navigation, that are, locomotion, wayfinding, and maintaining orientation as identified by environmental psychologists (e. g., APPLEYARD 1970; DOWNS & STEA 1973). It further describes what characterizes large-scale spatial behavior.

Locomotion and Wayfinding

According to MONTELLO (2005), navigation is a conglomerate of two types of behavioral patterns, locomotion and wayfinding.[2] Locomotion is "the movement of one's body around an environment, coordinated specifically to the local or proximal surroundings" (MONTELLO 2005, p. 259). Locomotion refers therefore to movement in the space that is immediately accessible to the body's sensory and motor system. Common problems that are solved by locomotion alone are avoiding obstacles and barriers (see e. g. CUTTING, VISHTON, & BRAREN 1995), going through doors or corridors without bumping into the walls, or directed movement towards visible landmarks which are important objects in the surrounding environment. Locomotion refers therefore to primary uses of spatial information.

Wayfinding is the "purposeful movement to a specific destination that is distal and, thus, cannot be perceived by the traveler" (ALLEN 1999b, p. 47). Wayfinding requires a goal (a destination) which should be reached beyond the visible local environment (see also GÄRLING, BÖÖK, & LINDBERG 1986; OTTOSSON 1987, for further elaboration). Wayfinding requires the individual to cognitively engage with the available spatial information and is more than merely locomoting the surrounding space. It requires the subject to successfully engage with the varying spatial relations between the self and objects despite certain uncertainty that might arise due to movement to a distant and probably unknown goal (ALLEN 1999b).

According to ALLEN (1999b), wayfinding can have three functional goals: goal-directed travel to familiar destinations, exploratory wayfinding, and travel with the goal to reach an unknown destination. All three types reflect typical wayfinding tasks

---

[1]LOBBEN (2004) uses the notion of navigational map-reading which emphasizes that the spatial product, the map, is in the focus of research. Since this study is concerned with spatial thought such as during navigation, which is mediated by maps, the term map-based navigation is used.

[2]WIENER ET AL. (2009) proposed a far more differentiated taxonomy (hierarchical structure) of unaided navigation on the basis of the study of MONTELLO (2005) and the landmark-route-configuration-model of SIEGEL AND WHITE (1975). As the current study focuses on map-based-navigation, their framework will not presented in detail here.

in everyday life, but travel to familiar destinations is probably the most encountered in society. Travel to familiar destinations refers to the travel between known places such as home and school or home and work. Exploratory travel is a common wayfinding activity for all mobile species and refers to the self-directed exploration of the environment with the goal to return to the starting point (the origin) after moving. Whereas the first two types of wayfinding, can be mastered without using additional information on the spatial information encountered during movement[3], wayfinding to unknown destinations involves the arrival at a goal state through consultation of any symbolic information such as maps or verbal directions (e. g., GALLISTEL 1990; MEILINGER & KNAUFF 2008).

Maintaining Orientation and a Sense of Direction
Wayfinding requires the individual to maintain orientation while moving, that is, "as we move, we maintain a sense of where we are relative to our goal, where places and objects we should avoid are located, and so on" (MONTELLO 2005, p. 264). Orientation has further been described as "our awareness of the space around us, including the location of important objects in the environment" (HUNT & WALLER 1999, p. 4). Being aware of one's position can therefore be considered as a prerequisite for successful wayfinding, which, in turn, further requires the abilities to determine one's heading and way-planning or decision-making how to reach a distant location in space (e. g., GÄRLING & GOLLEDGE 1989).

There are numerous objects and locations to which the individual can be oriented, and, consequently, disoriented. Being aware of a subset of the set of all possible orientations towards all objects and locations is sufficient for wayfinding. An individual is said to be oriented when knowing their own location relative to important features in the environment. Those might be subjectively chosen or objectively predefined by a task. Individuals that remain oriented are said to have a good sense of direction, that is, "knowledge of the location and orientation of an organism's body with respect to facing direction (heading) and the location of significant nearby or distantly perceived (or memorized) features (landmarks)" (GOLLEDGE 1999, p. 33). Whenever an individual suffers from knowing the own location in space, the individual is said to be disoriented and is getting lost.

---

[3] ALLEN (1999b) identified six possible means of accomplishing those two kinds of tasks (oriented search, following a marked trail, landmark-based piloting, path integration, habitual locomotion, using a cognitive map) that make, more or less, use of a particular spatial knowledge acquired from the spatial information that the individual engages with during movement (MONTELLO 2005). The spatial behavior to accomplish any of the first two wayfinding tasks by any of those means is denoted as unaided wayfinding. Since this study involves the use of maps, it is not further elaborated.

Using Maps in Navigation

Maps have been classified as external (LIBEN 1999), indirect (MONTELLO ET AL. 2004), and secondary (PRESSON & HAZELRIGG 1984) sources of spatial information (but see MONTELLO & FREUNDSCHUH 1995, for a complete review on different sources of spatial learning). Maps depict spatial information in a visual, simultaneous, highly symbolic, precise and metric manner (MONTELLO & FREUNDSCHUH 1995). They therefore familiarize the individual with an unknown or poorly known environment:

> Maps [...] eliminate the need for familiarity. They explicitly and immediately provide a survey representation of the global structure of an environment, and reveal spatial relationships that may not have been realized from direct experience. (MONTELLO ET AL. 2004, p. 273)

Those pieces of spatial information, however, become only meaningful in navigation if the individual relates the map to the depicted reference space, that is, when deriving "real-world meanings of the map that constitute objects of thought" (OTTOSSON 1987, p. 47). To successfully navigate large-scale spaces with the help of maps, an individual has therefore to refer to the space that is abstractly represented on the map. Spatial information acquired and processed while moving with the map in referent space is sequential, rather fragmented, visual, and three-dimensional and involves also vestibular and kinesthetic pieces of information (MONTELLO & FREUNDSCHUH 1995). All those pieces of information need to be related to the one acquired from the map. Consequently, the individual cannot perceive all the spatial information needed from one single vantage point but has to integrate pieces of information from multiple vantage points and visual sources. To successfully use maps in navigation, in particular to maintain orientation, the individual, then, has to interact with the space that is represented on the map.

Every large-scale spatial behavior reflects an individual's conception of an encountered situation or task, that is, his or her subjective understanding of aspects of these, thus involving aspects of the perceived spatial information of the map and the referent space. Since the individual moves in the environment and has to constantly assign a meaning to the spatial relations represented on the map, each consultation of it, and each processing of spatial information of the environment dynamically and continuously change the individual's conception of the encountered situation or task:

> The conception is continuously and dynamically developed during the accomplishment of the task, and the map-reader's consultations of the map and continuous encounter

with the terrain provide the basis for this development. Each delimitable action taken is an expression of the current conception [...] (OTTOSSON 1987, pp. 46–47)

Relating these insights to PERRIN-GLORIAN ET AL.'s (2013), one could argue that the particular characteristic of large-scale spatial behavior, and the underlying class of abilities, is that one part of the available information, the map, is situated in graphical space. The resulting behavior, however, requires the individual to directly interact with the space that is depicted when using large-scale spatial abilities.

Conclusion

Small-scale and large-scale spatial behavior does not only differ with respect to the way in which potential solutions to pre-defined spatial tasks might vary with respect to their application in day-to-day life, but also concerning the involvement of the physical space. Although at both scales of space, certain objects, configurations of those, or spatial situations are simultaneously depicted in an abstract format (objects, figures and maps) in graphical space, there are qualitative differences when using this spatial information. In the case of small-scale spatial behavior, no reference to the depicted physical space is necessary because all spatial information that is necessary for further inferences is presented. In the case of large-scale spatial behavior, individuals have to assign meaning to the map by relating it to the depicted referent space. By referring to the real-world meanings of the spatial relations depicted on the map and by additionally consulting sequential spatial information from the surrounding space, individuals are able to orient themselves and to navigate. Scale, then, affects the interaction of the individual with space and the extent to which an individual's conception of a spatial situation or task is static or dynamic.

3.2   Cognitive Analysis of Small-Scale Spatial Abilities

This section analyzes the cognitive processes that are involved in small-scale spatial abilities. Several researchers have proposed to study them using a bottom-up approach, thus trying to identify the cognitive components that may contribute to solving a particular task, and to study sources of individual differences that may emerge from these components (e. g., PELLEGRINO & KAIL 1982).

Sections 2.2 and 2.3 established a basic relationship between physical space containing spatial information, the corresponding spatial representation and spatial behavior. In line with HEGARTY ET AL. (2006), this study assumes that there are three sources of variance among individuals when processing small-scale spatial information. Individuals might vary concerning the abilities (1) to encode spatial information into a spatial representation; (2) to maintain the representation in working memory during the solution process; and (3) to perform mental transformations[4] that allow them to draw inferences from the spatial information to stimulate a certain spatial behavior (Figure 3.3).[5]

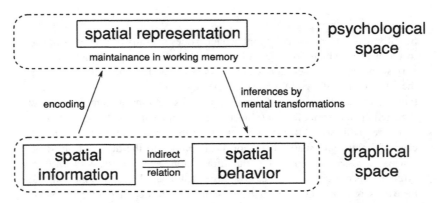

Figure 3.3 Three assumed sources of variance: encoding, maintenance in working memory, and making inferences by mental transformations

The first two sources of variances, encoding of spatial information in a spatial representation and maintenance of these in working memory, refer to unconscious use of spatial information, that is, LIBEN (1981) spatial storage. The third source of variance, making inferences by means of mental transformations, refers to conscious use of spatial information during reflection or mental manipulation of this kind of information, that is, LIBEN (1981) spatial thought. Both conscious and unconscious

---

[4]The notion of mental transformation is used to address cognitive processes that rely on the transformation of mental images.

[5]When taking a cognitive perspective on spatial abilities, one also has to examine representations and mental transformations. Both concepts have been described as being intertwined in a complex manner, thus being rather a single construct (e. g., J. R. ANDERSON 1978). A theoretical distinction between representations and transformations seems an adequate way to study cognitive processes (see also TVERSKY 2005). This approach is chosen in this study.

uses are important when an individual engages in spatial cognition (see Section 2.4). This section addresses all three sources, but since this study focuses on the conscious use of spatial information, this chapter summarizes findings concerning encoding and maintenance of spatial information and provides more nuanced insights concerning sources of individual differences when it comes to mentally transform spatial information.

This section is organized as follows. In a first step, it summarizes the basic idea of how small-scale spatial representations can be understood or modeled. In a second step, this section delves into the understanding of two types of mental transformations, namely transformations of objects and of the imagined self. It further describes how mental transformations relate to spatial updates of frames of reference. Since some of the terminology introduced here will be needed to describe the impact of working memory, the maintenance of spatial information in working memory is summarized in the third and last step.

3.2.1   Representation of Small-Scale Spatial Information

Humans represent spatial stimuli as spatial representations whenever the stimuli is absent or requires further processing using multiple frames of reference (e. g., ZACKS & MICHELON 2005). There is no doubt about the existence of this introspective aspect of human thought, mental images, and its importance in a wide range of spatial settings (PALMIERO ET AL. 2019). Despite the evidence for mental images and their involvement in spatial cognition, the format of mental images of adults is an ongoing subject of debate that has not been resolved until today (but see PALMIERO ET AL. 2019; TYE 2000, for an overview). Moreover, this issue is not well addressed for children (WIMMER, MARAS, ROBINSON, & THOMAS 2016). The traditional imagery debate has discussed two controversy ideas about the format of internalized small-scale spatial information of adults (PALMIERO ET AL. 2019). Whereas KOSSLYN (1980, 1994) and other researchers have suggested that mental images are depictive in format because they are the constituting parts of visual imagery (as cited in PALMIERO ET AL. 2019, p. 1), PYLYSHYN (1973, 1979, 1981, 2002) argued that mental images are descriptive representations that are closely related to abstract, language-like format (as cited in PALMIERO ET AL. 2019, p. 2).

KOSSLYN (1980, 1994, as cited in PALMIERO ET AL. 2019, p. 1) and co-workers argued for the existence of depictive representations by demonstrating an analogy of the mental transformation process for imagery and perception in a series of experiments on mental scanning and mental rotation processes (see TYE 2000, for further elaboration). They argued that mental rotation, for example, "corresponds to a physical rotation in the sense that the processes (whatever they may be) that occur in the subject's brain during a mental rotation have much in common with the processes that go on in the subject's brain when he or she is actually perceiving the same object physically rotating" (SHEPARD & PODGORNY 1978, p. 223). They concluded that mental images must be analogue, picture-like representations of the spatial stimuli. They have further assumed that images are stored in such a way that geometric properties and all other inherent properties of the visual stimuli are maintained (see PALMIERO ET AL. 2019, for further elaboration). These ideas were doubted by PYLYSHYN (1973, 1979, 1981, 2002, as cited in PALMIERO ET AL. 2019, p. 2) who assumed that cognitive structures representing a symbol system would 'translate' perceptual states into non-perceptual states. Those represented information, now not linked with perception any more, would become amodal and be integrated into a representation based on propositional relations.

WIMMER ET AL. (2016) recently argued that children might represent small-scale spatial information in a depictive format, thus preserving metric properties in their mental images. In their study, they found 5-year-old children to show effects of analogy between perception and mental imagery in a mental scanning experiment.

Although the classical imagery debate did not provide any definitive conclusions, recent studies, however, have provided other theoretical explanations of imagery, for example approaches based on embodied cognition that emphasized the close link between perception, cognition, and action (PALMIERO ET AL. 2019). Researchers stating that imagery is embodied emphasized that cognition always involves the brain and the body of the individual. When they interact with the environment, they make sensorimotor experiences and consequently, cognition itself should be grounded on the body. Based on these ideas, imagery could then be explained as simulated movement of the own body or other objects represented (see PALMIERO ET AL. 2019, for further elaboration on those approaches).

Individual Preferences in the Processing of Spatial Information
The literature on adults (e. g., KOZHEVNIKOV 2007) and children (XISTOURI & PITTA-PANTAZI 2011) has emphasized that the extent to which perceptual and motor components are integrated in the format of mental images does not only depend on an individual's abilities to form such mental images but, among others, a subject's individual preference in the processing of spatial information. Those can be described as a "psychological dimension representing consistencies in an individual's manner of cognitive functioning, particularly with respect to acquiring and processing information" (KOZHEVNIKOV 2007, p. 464). They can further be understood as 'stable attitudes' or 'habitual strategies' that govern the perception, encoding and thinking process (p. 464), and can therefore further be seen as "cog-

nitive principles that underlie complex behavior" (KOZHEVNIKOV ET AL. 2005, p. 710). Several studies have proposed that individuals may be classified according to a visualizer–verbalizer dimension.[6] Visualizers mainly rely on imagery when solving cognitive tasks whereas verbalizers use verbal-analytical strategies (see KOZHEVNIKOV ET AL. 2005, for further elaboration). Recent models of individual preferences in the processing of spatial information include a second dimension to that distinction to emphasize that there are two qualitative different types of visualizers; those using imagery to construct high-quality mental images of objects and their characteristics, object visualizers, and those using imagery to represent spatial relations, spatial visualizers (KOZHEVNIKOV ET AL. 2005). These findings are in line with the neurocognitive literature concerning the processing of visuospatial information that suggested two distinct neural processing systems; one for object features, the 'what'-stream, and one for localization and spatial information, the 'where'-stream (see KOZHEVNIKOV ET AL. 2005, for further elaboration).

In line with the findings of WIMMER ET AL. (2016) and in line with what has been proposed by K. CLEMENTS (1982) in the context of mathematics education (p. 36), this study assumes that small-scale spatial information is mentally represented in depictive format, and in line with some ideas presented by PALMIERO ET AL. (2019), considers mental imagery as a dynamic and probably embodied mental representation that adapts according to individual factors.

3.2.2   Mental Transformations of Small-Scale Spatial Information

There are many possible mental transformations to make inferences from mentally represented small-scale spatial information. Subjects may mentally alter size, shape and position (e. g., CARPENTER & JUST 1982), they may twist, rotate, fold and bend spatial stimuli (e. g., SHEPARD & COOPER 1982). Those myriad of possible transformations can be subsumed into two types of mental transformations of small-scale spatial information. On the one hand, subjects can mentally transform objects[7] by changing their orientation by rotation, changing the scale by shrinking or expanding, or folding objects. On the other hand, subjects can mentally transform their imagined self by taking a new perspective on the stimuli using rotation and translation (e. g.,

---

[6]Whether such individual differences exist or not remains, however, a subject to intense debate (e. g., PASHLER, MCDANIEL, ROHRER, & BJORK 2008).

[7]The expression 'objects' refers to flat, two-dimensional figures, and haptile, three-dimensional objects.

TVERSKY 2005). In both cases, transformations may be performed on the whole stimulus array (if consisting of different objects), or operations may be performed on parts of the stimulus array. Research on mental manipulation of objects dates back to the imagery debate (see TYE 2000), and research on perspective taking was substantially influenced by the research of PIAGET AND INHELDER (1948/1956). This section summarizes findings concerning both types of mental transformations from an experimental psychology point of view, thus referring to studies of reaction times and accuracy given a certain stimuli.

Transformations of Objects

SHEPARD AND COOPER (1982) proposed that there are three different types of transformations of objects: rigid, semi-rigid and nonrigid. According to them, rigid transformations refer to rotations, translations or reflections of visual stimuli, thus preserving the internal structure of the spatial information represented. Semi-rigid transformations alter some of the internal parts of the stimulus array by means of folding or (dis)assembling. Non-rigid transformations change the internal structure of the represented stimuli by bending, deformation, elongation, compression, or even melting (see also PLATH 2014; QUAISER-POHL 1998). This study refers to the first two types of transformations since those have been shown to be structurally similar and have been well addressed in the literature (HARRIS, HIRSH-PASEK, & NEWCOMBE 2013) but different to the third type of transformations (ATIT, SHIPLEY, & TIKOFF 2013).

SHEPARD AND METZLER (1971) studied mental rotation of small objects constructed of ten solid cubes in a sneak-like three-dimensional structure. Adults solved 1600 mental rotation tasks given in pictures that consisted of deciding whether or not two objects were identical ('same pair') or different ('different pair'). 'Same pair' items differed with respect to whether an object was rotated in the picture plane (picture plane pairs) or in the depth (depth pairs) (Figure 3.4, left). To emphasize that mental rotations can be interpreted as being continuous with respect to the rotation angle, SHEPARD AND METZLER (1971) included 'same pairs' with varying degrees of rotation from 0 to 180 degrees. Results showed that mean reaction times for 'same pairs' were linearly dependent with the increasing angle of rotation (Figure 3.4, right). The reaction time profiles were similar for the picture plane and depth conditions, which indicated that none of the two rotations took longer than the other one. SHEPARD AND METZLER (1971) interpreted the resulting linear relationship, also denoted as angular disparity, as an indicator that the mental transformations involved in mental rotation are analogous to the perception of a manual rotation of an object.

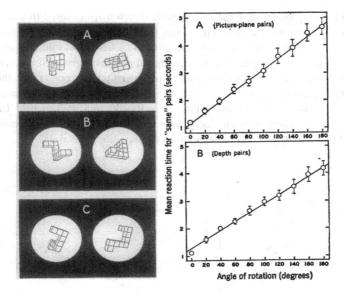

Figure 3.4 Shepard-Metzler-mental rotation experiment. (left) Examples of pairs: (A) picture plane pair, (B) depth pair, (C) different pair. (right) Linear increase of reaction times with increasing rotation angle with the same profile for depth and picture plane stimuli (in SHEPARD & METZLER 1971, p. 702). ©1971 AAAS. Reprinted with permission

Subsequent studies confirmed that angular disparity cannot only be found in mental rotation processes of three-dimensional stimuli but also two-dimensional stimuli (e. g., PELLEGRINO & KAIL 1982). Recent studies have confirmed that children solving a two-dimensional mental rotation task also show linear increasing reaction times with respect to the angular disparity (FRICK, DAUM, WALSER, & MAST 2009).

Angular disparity has also been found when it comes to mental scanning of an image. KOSSLYN, BALL, AND REISER (1978) studied mental scanning of an island image and emphasized that mental movement within the image is analogous to manual movement.[8] Adults were shown a map of an arbitrary island that involved seven salient landmarks[9] (Figure 3.5, p. 72, left). Subjects were first told to study

---

[8] Another interpretation of the experiment is that mental images preserve metric relationships that are given during perception.

[9] By now, landmarks can be understood as important features of the environment (e. g., SADALLA, BURROUGHS, & STAPLIN 1980).

and learn the map in detail. Experiments with sketch maps ensured that they had internalized the map with sufficient precision. Once the learning phase has finished, subjects were told to picture the map in their head. A first word told a location on the map which they had to imagine. Then, a second place was given. Subjects had to imagine to move along the straightest path from the first to the second place, thereby performing a process denoted as mental scanning. Reaction times were measured for each of the 84 word-pairs that were given. Results indicated that the time to mentally scan an image from one location to the other was linearly dependent on the actual physical distance of both locations (Figure 3.5, p. 72, right).

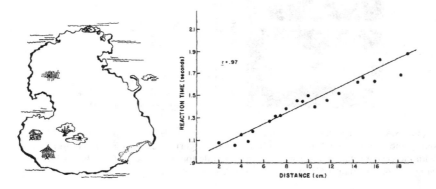

Figure 3.5 Mental scanning experiment. (left) Map with salient landmarks. (right) Linear correlation between actual distance of two locations and time needed for mental scanning between the two of them (in KOSSLYN, BALL, & REISER 1978, p. 51). ©1978 APA. Reprinted with permission

It has also been shown for children that reaction times increase in function of the mental distance to scan. WIMMER ET AL. (2016) presented a fictional map of an island with different landmarks and signposts to children. The children then completed a task asking them to mentally scan different distances. 5-year-old children showed a linear increase in reaction times in dependence of imagined distance.

A preliminary conclusion of the studies outlined is that mental manipulation of objects is a mentally equivalent process to the perception of manipulation of objects. Consequently, mental operation of depictive images resembles the actual physical activity (see also SHEPARD & PODGORNY 1978, for further elaboration). This isomorphism has been shown for other mental transformations of objects, for example scaling of objects, in which the time to compare two different rectangles has been shown to increase as a function of the the ratio of sizes between the two of them (BUNDESEN & LARSEN 1975). Moreover, it has been shown during paper-folding tasks that there is an equivalence between the physical operation of folding and the purely mental analogue activity. Indeed, reaction times of folding were shown to correlate linearly with the number of folds to be performed (SHEPARD & FENG 1972). The authors of that study, however, argued that mental paper folding does not only involve the mental simulation of a physical operation, the folding, but also analytical reasoning about the given stimulus. Paper folding has therefore been described as being a task that is cognitively more demanding than mental rotation (see also LINN & PETERSEN 1985). Linear increases in response times have also been replicated for 3D mental folding tasks in studies with adults (HARRIS, HIRSH-PASEK, & NEWCOMBE 2013).[10]

To sum up, mental transformations of objects are characterized by cognitive processes that are isomorphic to the perception of the corresponding physical manipulation of objects. Although past research on transformations of objects has dealt with a few measures of rigid mental transformations such as mental rotation, it has been shown that this assumption holds true even for semi-rigid transformations such as paper folding.

Mental Transformation of the Imagined Self
Studies that addressed mental transformations of the imagined self emphasized that those are governed by the individual's self-centerbeen observed in Figure 12.1 (p. 352) and Figure 12.2 (p. 355), both OB and EGO predicted the lowest extent of variance in SurveyMap (15.7 predicted about a quarter of variance in Prod (26.2 (52.2 According to the assumptions outlined in Sectioned imagery of motor processes. SAYEKI, COLE, ENGESTROM, AND VASQUEZ (1997) for example, studied mental rotation with objects similar to objects in SHEPARD AND METZLER (1971) study. Some subjects, however, were suggested a 'body analogy' that "was not explicitly stated; subjects were shown only a single slide of a block with a human head placed at a proper position" (SAYEKI ET AL. 1997, p. 98). Adding a head on the object made appear the figures similar to a schematic depiction of a human body with an outstreched arm. Subjects that were presented the 'body analogy' before solving 'normal' mental rotation tasks showed a significant reduction in reaction times. The corresponding reacting time-curve was found to be almost flat, and error rates remained stable during the whole experiment. One possible explanation for the

---

[10]Studies with children that addressed paper folding (e. g., HARRIS, NEWCOMBE & HIRSH-PASEK 2013) showed that the abilities to solve these tasks emerge quite early in children's development, but did not yet address whether similar reaction time-patterns may be found.

findings might be that subjects in the 'body analogy' group benefited from motor imagery, that is, the imagery of embodied self-centered movements.[11]

The finding that mental rotations might involve self-centered motor imagery was further supported by a series of studies by PARSONS (1987a) and PARSONS (1987b). PARSONS (1987b) studied reaction times in determining whether a given hand was a left or right one. Results showed that there was a correlation with the actual time taken to move a hand in the right position. The author concluded that "subjects imagine moving their own corresponding body part to the orientation of a stimulus to compare imagined and external body parts" (PARSONS 1987b, p. 220). In another study, PARSONS (1987a) investigated reaction times of bodies with an outstretched arm. Subjects had to determine whether the left or right arm was outstretched for various orientations of the body among the three canonical planes of mathematical space. Functions of reaction times did not show a linear increase in dependence of angular disparity for all conditions. For angular rotation around the vertical body axis, reaction times were independent of the varying orientation.

Results of PARSONS's (1987a) can be interpreted as first evidence that no patterns of angular disparity can be found when individuals solve tasks that require them to manipulate the imagined self. Indeed, reaction times for tasks requiring the subject to take another perspective have been reported to remain constant at small angles of transformations of the self with a sudden increase occurring at turning angles of about 60 to 90 degrees. At this point, angular disparity can be observed, thus reaction times start to linearly increase with increasing self-rotation angle (e.,g., KEEHNER, GUERIN, MILLER, TURK, & HEGARTY 2006; MICHELON & ZACKS 2006).

Consequently, in contrast to transformations of objects, transformation of the self may not be appropriately described as an isomorphism between perceiving the manipulation of an object and imaging the scenario (KESSLER & THOMSON 2010). Instead, other theories have been proposed to explain which cognitive mechanisms might underlie transformations of the self. KESSLER AND THOMSON (2010), for example, suggested that individuals solve perspective tasks involving low angles by applying a simple visual matching process since the perspective to take is close to the egocentric one. In contrast, they proposed that tasks with angles above 60 to 90 degrees are solved by mental self-rotations based on endogenous emulation of a body, which can be understood as emulated movements that the subject has experienced in real space and subsequently internalized.

---

[11] Another explanation would be that the 'body analogy'-group benefited from the additional spatial cue that might have helped during mental rotation. This interpretation was, however, not in line with research on additional spatial cues in mental rotation presented at this time (see AMORIM, ISABLEU, & JARRAYA 2006, p. 329, fur further discussion)

Problems of taking another perspective typically require individuals to master conflicting frames of reference (e. g., HUTTENLOCHER & PRESSON 1973; NEWCOMBE & FRICK 2010; YU & ZACKS 2017). When solving a task that requires them to take another perspective, there is a conflict between their physical egocentric frame of reference, that is, the initial, primary egocentric reference frame that a subject takes when being presented such a task, and the projected egocentric reference frame, that is, the egocentric reference frame that corresponds to the one of the imagined transformation of the self in the tasks. To do so, individuals have to reduce the sensorimotor information that is given by their primary physical orientation in space to imagine another perspective (see also PRESSON 1987). The conscious control of conflicting frames of reference has been demonstrated to involve higher-order cognitive processing that guide controlled adjustment of the initially taken egocentric perspective in the respective task. EPLEY, MOREWEDGE, AND KEYSAR (2004) suggested that egocentric biases occur whenever there are not sufficient cognitive capacities to fulfill this task:

> Because people view the world through their own sensory organs, one's own perspective is readily accessible and generated automatically while another's perspective requires additional controlled processing. Adults and children do not appear to differ in the automatic processing of their own perspective but do differ in the controlled adjustment required to accommodate another's differing perspective. Egocentric biases diminish but do not disappear because such corrective procedures are not always activated, and when they are, tend to be terminated pre-maturely. (p. 765, emphasis in original)

FRANKLIN AND TVERSKY (1990) proposed that individuals manage this cognitively effortful activity with the help of spatial frameworks. Spatial frameworks are particular types of spatial representations based on egocentric frames of reference. According to the framework, the speed to locate an object with respect to the egocentric framework is based on schematic knowledge about canonical axes (dimensions) of a person interacting with space (YU & ZACKS 2017). In particular, the ease to locate an object is dependent on which of the axes the individual has to reason by their current body posture. That is, for an upright subject, the basic position is being upright, thus making the head-feet axis most accessible. This might be explained by the gravity that has an effect on the subject, but also the head-to-foot asymmetry that is important for interaction with the environment, for example when grasping objects (FRANKLIN & TVERSKY 1990; YU & ZACKS 2017). The second accessible one is the front-back axis. It is linked to the front-back asymmetry of the human body that determines which actions are easier for subjects: objects in front of the body are easier to grasp than behind the body (FRANKLIN & TVERSKY 1990; YU & ZACKS 2017). Finally, the least accessible axis is the left-right one. It is not subject to a stable asymmetry such as gravity or front-back direction of the body because the latter can be modeled to be close to symmetric. The accessibility of the axes changes whenever a person has another position in space, for example, when lying down. According to FRANKLIN AND TVERSKY (1990) theory, the abilities to transform the imagined self to make inferences on the perspective of another person on an object configuration refers to the cognitive process of the formation of a new, emulated spatial framework that represents the projected perspective of the other person (see also YU & ZACKS 2017). Based on this framework and the corresponding canonical body axis, the individual is able to make inferences about the location of objects from another person's viewpoint.

Mental Transformations as Updates of Frames of References

This section explains the relation between mental transformations and frames of reference. As introduced in Section 2.3, individuals encode spatial stimuli in three types of frames of reference that are egocentric, intrinsic, and environmental ones.

According to ZACKS AND MICHELON (2005), mental transformations during spatial reasoning imply a permanent spatial update of the three frames of reference and their interrelations during mental transformations. According to their multiple systems framework, individuals encode spatial relations in multiple frameworks, and the pairwise relation between those frameworks constantly change when performing a mental transformation. Given that individuals can encode spatial stimuli such as an object array in three different frameworks, there are three different relations that might change; first, the relation between the egocentric and the environmental frame of reference, that is, the subject's heading in the environmental space. Second, the relation between the intrinsic and the environmental frame of reference, that is, the object array's position and orientation in the environmental space. Third, the relation between the egocentric and the intrinsic frame of reference, that is, the spatial relation between the subject and the object array.

According to their framework, transformations of objects, also denoted as object-based transformations or as object manipulation, require the subject to update each of the relations with the intrinsic frame of reference since they represent imagined rotations of objects. Their own perspective on the scenario remains unchanged, and consequently the relationship between the environmental and egocentric frame of reference remains unchanged. Transformations of the imagined self, also denoted as transformation of the egocentric perspective, require the individual to mentally update each of the relations with the egocentric frame of reference since they are rotations or translations of one's perspective. They require the subject to change the imagined perspective on a spatially invariant object array, and consequently the relationships between the environmental frame of reference and intrinsic one of objects in the environment remain unchanged (Figure 3.6).

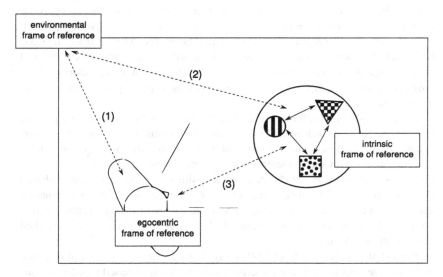

Figure 3.6 Update of relations between frames of reference. Object-based transformation: (1) remains fixed; (2), (3) are updated. Transformation of the egocentric perspective: (2) remains fixed; (1), (3) are updated.

YU AND ZACKS (2017) provided evidence that spatial transformation use is cognitively equivalent to spatial representation use when it comes to solving perspective taking tasks. More specifically, subjects draw upon cognitive processes relying on a task-specific spatial representation (the spatial framework) and a task-specific mental transformation (an egocentric perspective transformation). By considering reaction time profiles, the authors were able to show that whenever subjects used a perspective transformation to solve a given task, they build up a spatial framework. Conversely, whenever they build up a spatial framework to solve a task, they were using egocentric perspective transformations. Taking another person's perspective can therefore be interpreted to be a powerful cognitive computational mechanisms that allows individuals to rely on a schematic representation of the space around their bodies that is subsequently updated to take the desired point of view and make spatial inferences (YU & ZACKS 2017).

Towards Two Distinct Types of Mental Transformations

So far, this section has argued that object-based transformations seem to be qualitatively different from transformations of the self. Findings from the experimental (AMORIM & STUCCHI 1997; CREEM, WRAGA, & PROFFITT 2001; PRESSON 1982b; WRAGA ET AL. 2000; WRAGA, CREEM-REGEHR, & PROFFITT 2004; ZACKS ET AL. 2000), neuropsychological (WRAGA, SHEPHARD, CHURCH, INATI, & KOSSLYN 2005; ZACKS 2008; ZACKS & MICHELON 2005; ZACKS, RYPMA, GABRIELI, TVERSKY, & GLOVER 1999; ZACKS, VETTEL & MICHELON 2003) and object recognition literature (SIMONS & WANG 1998; R. F. WANG & SIMONS 1999) provided further evidence for the dissociation of these two types of mental transformations.

ZACKS ET AL. (2000), for example, directly compared the chronometric profiles of adults performing each type of mental transformations. They argued that differences in the neural processing of both transformations would imply differences in the efficacy curve for different orientation angles. Adults completed two tasks involving representations on human bodies: the same-different tasks tapped object-based transformation and the left-right task tapped egocentric viewpoint transformation abilities. In a first task, subjects had to determine whether both representations of a body were the same. In a second one, they had to determine whether the outstretched arm was the left or right.

Results of ZACKS ET AL.'s (2000) study showed a typical increase of response times with respect to increasing orientation angle in object-based transformations which is in line with classic behavioral data on mental rotation tasks outlined in Section 3.2.2. In contrast, the left-right task did not show effects of angular disparity (see Figure 3.7).

Those results have been confirmed when using other visual stimuli involving objects rather than humans in both tasks (SHELTON & ZACKS 2015). There is, however, a stimulus-dependence when providing figures of small objects versus pictures of the body. In the latter case, subjects tend to solve object-related tasks with object-based transformations whereas body-related tasks are solved by perspective transformations (ZACKS & TVERSKY 2005).

The Neuropsychological Literature   A range of studies examined brain activities of individuals engaging with both types of mental transformations. They showed that there are different processing resources involved at the neural level when individuals engage with either object-based mental transformations or egocentric perspective transformations (WRAGA ET AL. 2005; ZACKS & MICHELON 2005; ZACKS 2008), thereby using different stimuli for each of the transformations. In addition, ZACKS ET AL. (1999) investigated both types of mental transformations using the same mental stimulus to derive evidence for distinct underlying neural mechanisms.

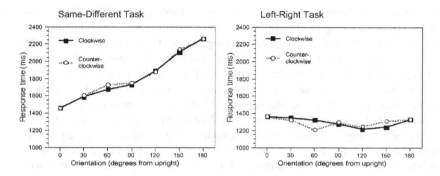

Figure 3.7 Chronometric profiles for both (left) object-based transformations and (right) egocentric perspective transformations (in ZACKS ET AL. 2000, p. 324). ©2000 Springer Nature. Reprinted with permission

Functional neuroimaging data analysis of an experiment in which subjects had to decide about 'Whose hand is it?'- questions on schematized human bodies emphasized that there are specialized cortical regions for each transformation. ZACKS ET AL. (1999) found that egocentric transformations led to an activation of the areas near the parietal-temporal-occipital (PTO) junction with a strong lateralization to the left, whereas object-based transformations led to activation in the posterior areas. A frontal activation in all tasks indicated that there is a set of processes, which is common for both mental transformations. This dissociation was also found when individuals reasoned about non-human visual stimuli (ZACKS ET AL. 2003).

In summary, this section showed that during processing of small-scale spatial information, the individual has to master two distinct types of mental transformations. Although both might be used interchangeably to solve the same kind of tasks, they are different in terms of the underlying processing resources, which leads to qualitative different behaviors in reaction time experiments.

3.2.3   Maintenance of Small-Scale Spatial Information in Working Memory

Process analysis of spatial abilities has shown that using them to solve a spatial task entails the encoding of the visuospatial stimulus and the maintenance in visual memory. PELLEGRINO AND KAIL (1982) suggested that "individual differences in spatial ability may emanate from fundamental representational and 'visual memory' skills that affect the total time and course of item processing" (p. 325). This section summarizes some empirical findings concerning the influence of working memory (WM), or more precisely, visuospatial working memory (VSWM, see Section 2.3.3).

VSWM capacity has been demonstrated to be related to spatial abilities using two approaches; dual-task and correlational studies (e. g., SHOLL & FRAONE 2004). In dual-task approaches, subjects perform two tasks that are intended to rely both on the maintenance of spatial information in VSWM and a comparison condition in which one task is intended to load on VSWM and one on another component, such as verbal memory. Whenever the corresponding analysis shows impaired performance in the first condition, results can be interpreted as an evidence for the same subcomponent of VSWM being involved (e. g., LABATE, PAZZAGLIA, & HEGARTY 2014). Correlational studies provide evidence by finding significant correlations between performances in the tasks to compare. They require, however, a precise definition of the constructs measured. This minimizes the potential overlap between the constructs and helps to interpret the correlational patterns carefully (e. g., PURSER ET AL. 2012).

The findings of studies using both dual-task and correlational approaches are converging; there is a strong evidence that individual capacity of VSWM is an important predictor of performances in solving small-scale spatial tasks (ZIMMER ET AL. 2010). There is a general agreement that VSWM is related to mental rotation tasks (e. g., LORING-MEIER & HALPERN 1999) and to visualization tasks such a paper folding (see the literature review in PLORAN, ROVIRA, THOMPSON, & PARASURAMAN 2015).

A positive relationship between perspective taking abilities and working memory capacity has been identified in children. VANDER HEYDEN, HUIZINGA, RAIJMAKERS, AND JOLLES (2017) found that WM significantly predicted perspective taking performances after controlling for age and intelligence. These findings, taken together with the literature on cognitive processes in solving perspective taking tasks (see Section 3.2.2), suggest that children's performances when controlling for conflicting frames of reference in perspective tasks are strongly related to their individual working memory capacity.

Focusing on adults, MIYAKE, FRIEDMAN, RETTINGER, SHAH, AND HEGARTY (2001) studied the relation between different subsystems of working memory and a set of spatial tasks both for quantifying the extent to which they relate and to determine important subcomponents of working memory that account for individual differences in different subclasses of small-scale spatial abilities. Results indicated that the abilities to maintain visuospatial representations in the mental workplace, VSWM capacity, related to three subclasses of small-scale spatial abilities, namely spatial visualization (as measured by paper folding tasks), spatial relations (rotation of 2D figures), and perceptual speed (identify identical figures or hidden graphical patterns) with correlations ranging from .54 to .69. However, correlations with the central executive (CE) were weight higher (.71 to .90), thereby indicating a strong involvement of executive functioning and controlled attention. When controlling for the correlation between CE and VSWM, CE emerged as an significant predictor of all three subclasses of small-scale spatial abilities, but to a different extent. Moreover, only in the subclass perceptual speed, VSWM made a significant extra contribution in explaining variance. The authors interpreted their findings as a working memory-based description of different subclasses of spatial abilities. According to them, all of these rely on executive functioning and maintenance of visuospatial information, but the tests that reflect those subclasses differ in their respective demands concerning the involvement of the central executive. Indeed, individual differences in spatial tasks, reflecting the subclass visualization typically involving multistep solution processes on complex stimuli (e. g., LINN & PETERSEN 1985), were highly predicted by differences in controlled attention, whereas tasks requiring apprehending and identifying a visual pattern were less well predicted by executive functioning. Since all spatial tasks involve the storage and maintenance of spatial information, the degree of involvement of CE seems to differentiate among the tasks. Consequently, at its broadest sense, these results suggested that spatial tasks may be aligned along continuum representing the involvement of CE.

The idea that spatial tasks might be classified along such a continuum dates back to studies in which researchers identified basic cognitive processes (denoted as components) that were involved when solving spatial tasks. After analyzing those tasks from this perspective, researchers separate the involved cognitive processes and analyzed their particular impact to account for individual differences in the whole tasks. As a result of these component analyses, a series of studies (see MUMAW, PELLEGRINO, KAIL, & CARTER 1984; JUST & CARPENTER 1985; PELLEGRINO & KAIL 1982) have demonstrated that the speed of processing spatial information accounted for individual differences in tasks involving 'basic' spatial transformation such as mental rotation with simple stimuli (all cited in HEGARTY & WALLER 2005, pp. 137–139). This was, however, not the case for more complex tasks that involved multistep solution processes and complex three-dimensional visual stimuli (LOHMAN 1982; MUMAW & PELLEGRINO 1984).

Taken together, those componential approaches have led to the identification of a 'speed-power'-continuum along which different spatial tests can be arranged after having performed an analysis of the constituting cognitive processes (see HEGARTY & WALLER 2005, p. 138, for further elaboration). Tasks that are on the 'speed'-side of this continuum are simple for most people, and to measure individual differences, those tasks need to involve an imposed time limit since they reflect the individual speed of processing. In contrast, items that are on the 'power'-side, that is, items involving multistep solution processes and complex visual stimuli measure individual differences in cognitive resources available. Those items cannot be solved by all individuals equally well, even though no general time limit might be given, and individual differences might be due to the 'power', that is, the level of the individual spatial abilities (see also HEGARTY & WALLER 2005).

One interpretation of these componential analyses is that for solving complex spatial tasks in particular, individuals might differ in terms of their capacity to encode the visual stimuli accurately and to continue to do so even after performing various transformations of these. Individual differences in tasks on the 'power'-side of the continuum have therefore been assigned to differences in working memory resources in a way that low-spatial individuals have less resources for storing and maintaining spatial information in memory than high-spatial individuals (see HEGARTY & WALLER 2005, for further elaboration).

Conclusion
Small-scale spatial abilities allow individuals to mentally represent a spatial stimulus in depictive format, to maintain it in visuospatial working memory, and to make inferences from the stored information by means of mental transformations that require the individual to reflect and update relations of frames of reference. Encoding and maintaining spatial information in working memory, and making inferences from the represented information, are sources of variance that might result in individual differences when solving a pre-defined small-scale spatial task. The cognitive analyses revealed two distinct types of mental transformations: mental manipulations of imagined objects (object-based transformations) and coordination perspectives of mental movement of the imagined self (egocentric perspective transformations). Whereas there is a structural isomorphism to perceptions of the physical operation for the former type of transformation, this is not the case for the latter one. Mental transformations of the imagined self correspond to the maintenance of an egocentric representation of the space around one's body during subsequent spatial updates of it, probably based on the mental emulation of previously experienced and embodied movements. Since both types of transformation require the individual to consciously manipulate and update frames of reference, executive functioning might allow them to control attention, in particular when solving more complex spatial tasks that require multistep solution processes.

3.3   Cognitive Analysis of Large-Scale Spatial Abilities

Large-scale spatial abilities enable spatial behavior such as map-based navigation, a conglomerate of map-reading and navigation, and therefore a complex
information-processing task (MACEACHREN 1995). CRAMPTON (1992)[12] emphasized that wayfinding with the help of maps is "a profoundly psychological activity, and that in order for researchers to grasp the essence of what happens [...]
the cognitive side of the story is crucial" (p. 46). Although researchers such as
OTTOSSON (1987) provided models on how to conceptualize map-based navigation,
large-scale spatial abilities have been less systematically addressed than small-scale
ones (e. g., LOBBEN 2004). Recent studies have investigated map-based navigation
to understand how large-scale spatial information might be stored in memory (e. g.,
MEILINGER, FRANKENSTEIN, WATANABE, BÜLTHOFF, & HÖLSCHER 2015).

   This section addresses large-scale spatial abilities as a cognitive phenomenom. It
investigates them using a top-down approach by delving systematically into possible cognitive processes with a visuospatial basis that govern map-based navigation.
A starting point for this kind of analysis might be the classification suggested by
LOBBEN (2004). She identified five major processes that are important during map-
based navigation: (1) survey mapping, (2) object rotation, (3) symbol identification,
(4) path integration, and (5) map-environment interaction. According to her, survey
mapping can be understood as the process of encoding knowledge from the map-
mediated direct experience of real space. Object rotation refers to the cognitive
abilities that allow an individual to align the map with the referent space. Symbol
identification refers to the process of encoding and understanding that symbols represent objects in the environment. Path integration is an unconscious cognitive process
that allows individuals to maintain a sense of orientation, whereas map-environment
interaction refers to the constantly changing relationship that the map-reader has
to establish between the self, the map, and the environment. This study addresses
these processes and related ones, thereby classifying them into the different types
of spatial representations as proposed by LIBEN (1981).

   In a first step, this section addresses large-scale spatial abilities by considering an
individual's understanding of the map itself (i. e., spatial product according to LIBEN
1981). In a second step, theoretical insights are provided concerning an individual's
representation of large-scale spatial information in survey maps (i. e., spatial storage
according to LIBEN 1981). Then, this section summarizes the unconscious and con-

---

[12] A key finding of his study is that cognitive processes of wayfinding differ for novices and
experts. Since this study addresses children, the literature concerning experts has not been
considered in this section.

scious cognitive processes required to stay oriented during map-based wayfinding in real space, thus elaborating on object rotation and map-environment-interaction (i. e., spatial thought according to LIBEN 1981). In a last step, the impact of working memory is described.

3.3.1 Understanding Maps

In line with the definition proposed by LIBEN AND DOWNS (1989), this study understands maps as abstract, reduced, scale-dependent spatial representation of a chosen set of spatial relations of the referent space that makes use of symbols. The introduction argued that maps help the individuals to conceptualize the space they interact with, since they depict spatial relations simultaneously, which might influence how the map-using individuals understand the aspects of their surrounding environment. Consequently, there is no doubt that "[s]patial representations help us think" (LIBEN 2003, p. 45), even when used during complex behavior in large-scale space such as navigation.

Different propositions have been made about what individuals need to understand about maps. According to LIBEN (2003, 2006), individuals need to understand that maps comprise three general principles; the duality, the spatialization, and the purpose principle. The duality principle refers to the fact that maps do not only represent a referent space by standing for it using symbols. Moreover, maps are something haptic, most simply a flat sheet of paper, which does not mean that the referent space is also flat (LIBEN 2006).[13] The spatialization principle refers to the double spatial essence of maps in a way that they do not only represent the referent space but are also spatial in a sense of being in relation to space (LIBEN 2006).[14] The purpose principle refers to the fact that maps are not only representations of something but are made by the map maker with the purpose of showing something (LIBEN 2006). Maps may be used for wayfinding but can also be illustrative by depicting complex socio-geographic data (DAVIES & UTTAL 2007).[15]

---

[13]The duality principle is not only inherent to maps but all kinds of LIBEN's (1981) spatial products since it relates to the dual existence of those.

[14]The latter aspects refer to three characteristic dimensions of a map the point of view of the map maker (see LIBEN 2006). These include viewing distance (distance from which referent space is represented, implying the notion of scale), viewing angle ('bird's eye' vs. opaque views), and viewing azimuth (angular disparity from north).

[15]In its broadest sense, the purpose principle therefore refers to a map as a tool for communication between the map maker and the map user (OTTOSSON 1987). The intended purpose

According to MUIR AND CHEEK (1986), an individual needs the following abilities to successfully deal with maps: (1) interpreting symbols, (2) viewing perspective, (3) finding location, (4) determining direction, (5) calculating distance, (6)
computing elevation, (7) imagining relief and (8) understanding scale. There is
general agreement in the literature that interpreting symbols is a pre-requisite for
map use (e. g., LIBEN 2006; LIBEN & DOWNS 1989; OTTOSSON 1987). Interpreting
symbols refer to the individual's abilities to understand that symbols on the map represent non-flat three-dimensional objects in the environment. Viewing perspective,
that ist, finding both the one's location and those of landmarks, can be summarized
as spatial aspects of map use that require the individual to use the map to reason on
the changing spatial relations in the referent space, in particular during movement.
Calculating distance and understanding scale refer to geometric aspects of map use,
since they require the individual to use the map on time-invariant aspects of the
environment. Computing elevation and imagining relief refer to geographic aspects
of map use if the chosen map allows for that.

Maps have often been described as representing the perceived three-dimensional
referent space in a 'bird's eye' view (e. g., KASTENS & LIBEN 2007) or as a picture from above. Both expressions rely on the notion of perspective since they
imply that maps provide an aerial view on the referent space. To understand a map,
however, individuals have to understand that the map itself lacks perspective and
that the depicted spatial relations are independent of the existence of a perspective
(see OTTOSSON 1987, pp. 137–138, for further elaboration). However, maps may be
viewed from a certain perspective. In particular when an individual is located in the
map's corresponding referent space, the issue becomes relevent. When understanding this fundamental property of maps in addition to the symbol system, individuals
can start to reason about the spatial dualism of maps as outlined by LIBEN (2003)
as well as spatial or geometric aspects as proposed by MUIR AND CHEEK (1986).

3.3.2   Representation of Large-Scale Spatial Information

During map-based navigation, individuals encounter spatial information not only
from direct experience with the environment, but also from the map. That information needs to be mentally represented in a corresponding spatial representation.
Although much research has been devoted to studies of spatial representations focusing on the acquisition of large-scale information during unaided navigation, there

---

of a map may influence the spatial characteristics of the map that are articulated by the map
maker.

is little agreement on how to model the spatial representation resulting from spatial experience involving map use (e. g., LOBBEN 2004; MEILINGER ET AL. 2015). In line with what has been proposed by LOBBEN (2004), this study uses the terminological distinction between cognitive maps for the former spatial representation and the notion of survey maps when addressing the latter spatial representation. Since survey maps are particular cognitive maps, this section describes some general properties of cognitive maps at first and summarizes the findings when it comes to experiences in large-scale space involving maps in a second step.[16]

Cognitive Mapping

Multimodal stimuli are encoded in memory whenever an individual encounters large-scale spatial information. Besides visual information, the individual encodes information from the vestibular senses (acceleration information) and kinesthetic information (limb position, force, and movement) that results from movement (MONTELLO ET AL. 2004). Whereas the influence of vestibular information on the encoding of spatial information is not yet clear (WALLER, LOOMIS, & STECK 2003), kinesthetic experience has been shown to be influential (WALLER, LOOMIS, & HAUN 2004).

Encoding multimodal stimuli of the environment encountered during experiencing it has been denoted as cognitive mapping.[17] According to DOWNS AND STEA (1973), cognitive mapping can be understood as an ongoing process:

> Cognitive mapping is a process composed of a series of psychological transformations by which an individual acquires, codes, stores, recalls, and decodes information about the relative locations and attributes of phenomena in his everyday spatial environment. (DOWNS & STEA 1973, p. 9)

Cognitive mapping is a process that allows individuals to represent the subjective understanding of the spatial world that they experience during interaction (DOWNS & STEA 1977). A cognitive map[18] then is a cross section of the individual's representation of an environment at one particular instant of time. It is a "product in the

---

[16]In both cases, it can be assumed that individual preference in the processing of spatial information contribute to which extent and how individuals represent knowledge (e. g., KYRITSIS, M., GULLIVER, S. R., MORAR, S., & MACREDIE, R., 2009; TASCÓN, BOCCIA, PICCARDI, & CIMADEVILLA 2017).

[17]This process has further been denoted as spatial learning, spatial layout learning or environmental learning. Throughout this study, the term (spatial) layout learning may be used as a synonym to emphasize that cognitive mapping is a process.

[18]Other notions are imagery maps, mental maps, environmental image, spatial image and spatial scheme (SIEGEL & WHITE 1975). Cognitive maps have further been denoted as topo-

sense of transitory stages in an ongoing process" (DOWNS & STEA 1981, p. 149) and reflects the subject's conception of the environment as well as their understanding of subjectively important features of the environment.

The notion cognitive map was first introduced in the study of TOLMAN (1948). However, the metaphor of a 'map' is misleading.[19] Since learning the layout of the environment takes time and is not straightforward, cognitive maps as spatial representations of large-scale spatial information are highly schematic, distorted, fragmented and do not reflect metrical and topological relations of the corresponding physical space (e. g., DOWNS & STEA 1977, 1973; HARDWICK, McINTYRE, & PICK 1976; ISHIKAWA & MONTELLO 2006; MEILINGER 2007; MONTELLO 1992; SIEGEL & WHITE 1975; UTTAL 2000b). Consequently, they do not represent an isomorphic mental representation of the environment (e. g., S. L. COHEN & COHEN 1985). However, since cognitive maps are built while taking multiple vantage points in space, they are highly flexible, orientation-free spatial representations (EVANS & PEZDEK 1980; RICHARDSON, MONTELLO, & HEGARTY 1999). They are thus encoded in a way that allows the individual to make inferences on directions to subjectively important features in the environment in an easy way from any possible point of orientation in space (WALLER, MONTELLO, RICHARDSON, & HEGARTY 2002), encoded in several frames of reference (e. g., SHOLL 1992).

Landmark, Route, and Survey Knowledge   Individuals represent different kinds of knowledge when interacting with the environment. According to SIEGEL AND WHITE (1975), these are landmark, route, and survey knowledge. Landmarks are "unique patterns of perceptual events at a specific location, they are predominantly visual for human adults, they are the strategic foci to and from which one travels, and they are used as proximate or intermediate course-maintaining devices" (p. 23). Whenever individuals mentally represent landmarks, they acquire landmark knowledge. Landmarks have further been defined as "simple, declarative, and discrete form of environmental knowledge" (WALLER 1999, p. 3), as visually "distinct object or feature that is noticed and remembered" (PRESSON & MONTELLO 1988, p. 378) and have been emphasized to be important in navigation as they are "relatively better known and [...] serve to define the location of adjacent points" (SADALLA ET AL. 1980, p. 517). In summary, the notion of a landmark has been used in three ways. Landmarks can be (1) distinguishable features in the environment that are important

---

graphical schema (PIAGET & INHELDER 1948/1956) and topographical representation (HART & MOORE 1973).

[19]DOWNS AND STEA (1981) further criticized that the metaphor of a map implies that it must be some kind of fixed product of an internalization process rather than a process.

for navigational decisions; (2) distinguishable features in the environment that allow an individual to maintain orientation; and (3) salient information in memory tasks (see SADALLA ET AL. 1980; WALLER 1999, for further elaboration).

When connecting single landmarks by action, the resulting sequences of land-marks, routes, are "sensorimotor routines for which one has expectations about landmarks and other decision points" (SIEGEL & WHITE 1975, p. 24). The inter-nal representation of routes, denoted as route knowledge, can also be described as "an internal representation of the processes necessary for finding one's way from place to place" (MONTELLO ET AL. 2004, p. 270) which is also sequential by defini-tion. When goal-directed integration of route and landmark knowledge takes place, SIEGEL AND WHITE (1975) speak of survey knowledge or configurational knowl-edge that is formed. Survey knowledge is "a more flexible form of spatial knowledge […].[It allows] direct access to quantitative spatial relationships, such as distances and directions between arbitrary locations in an environment - not solely those locations between which one has traveled" (MONTELLO ET AL. 2004, p. 270). Both, route and survey knowledge are central for navigation without maps and support the navigation with maps. In contrast, landmark knowledge alone is not sufficient for successful wayfinding, as the individual does not know where to head after reaching the known landmark to reach another known one. The only advantage of landmarks is to avoid circular locomotion in an unknown environment as one would recognize the landmark in the case of moving in circles (MEILINGER 2007). Naviga-tion processes therefore require knowledge of at least some spatial relations among landmarks.

Models of Cognitive Maps From Directly Experienced Environments   Cognitive maps, whether or not being a metaphorical concept or a genuine cognitive entity, have been controversially debated in the scientific literature until today. Although there is general agreement that cognitive maps represent landmarks, routes, and configurations, there still remains a matter of debate how to describe cognitive map (thus referring to PALMER (1978) models of spatial representation), that is, how to model which knowledge is acquired and how knowledge about landmarks and routes is interconnected (e. g., MEILINGER 2008; WERNER & SCHMIDT 1999). Since cognitive maps have been emphasized to organize navigation in real space (e. g., DOWNS & STEA 1977; GALLISTEL 1990; MEILINGER 2008), there is general agreement that there must be some internal organization allowing the individual to adaptively behave in real space. There are three theoretical models of how knowledge could be represented in the brain (Figure 3.8).[20] It might be organized in a single coordinate system, in a graph-like representation or in an integrated network representation (MEILINGER 2008; WOLBERS & WIENER 2014).

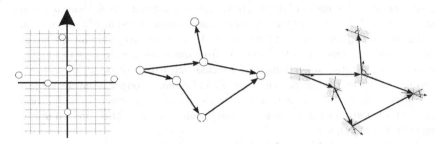

Figure 3.8 Three theories on the architecture of cognitive maps: (left) coordinate system (middle) graph-like (right) NRFT (in MEILINGER 2008, pp. 345 and 348). ©2008 Springer. Reprinted with permission

Theories that have supported the existence of one single coordinate system assume that individuals encode metric and spatial-relational information between different locations in a global coordinate system. This process is supposed to be supported by neural structures in the brain (GALLISTEL 1990; O'KEEFE & NADEL 1978).[21] Graph-like representations (e. g., MALLOT 1999; POUCET 1993) assume a set of independent units that represent landmark knowledge (MEILINGER, RIECKE, & BÜLTHOFF 2014). Each node represents a landmark in space. An edge represents the connection between those locations and can be interpreted as the action that is necessary to move from location A to location B (MEILINGER 2008; WOLBERS & WIENER 2014).[22]

---

[20]Discussing the architecture of cognitive maps returns to the question of which frames of reference are involved during encoding of large-scale spatial information (FILIMON 2015). The idea of a single coordinate system, for example, reflects the idea of an environmental frame of reference underlying the mental representation.

[21]O'KEEFE AND NADEL (1978) assumed the existence of place and head cells. Place cells fire differently, depending on the location of an animal whereas head cells represent the heading of the animal in a particular location. Both cell types are part of the hippocampus. They concluded that the hippocampus functions as a cognitive map (see HARTLEY & BURGESS 2002, for further discussion).

[22]This is actually a transfer of the idea of a network of intrinsic frames to the representational level (see the developmental theory of HART & MOORE 1973, on fixed and coordinated frames of reference).

Network of Reference Frame Theory (NRFT) (MEILINGER 2008, 2007) proposes that the "environment is encoded in multiple interconnected reference frames" (MEILINGER 2008, p. 347). Each frame of reference is a coordinate system in the sense of O'KEEFE AND NADEL (1978) with a particular orientation, and all of them form a graph.[23] The nodes of the graph are representations of single vista spaces in the sense of MONTELLO (1993)[24]. They contain landmarks and metric information and constitute the basic units of the network. Edges in the graph are interpreted as the necessary perspective shifts[25] to move from one frame of reference to an adjacent one. Perspective shifts consist of directed translations (egocentric 'vectors') and rotations towards another vista space and become stronger when the subject becomes familiar with an environment. Vista spaces and perspective shifts form abstract relations that are useful for planning paths and guide goal-directed movement in the environment.

In summary, graph-like and NRFT-like representations of cognitive maps explicitly explain the distorted and fragmented nature of large-scale spatial knowledge. NRFT states that there is, to some account, metric information stored in memory not relying on the argument of having a metric coordinate system in the mind. NRFT specifies both, route knowledge and metric information. Metric information is gained and stored out of vista spaces by perception processes that allow the individual to perceive depth and structure of a particular scene. NRFT has been empirically validated (MEILINGER, KNAUFF, & BÜLTHOFF 2008) and can be interpreted as a modern model for cognitive mapping. This study therefore assumes that the architecture of a cognitive map can be modeled in this way, with vista spaces, as they can be perceived from one single vantage point, being encoded similar to small-scale spaces.[26]

Survey Mapping From Map-Based Experiences Only   Although survey knowledge acquired by map experience (the survey map) could be thought of a replacement for configurational knowledge gained by direct experience, it has been shown that

---

[23]NRFT is therefore closely related to POUCET's (1993) idea of local charts (representing vista spaces) that are connected by polar coordinate vectors in a graph.

[24]There is empirical evidence that humans are able to encode vista spaces, the largest entities that can be perceived directly from one single vantage point, within one egocentric frame of reference in the cognitive map (see MEILINGER 2007, for further elaboration).

[25]This shift is different from the abilities to imagine shifts of the own perspective. It is stored in memory but is not a result of mental inferences. It connects encoded vista spaces and does not only operate on one visual stimuli (MEILINGER 2008).

[26]Again, individual preference in the processing of spatial information might affect how those are encoded by the individuals.

there are considerable qualitative differences between both spatial representations (FIELDS & SHELTON 2006; MONTELLO ET AL. 2004; RICHARDSON ET AL. 1999; SHELTON & MCNAMARA 2004; SHELTON & PIPPITT 2007; THORNDYKE & HAYES-ROTH 1982). Indeed, direct experience of the environment has been described as being dominated by the visual-perceptual experiences that the observer makes from a ground-level view. That is, all perceived orientation changes with respect to the canonical axes of the observer (in particular left/right and front back). There is converging agreement in the literature that these experiences are at least partly encoded with respect to the observer's body by using egocentric frames of reference. Consequently, encoding of subsequent ground-level information has been shown to activate brain areas that are associated with the updating of egocentric orientations. In contrast, encoding of survey information, such as those presented in a map, has been shown to be related to brain areas that are representatives for object-processing (see also FIELDS & SHELTON 2006, for detailed discussion).[27]

Important behavioral findings have demonstrated that map learners outperform navigation learners when giving straight-line distance estimates to a certain landmark after short (RICHARDSON ET AL. 1999) and long (THORNDYKE & HAYES-ROTH 1982) exposure times to the map, an effect that might result from the fact that learning from a map might result in a survey-like (analogue) image, whereas navigation produces a sequence of interconnected vista spaces from ground level view (see THORNDYKE & HAYES-ROTH 1982, for further discussion). RICHARDSON ET AL. (1999) argued that map learning results in an orientation-specific, inflexible representation, whereas navigation results in an orientation-free, flexible representation. It has been widely reported in the psychological literature that survey maps are indeed encoded in a preferred direction (EVANS & PEZDEK 1980; M. LEVINE 1982; M. LEVINE, MARCHON, & HANLEY 1984; MACEACHREN 1992; PRESSON 1987; PRESSON & HAZELRIGG 1984; PRESSON, DELANGE, & HAZELRIGG 1989; WALLER ET AL. 2002).[28] Consequently, map learners show orientation specific knowledge when performing a pointing task to unseen landmarks that is remarkably accurate

---

[27] Although one might hypothesize that there is a relationship between the way information is presented and encoded and between the subsequent type of knowledge, the corresponding relations are not straight forward. There is converging agreement that there is no one-to-one correspondence between the mode of experience in the environment and the type of knowledge (FIELDS & SHELTON 2006; SHELTON & PIPPITT 2007). The acquisition of route and survey knowledge is therefore not bound to the mode of experience. In particular, both types of knowledge can be acquired from direct experience without a map, thus involving permanent changing ground-level viewpoints or be acquired from map-based experiences involving areal and ground-level perspectives.

[28] The map alignment effect has been particularly emphasized in the literature concerning 'You-Are-Here'(YAH) maps for public spaces. When the posted map is not in alignment with

when testing in an aligned condition but decreases in contra-aligned condition. As a result, map learners have been shown to be outperformed by navigation learners when estimating route estimates or when pointing to unseen locations (RICHARDSON ET AL. 1999; THORNDYKE & HAYES-ROTH 1982). Although orientation specificy of survey maps has been shown to disappear when presenting and experiencing multiple viewpoints of a map (MACEACHREN 1992), the initial viewpoint remains dominant in memory even after consulting other viewpoints (see MEILINGER 2007, for further discussion).

Survey Mapping From Map-Mediated Direct Experience The studies outlined were conducted in very limited experimental conditions, addressing survey mapping mainly in comparison to cognitive mapping from direct experience. Subjects were tested when learning either from a map or from direct experience. In practice, subjects use maps for navigation and encode multimodal information from the map. While directly experiencing space, they encode both types of information in a survey map (LOBBEN 2004). LOBBEN (2004) emphasized that a survey map is not a product, but a constant process involving updates of knowledge whenever a subject navigates in the environment. According to her, one possible model of this process is that every subject encodes an "original cognitive map" (LOBBEN 2004, p. 275) either by direct experience or by consulting a map. This original cognitive map is then constantly transformed during the map-based navigation process. She speculated that "[e]ventually, both processes lead to the development of a single cognitive map" (p. 275), the survey map.

MEILINGER ET AL. (2015) compared cognitive and survey mapping in a virtual reality setup. Results indicated that participants preferred the mental representation that was encoded and experienced first. Map learners relied on the mental representation of the map during measurement of spatial knowledge even after navigating the space:

> When learning from maps before and during navigation, participants employed a map-based reference frame which was different from the reference frame used by navigation-only learners. Consequently, map experience caused the employment of a map-based reference frame. [...] Consequently, also learning from different sources like maps and navigation seems to follow the more general rule in spatial learning that established reference frames form the basis to integrate later experienced information (unless highly salient later information overrides earlier experiences). (MEILINGER ET AL. 2015, p. 1006)

---

the actual space, people tend to move towards the wrong direction by interpreting the 'up' on the map as 'forward direction' (e. g., M. LEVINE 1982; M. LEVINE ET AL. 1984).

According to the authors, empirical data can best be explained by a mechanism that allows the individual to structure further experience based on the primary experience. Map readers that consult the map during navigation, structure their navigation experience with the help of the map, but they do not necessarily develop a second, navigation-based cognitive map that can adaptively be used in different settings of the pointing task.

It is not clear that individuals actually rely on the primary experience of real space during survey mapping, that is, the map. MEILINGER ET AL. (2015) studied subject's performances after encountering a simple spatial layout in virtual space that did not reflect the complexity of buildings or neighborhoods. Spatial learning in complex spatial layouts may lead to an intertwined representation based on the primary experience (consulting the map) and secondary experience (direct experience during navigation). Moreover, although learning spatial layouts from learning in real and virtual environments has been proven to rely on common processes (RICHARDSON ET AL. 1999), it is not clear whether the empirical results hold for multimodal learning in real environments. Consequently, the phenomena of survey mapping is still not well known.

3.3.3   Cognitive Processes of Map-Based Orientation

During map-based navigation, the individual has to engage in orientation as a pre-requisite for further wayfinding. Being able to maintain a sense of the own location in space and that of important landmarks, in particular during movement, is an important cognitive process with visuospatial basis since it requires the individual to reason about the varying relations between the self, the map and the environment. Once established, the map cannot only help to locate oneself precisely in space but allows the individuals to extract important information on the referent space from the map, or to move towards a pre-defined goal. This section provides theoretical insights concerning map-based orientation processes during navigation, one important aspect of large-scale spatial thought.

During movement in real space, for example when navigating with the help of maps, the subject must be aware of the changing locations in space, must be able to track and update them. Using or reading a map thereby refers to a "means to 'consult' it to answer more or less articulated questions about characteristics of some part of the physical environment" (OTTOSSON 1987, p.36). According to LIBEN AND DOWNS (1993), individuals have to understand three different relations between themselves (self), the environment, and the symbolic representation of the environment (map) when they consult maps in real space. These three relations are perception and reflecting on the own location in space (self-environment relation), understanding and establishing links between the space and its representation (map-environment relation), and understanding their own location on the map (self-map relation). Figure 3.9 details the three relations.

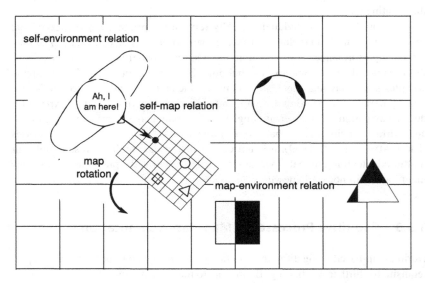

Figure 3.9 The map-environment-self relation (adopted from LIBEN 2006, p. 224). ©2006 L. S. Liben, John Wiley & Sons. Adopted with permission

The Self-Environment Relation

Understanding the self-environment relation refers to the conscious reflection about the own position and heading direction in space with respect to important landmarks. Individuals may develop an approximate understanding of where they are in space using path integration. Path integration allows an individual to update the own position in space during movement. It is "an orientation process in which self-motion is integrated over time to obtain an estimate of one's current position" (MEILINGER 2007, p. 32). Path integration refers therefore to an effective monitoring of self-movement and allows individuals to return to a starting point without the use of memorized landmarks or external features of the landscape (MONTELLO 2005). Moreover, LOOMIS, KLATZKY, GOLLEDGE, AND PHILBECK (1999) emphasized that path integration is a subjective process bound to a certain location (landmark) that typically functions as a means of orientation in the larger, surrounding environment:

To generalize, path integration is the process of navigation by which the traveler's local translations and rotations, whether continuous or discrete, are integrated to provide a current estimate of position and orientation within a larger spatial framework. (p. 129)

Path integration can further be subdivided into dead reckoning, that is, monitoring on the base of velocity information, and inertial navigation, that is, monitoring on the basis of acceleration information. According to GALLISTEL (1990), dead reckoning can be defined as follows:

It refers to the process of updating one's estimate of one's position on the basis of knowledge of how fast one has been moving, in what direction, for how long. (p. 4)

Confusingly, the notions of path integration and dead reckoning have been used as synonyms for referring to the abilities to keep track of 'where you are' with respect to the starting point or another important landmark in the environment (MAY & KLATZKY 2000; MONTELLO 2005). It has been shown that both humans and animals use those abilities (MÜLLER & WEHNER 1988). Moreover, humans are even able to do so without vision (LOOMIS ET AL. 1999; RIESER, GUTH, & HILL 1986). Path integration typically involves only working memory and does not make use of long-term memory (MALLOT 1997; SHOLL & FRAONE 2004).

Spatial updating by path integration happens unconsciously. Individuals cannot resist performing path integration while they move in space as "representations in spatial working memory are naturally and easily transformed as a result of moving" (WALLER ET AL. 2002, p. 1051). However, path integration suffers from the accumulation of errors over time and is therefore not a stand-alone means of navigation (without map). It is limited to rather short distances and not to long-term exploration of an environment (MEILINGER 2007, pp. 34–35).

Path integration is based on information gained from active experience within a route. However, it is distinct from route knowledge referring to represented spatial knowledge containing information about landmarks and their connection in form of internal directional advices in long-term memory. The underlying representation of path integration, in contrast, "is a constantly updated abstraction derived from computations of route information" (LOOMIS ET AL. 1999, p. 132) that contains updates on location and orientation with respect to the origin in working memory. The representation gained from path integration is therefore not suitable for retracing routes because it does not contain route knowledge but it allows individuals to have a sense of their own location in space.

When individuals additionally consult a map to determine their position in space, this requires them, as outlined before, not only to understand representational correspondences, that is, the 'stand for'-relation between the map and the referent but also to reason about configurational correspondences. Establishing these requires the individual to understand that spatial relations among the depicted symbols on a map are consistent with the corresponding locations in the referent space. Establishing a reference between the map and the referent space is an important pre-requisite when using maps, for example, during wayfinding, which has been emphasized by OTTOSSON (1987):

> To make possible the use of a map as an aid in wayfinding, we must first establish our real-world position on the map through inference from what is observed in the terrain and from what is presented graphically on the map. The first might, I suggest, be called a problem of *general reference*, namely, that symbols on the flat map refer to physical counterparts in a three-dimensional reality. The second problem is to relate specific features on the map to specific counterparts in the environment, which consequently can be called a problem of *specific reference*. (p. 39, emphasis in original)

To use geometric correspondences, for example, for locating oneself in space or on the map, or for extracting information about the referent space from the map, the individual has to complete two intertwined cognitive processes. First, the individual has to reason about the map-environment relation by identifying whether the map correctly aligns with the environment. Second, the individual has to apply projective reasoning on the same relation for the purpose of establishing the self-map relation or extracting information from the map.

The Map-Environment Relation
To maintain the map-environment relations, individuals engage in object rotation, "a single cognitive ability, as it is associated with a particular map-reading task" (LOBBEN 2004, p. 275). Object rotation has been studied extensively in the literature on map use since it is the cognitive process associated with keeping map and environment in alignment (e. g., EVANS & PEZDEK 1980; OTTOSSON 1987; SHEPARD & HURWITZ 1984). A map is denoted to be aligned with the environment if the 'up' on the map corresponds to the relation in front, or 'forward' of the map user, and the left-right directions on the map correspond to left-right directions of the map-reader (OTTOSSON 1987; SHEPARD & HURWITZ 1984).

When an individual uses a map in real space, three different frames of reference are involved (e. g., SHEPARD & HURWITZ 1984; OTTOSSON 1987); first, the egocentric system of spatial relations that are perceivable at a specific moment for the map-reader. Forward, left, and right are typically encoded with respect to the map-reader's body. Second, the environmental frame of reference including all locations, and the map-reader being defined to one global feature and third, the intrinsic frame of reference, containing a subset of map features that correspond to perceivable features in the actual situation. To align a map, the individual has to become aware that those three frames of reference need to be physically aligned. Only in this case, "all directions are then 'immediately given.' [...] Thus, all directions can be *egocentrically determined*" (OTTOSSON 1987, p. 42, emphasis in original). Consequently, only in alignment condition, the spatial relations are in "a natural correspondence" (SHEPARD & HURWITZ 1984, p. 166) that is inherent to human beings (but see also FRANKLIN & TVERSKY 1990, for further elaboration).

Whenever a map is misaligned, the map-reader has to compensate the misalignment (MONTELLO 2005) by either physically turning the map or by performing the compensation process mentally (Figure 3.9, p. 94). This can be done by either mentally aligning the map or by mentally reorienting the imagined self. SHEPARD AND HURWITZ (1984) studied chronometric profiles for decision on left-right turns in a series of experiments on computer screens. Their study had three major findings. First, the results indicated that subjects identified that maps were misaligned but it took them additional mental computations. Second, chronometric profiles showed effects of angular deviance, therefore indicating that the additional computation for compensation could be a 'mental rotation'-type like mental transformation in the sense of SHEPARD AND METZLER (1971). Third, mentally rotating the map stimulus was found to be more effective in terms of rotation speed than mentally transforming the own perspective.

The Self-Map Relation
When the map-environment relation is established, that is, the map is in alignment with the environment, individuals can use configurational correspondences to extract information from the map, or vice versa. Doing so has been denoted as projective reasoning. It can be further differentiated into two processes, visualization and self-location (LOBBEN 2004). According to her, visualization refers to the abilities to "mentally transform the two-dimensional map into a three-dimensional form and 'see' its characteristics and objects [...] This act of seeing with the mind's eye, or developing a mental representation as a result of seeing a visual image [...] allows map-readers to 'see themselves' on the map and place themselves in the real-world environment" (pp. 276–277).[29] When using visualization, the map user creates a particular spatial representation using the information from the map, which may

---

[29]The notion of visualization in this context has to be distinguished from the factor-analytical visualization-factor (see e. g., ELIOT 1987).

subsequently be compared to the perceived features in the environment (Figure 3.10).

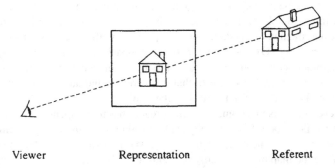

| Viewer | Representation | Referent |

Figure 3.10 Visualization can be interpreted as stretching out the 2D visual information from the map into 3D visual information of the environment (in LIBEN 1999, p. 304). ©1999 L. S. Liben, Lawrence Erlbaum Associates. Reprinted with permission

LOBBEN (2004) further refers to the sub-process of self-location as "locating oneself on a map using the clues presented in the surrounding environment" (p. 277). According to her, self-location is a directed process from the environment to the map, which is a problem solving process that involves both determining salient landmarks and relationships among them in the environment and comparing their relations to relations on a map. Knowing where one is 'on the map' is important for navigation with the help of maps (e. g., LIBEN & DOWNS 1993; LIBEN, MYERS, & KASTENS 2008; OTTOSSON 1987).[30]

There is doubt about what mental processes underlie visualization and self-location. LOBBEN (2004) hypothesized that self-location and visualization are similar processes that demand for a dimensional translation between map and environment. Moreover, she assumed that self-location can be considered a discrete process, whereas visualization can be seen as a continuous cognitive process:

---

[30] LOBBEN (2007) investigated the role of self-location and other cognitive processes on the performance in map-based navigation. She concluded that besides map rotation, self-location abilities (the abilities to perform the process of self-location) are an important predictor, whereas route memory (an individual's capacity to store a self-prescribed route in memory) was not.

> Unlike visualization, which is presumably a continuous process occurring while the map-reader is *in the process* of navigating, self-location is discrete and will take place at the start of the navigation when locating oneself on the map, at the end of navigation, and in between at critical junctures such as when 'double-checking' location or correcting erroneous decisions (getting lost). (LOBBEN 2004, pp. 277–278, emphasis in original)

ARETZ AND WICKENS (1992) hypothesized that besides the mental rotation that is necessary for aligning the map, there could be a second mental rotation that brings the flat, two-dimensional map into the three-dimensional line of sight of the map-reader. Participants were shown pairs of simple maps (2D-condition) or a map (2D) and a forward view (suggested 3D-view). The subjects had then to determine whether the pairs represented a 'same' (can be brought into alignment or corresponds to the forward view) or 'different'-pair and reaction times were measured. The result showed a linear increase of reaction times with respect to increase in angular disparity and significantly higher reaction times for pairs that involve 2D/3D-stimuli (3D-condition) than for pairs that involved only 2D-stimuli. ARETZ AND WICKENS (1992) concluded that the additive reaction times in the 3D-condition are due to a second mental rotation. However, both authors also admitted that "[t]hese data do not, of course, eliminate other possible explanations. The additive increase may be explained by some other cognitive transformation used in creating a 3-D mental representation of a 2-D stimulus" (p. 315). Consequently, mental transformations underlying visualization may be other types of transformations, such as mental transformations of the imagined self. Applying those during visualization would allow individuals to establish a spatial framework out of the information perceived on the map, thereby being able to anticipate possible directions of important landmarks with respect to the egocentric body-schema.

Updating the Map-Environment-Self Relation During Movement
When individuals navigate real space with the help of maps, they have to engage in a time-variant network of relations that is inseparably composed of (a) the individual's present position in the environment, and consequently, on the map; (b) the target location; and (c) features and relations between characteristics as derived from the map (GERBER & KWAN 1994). While moving in space, the individual's subjective interpretation of this network results in a specification of the so-called navigational plan. This navigational plan includes all actions and strategies that the map-reader needs to stay oriented while moving from one location to another (OTTOSSON 1987, p. 44). This plan is encoded into working memory, where it is further processed dur-

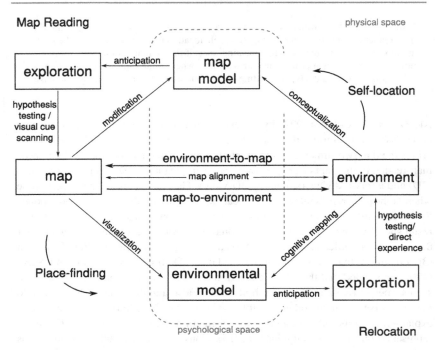

Figure 3.11 A cognitive process model of updating the map-environment-self relation during movement. (Adopted from SEILER 1996, p. 62)

ing completion. A process model of how individuals might complete a navigational plan has been described by OTTOSSON (1986, 1987) and combined with NEISSER's (1976) concept of schemata in a study by SEILER (1996). Figure 3.11 shows a modified version of this process model. It integrates the cognitive processes that were identified in this chapter and takes account of the multiple facets of staying oriented in real space[31].

---

[31] In the process model, the term 'model' is used instead of 'representation' to emphasize that the process model, as originally introduced, is based on internalized information and schemata rather than spatial representations alone. Both OTTOSSON (1986) and SEILER (1996) assumed that the process is enriched by an individual's schemata both for map and environment (not depicted in Figure 3.11). Map schemata might, for example, represent prior experiences of the map-reader while environmental schemata might represent prior spatial experiences in large environments during activities. During completion of the navigation plan, they might

This study uses the process model to explain three different phenomena. First, it explains how individuals use maps to establish their self-environment relation. Second, it describes the cognitive processes when it comes to relating information from the map to the environment and vice versa. Third, it points to a possible mechanism how survey maps may be encoded during map-based navigation.

Using Maps to Locate Oneself in Space   In line with OTTOSSON (1986) and SEILER (1996), one way to interpret the model is that map-based orientation during movement in space is an intertwined process of 'map-reading' and 'relocation'. SEILER (1996) used the term map-reading to refer to a process similar to LOBBEN's (2004) notion of visualization:

> [m]ap reading is the process of constructing a model from the map symbols. This model, or plan, is developed in advance and predicts how the terrain will be when arriving at a certain point in the model (route planning). The model may of course be a preliminary one and it is far not as detailed as reality. The construction of the model is based on the information the map contains, but is also affected by expectations and prior knowledge. (SEILER 1996, p. 60)

Relocation is a process of consciously establishing the self-environment relation:

> Once an orienteer becomes aware that he or she has lost contact between his or her map model and the terrain, the so-called process of relocation starts. It is similar to photo-orienteering: Information extracted from the terrain has to be matched with the map at hypothetical positions. The lost orienteer looks around or even walks a few steps to explore more of the relevant terrain features that might help to redefine the position. (SEILER 1996, p. 60)

Both processes can be integrated into the process model (Figure 3.11), that explains how map-reading individuals may conceptualize the challenges they face.

According to OTTOSSON (1986) and SEILER (1996), the process of updating the map-environment-self relation during movement is based on two spatial representations that are based on the two sources of information involved in large-scale spatial thought: the environment and the map. During movement, the map-reader uses the information from the map to conceptualize a model of the environment by means of mentally visualizing how the two-dimensional information would look like in

---

direct the initial attention of the individual to relevant information in the environment and on the map,and are therefore particularly relevant during anticipation processes.

three dimensions, which, in turn, allows the individual to anticipate landmarks that may be perceived along the route in the environment. This anticipation leads to the formulation of a spatial hypothesis that is continuously tested against the spatial information perceived during exploration of the environment. The information acquired during movement is further encoded by cognitive mapping and refines the individual's spatial representation of the environment. This might be an advantage when deriving further spatial hypotheses (see also OTTOSSON 1986, p. 81).

Conversely, a preliminary representation of the map might be conceptualized from the landmarks observed in the environment. This representation allows individuals to anticipate how some particular part of the map will look like, for example, when they aim to determine their position on the map. When visually exploring the map, spatial cues that are picked up during perception of the map are used to test the hypothesis. The more often individuals consult the map, the better their spatial representation of it may become.

OTTOSSON (1987) emphasized that both approaches are intrinsically linked and that "these two processes interact in a complex interplay where different phases can be very closely interwoven" (p. 82). He further pointed out that this might lead to a continuous development or change of the individual's conception of a given situation or spatial task (see also 3.1.2). Indeed, whenever an individual consults the map, it can be interpreted as an expression of maintaining an "internally consistent conception of the task" (OTTOSSON 1987, p. 129). This involves the map-mediated conception of the environment, the perceived environment and the individual's conception of the self-environment relation, that is, the conception of his or her movement in relation to the (map-mediated or not) environment, in short, an updated map-environment-self relation.

Using Map-Environment Relations   The idea of deriving spatial hypotheses that are then tested against the actual perceived information is also central when using the model to explain the cognitive processes of reasoning about the map-environment relation, and when 'translating' information from the map to the environment and vice versa. In line with KASTENS AND LIBEN (2010), the model implies that both translation processes are a conglomerate of non-linear cognitive processes that involve the development and testing of subsequent spatial hypotheses.

The corresponding hypotheses do, however, differ. When translating information from the environment to the map, individuals develop a spatial hypothesis concerning cues they expect to find on a map. When translating information from the map to the environment, individuals develop a spatial hypothesis concerning landmarks they expect to find in the surrounding environment.

Self-location, as introduced by LOBBEN (2004), usually does not only require the subject to establish one hypothesis but several ones that are more and more refined and tested until the position on the map is sufficiently constrained. This process does not only involve transformation processes from the environment to the map but also translation processes from the map to the environment.

Visualization, as introduced by LOBBEN (2004), is the process that allows an individual to derive an environmental model from the information depicted on the map, thus leading to a single spatial hypothesis about the landmarks to be anticipated. The cognitive process of deriving several spatial hypotheses that are then subsequently tested against environmental features perceived during exploration of the environment can be denoted as place-identification. Place-identification is the process of identifying pre-defined visual cues on the map as landmarks in the environment (LOBBEN 2007). It is the conceptual analague to self-location in the environment. Consequently, similar to map-reading and relocation as proposed by SEILER (1996), self-location and place-finding may be understood as intertwined in different phases during map-based orientation because the individual has to constantly translate information from the map to the environment and vice versa.

To develop spatial hypotheses by means of vizualization during self-location and place-findings requires the map to be aligned with the environment. Consequently, in every translation, a mental rotation process can be assumed to be involved. However, the mental transformations underlying visualization and conceptualization of a two-dimensional representation of what has been perceived are not fully understood. Since both processes involve establishing or collapsing a spatial framework, either to determine locations of anticipated landmarks with respect to the egocentric frame of reference or to determine locations of symbols of landmarks that have been perceived in vista space (thus involving an egocentric frame of reference). As a consequence, those processes might be governed by mental transformations of the imagined self.

Understanding Survey Mapping   Although OTTOSSON (1986) and SEILER (1996) refer to two kinds of models, neither the map and the environment model explain how a spatial representation of the environment is encoded nor do they elaborate on the differences and similarities between knowledge acquired from maps and the direct experience of space (see also CRAMPTON 1992). However, based on the findings outlined above, this study uses the process model to provide a theoretical explanation of survey mapping.

In line with LOBBEN (2004) and MEILINGER ET AL. (2015), individuals might encode an initial representation of the environment from the map. This representation is oriented and provides a metric scaffolding of possible landmarks and perspective shifts in an 'empty' network of references frames (MEILINGER 2008, 2007). When individuals move in space, this network is further developed, and information from direct experiences is assimilated. It is subsequently updated by undergoing the intertwined cognitive processes of self-location during map-reading and in real space. Vista spaces are assimilated into the network by visualization and subsequent hypothesis testing and are stored in depictive format since they can be considered to be small-scale spatial information. The map thereby mediates of how subjects perceive and conceptualize the surrounding environment. Direct experience and embodied movement during exploring the environment lead to perspective shifts, which, in turn, deepen the individual's knowledge in the edge of the network. Their encoding is mediated by the visualized change of locations on the map. Besides encoding a survey map using these processes, individuals probably encode a second internal representation of the map. Both the survey map and the map model are elements of psychological space.

3.3.4   Maintenance of Large-Scale Spatial Information in Working Memory

Representing information from the explored environment requires the individual to maintain different pieces of spatial information in working memory. This might allow individuals not only to acquire spatial knowledge stored in spatial representation and to transfer it to long-term memory (HEGARTY & WALLER 2005) but also to make inferences on directions to landmarks, routes, or distances from this representation. Consequently, working memory has been empirically demonstrated to play an important role in cognitive and survey mapping. In particular, VSWM capacity emerged as an important predictor of survey knowledge (see e. g., LABATE ET AL. 2014, for further elaboration) both in dual task (e. g., LABATE ET AL. 2014) and correlational studies (e. g., ALLEN, KIRASIC, DOBSON, LONG, & BECK 1996; HEGARTY ET AL. 2006) and in studies involving both approaches (COLUCCIA, BOSCO, & BRANDIMONTE 2007). This holds true for direct experience in unknown environments (LABATE ET AL. 2014) but also in virtual environments (HEGARTY ET AL. 2006), and when learning from a map (COLUCCIA ET AL. 2007). Moreover, VSWM proved to be involved not only in acquiring survey knowledge but also route knowledge (see MEILINGER 2007; PURSER ET AL. 2012, for overviews in adults and children, respectively). Again, this was shown using both dual-task approaches (e. g., MEILINGER 2007) and correlational approaches (e. g., PURSER ET AL. 2012), and for different learning conditions. Finally, VSWM proved to be an important predictor when it comes to map-based navigation with topographical maps in large-scale unknown complex natural environments in adults (PLORAN ET AL. 2015).

Conclusion

When individuals navigate their environment using a map, a subclass of large-scale spatial abilities allow them to mentally represent spatial stimuli encountered during movement in a spatial representation, denoted as the survey map. Since individuals perceive pieces of large-scale spatial information at multiple vantage points, they have to be assimilated in a map-mediated conceptualization of the environment. Consequently, spatial knowledge is stored in a fragmented and distorted way in working memory, where it is maintained for further cognitive processing. When individuals familiarize themselves with the environment, this knowledge is integrated into long-term knowledge structures.

Large-scale spatial abilities allow individuals to understand and keep track of the varying relations between the self, the map, and the environment, especially during movement. By doing so, individuals engage in a range of intertwined cognitive processes that allow them to determine their position on the map and in real space, continuously updating the map-environment-self relation. Individuals constantly relate information provided on the map to the environment and vice versa, thereby establishing a sequence of spatial hypotheses that allow them to anticipate spatial cues in the environment and on the map. A constant refinement of these hypotheses means a process of visual validation and re-formulation that allows individuals to track both their own location on the map (self-location) and that of landmarks in the environment (place-finding) will subsequently be described as two other subclasses of large-scale spatial abilities. Updating their position in space, but also translating information from the map and vice versa require the individuals to reason and update relations of frames of reference, for example as it is the case during map alignment which is an important pre-requisite of map use in real space.

3.4   Comparison of Spatial Abilities at Different Scales of Space

Sections 3.2 and 3.3 described small-scale and large-scale spatial abilities as cognitive phenomena. Using a bottom-up approach, and in line with HEGARTY ET AL.'s (2006) argumentation, Section 3.2 provided theoretical insights regarding three possible sources of individual differences concerning small-scale spatial abilities, namely encoding, maintenance in working memory, and making inferences (Figure 3.12). Concerning mental transformations used to make inferences on information stored in a spatial representation, the analysis revealed that there are at least two types of qualitatively different mental transformations; namely mental transformations of objects and of the imagined self.

One important implication of this theoretical insight is that small-scale spatial abilities should not be understood as a unitary construct, since the individual's abilities to perform mental transformations of objects could vary from those needed to perform mental transformations of the imagined self. If one therefore assumes that individuals do not differ with respect to their capacity to encode and maintain information in working memory but only with respect to performing those two types of

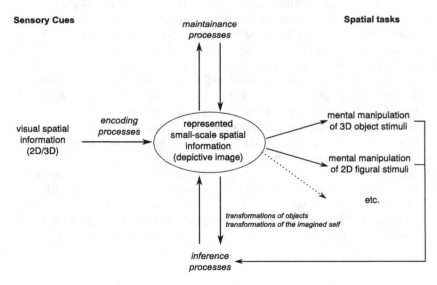

Figure 3.12 An information-processing description of small-scale spatial abilities (adopted from HEGARTY ET AL. 2006, p. 155). ©2006 Elsevier. Adopted with permission

mental transformations, a possible model of small-scale spatial abilities, one among potentially others, could include two subclasses. The first subclass could be defined as a subset of small-scale spatial abilities allowing an individual to perform mental transformations of objects. The other one would be defined as a subset of small-scale spatial abilities allowing an individual to perform mental transformations of the imagined self.

Section 3.3 provided theoretical insights concerning possible cognitive processes that underlie map-based navigation and that might be possible sources of individual differences regarding large-scale spatial abilities (Figure 3.13). Besides encoding of spatial information in a survey map and maintaining this information in working memory, the analyses revealed that large-scale spatial abilities are most likely a conglomerate of cognitive abilities. These allow individuals to be aware of both their own position in space and that of important landmarks, in particular during movement. To determine those positions, the individual has to cognitively establish a map-environment-self relation that is constantly updated during movement in space and the basis for making inferences on the environment from information gained from the map. An analysis of cognitive processes that underlie successful updates of this relation revealed that there seem to be two subclasses of transformations that

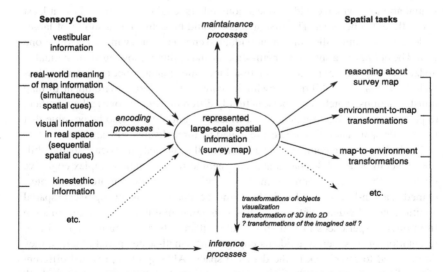

Figure 3.13 An information-processing description of large-scale spatial abilities (adopted from HEGARTY ET AL. 2006, p. 155). ©2006 Elsevier. Adopted with permission

differ in their kind of the established spatial hypotheses, transformations of spatial information from the map to the environment, and vice versa. Although mental alignment is a prerequisite for both, the former subclass relies on subsequent visualization processes that allow an individual to derive hypotheses about landmarks to be encountered in the environment, thereby also relying on knowledge stored in the survey map. The latter subclass relies on subsequent processes that allow an individual to derive hypotheses concerning visual cues on a map during a map-reading process. Map alignment can be assumed to be a process that is isomorphic to an object-based transformation (mental rotation). It is not clear, however, if visualization or the transformation of 3D information to 2D information share a structural isomorphism to a particular mental transformation, such as the transformation of the imagined self.

One important implication of these theoretical insights is that large-scale spatial abilities should also not be understood as a unitary construct. Even if one assumes that individuals do not differ with respect to their abilities to encode information of vista spaces and to maintain all perceived large-scale information in working memory, the abilities to integrate multiple pieces of information encountered from different vantage points into a survey map seem to differ from those ones to master the update of the map-environment-self relation. One possible model of large-scale spatial abilities, which could be one among others, could therefore include at least two different subclasses. The first subclass could be defined as a subset of large-scale spatial abilities allowing an individual to encode a survey map. The other one would be defined as a subset of small-scale spatial abilities allowing an individual to update the map-environment-self relation during movement. Since the latter subclass includes not only the abilities allowing individuals to perform environment-to-map transformations, which could be structurally different from the ones allowing them to perform map-to-environment transformations, but it could also be possible that large-scale spatial abilities are best modeled by three distinct subclasses.

Table 3.1 shows a comparison between small-scale and large-scale spatial abilities. Overall, large-scale spatial abilities seem to involve more complex cognitive processes, that is, they seem to involve cognitive processes that are highly intertwined in a non-linear way. This is due to the fact that when using large-scale spatial abilities, the individual continuously has to integrate multiple pieces of information from direct experiences in real space, rather than only referring to a spatial representation of it. When individuals engage with small-scale spatial abilities, they do not need to refer back to the depicted space. Although the provided different kinds of sensory cues result in qualitatively different spatial representations, both classes of spatial abilities share the need to maintain them in working memory for further processing. This might involve the visual component, the visuospatial working memory as well as the central executive probably to a different degree, but maintaining information for making inferences and storage processes is likely to be important for both classes of spatial abilities.

The analyses revealed that both classes of spatial abilities do not only share structural similarities at the level of maintenance in working memory, but also when it comes to making inferences from the represented information. Although the involvement in both classes remains to be clarified for some of the cognitive processes, it is safe to assume that transformations of objects are involved in both classes of spatial abilities.

Table 3.1   A Comparison of Small-Scale and Large-Scale Spatial Abilities

	Small-Scale Spatial Abilities	Large-Scale Spatial Abilities
Spatial behavior	mental manipulation of objects and figures in graphical space	map-based orientation during navigation in physical space
Sensory Cues	abstract depictions of physical space in graphical space	maps in graphical graphical space and direct experience in physical space
Spatial information	visual	visual, kinesthetic, vestibular
Vantage point	single	multiple
Modus	simultaneous	simultaneous (map) + sequential (physical sp.)
assign real-world meaning to spatial information	not necessary	necessary
Spatial memory		
Spatial representation	depictive, metric, exact	distorted, fragmented and non-metric survey map
Involvement of WM	yes	yes
Storage in LTM	probably not	yes, when familiarizing
Involvement of schemata	probably yes	probably yes
Spatial inferences	probably linear	non-linear, intertwined
Conflicting frames of reference	yes	yes
Transformation of objects	yes	yes
Transformation of the self	yes	probably
Visualization	probably	yes
Self-location	probably	yes
Place-identification	probably	yes

Conclusion

Scale influences spatial behavior indirectly because the scale of spatial information imposes different demands on the corresponding information processing when it comes to the representation and maintenance of spatial information and making inferences from them. The analyses showed that small-scale and large-scale spatial abilities do, however, not only show structural similarities when it comes to their definition, but also at the level of cognitive processes. Similar cognitive processes can be found at both scales of space, indicating that scale as a psychological construct might be used to conceptually differentiate spatial abilities, but does not lead to a total dissociation of those.

There is a wide range of sources of individual differences, both when it comes to examining spatial abilities at both scales of space, and their possible relation. Consequently, spatial abilities should not be understood as an undifferentiated construct. Indeed, there seem to be two classes of spatial abilities sharing some but not all structural similarities. In addition, these classes are multidimensional constructs themselves, that is, involving two or more subclasses.

Reviewing Empirical Findings of Spatial Abilities at Different Scales of Space

Chapter 3 described how scale affects the processing of spatial information. It identified possible similarities and qualitative differences in the cognitive processes and operations of small-scale and large-scale spatial abilities. These analyses yielded three important assumptions (see Section 3.4). First, concerning small-scale spatial abilities, there seem to be two distinctive types of mental transformations, namely object-based transformations and egocentric perspective transformations. Second, concerning large-scale spatial abilities, there seem to be at least two distinctive types of cognitive processes, namely the ones of encoding a survey map, and the ones related to stay oriented with a map, thereby maintaining the map-environment-self relation while moving. It can be assumed that the latter one, in turn, involves two types of processes, namely map-to-environment transformations and environment-to-map transformations. Third, there seems to be a partial dissociation between the cognitive processes at different scales of space.

This chapter reviews studies that have investigated whether an individual's abilities to use one (sub)class of spatial abilities and his or her abilities to use another (sub)class of spatial abilities can indeed be understood as separable cognitive abilities. In other words, although different cognitive processes have been identified in Chapter 3, which led to the assumptions above, these analyses do not necessarily imply a distinction from the perspective of individual differences. Consequently, although separate cognitive processes may be involved in these (sub)classes of spatial abilities, they could tap the same underlying abilities in the sense that when an individual performs well in tasks requiring one (sub)class of spatial abilities, he or she is also good in the other (sub)class of spatial abilities, and vice versa (e. g., KOZHEVNIKOV & HEGARTY 2001).

Chapter 4 reviews the literature providing empirical evidence that there are indeed individual differences when it comes to solving tasks reflecting those different (sub)classes of spatial abilities. Moreover, it identifies gaps in the literature. It subsequently addresses the three possible dissociations outlined above from three different perspectives. First, it summarizes, if available, findings from the psychometric literature. Psychometric researchers tried to specify possible models of spatial abilities by mapping groups of measures provided by spatial tests in homogenous dimensions, denoted as factors, using mainly exploratory factor analytical approaches. Those factors represented weighted sums of each of the variables and contained full information given by the correlations in a set of interpretable latent variables that stand for a certain subclass of abilities (e. g., HEGARTY & WALLER 2005). This chapter summarizes models of spatial abilities, mainly small-scale spatial abilities. It shows that a range of these models provided evidence that there might be two distinct factors that, in turn, are conceptually equivalent to the two subclasses of small-scale spatial abilities proposed here. A review of psychometric studies concerning the modeling of large-scale spatial abilities is, however, less fruitful.

This chapter further addresses findings from the individual differences literature. Similar to the psychometric literature, these studies aimed to identify patterns of individual differences when it comes to solving a set of pre-defined tasks. In contrast to applying exploratory factor analyses searching for a factor structure that best reflects the underlying data, those studies used confirmatory approaches to test theoretically based, hypothesized factor models given the empirical correlations of the underlying data set (HEGARTY & WALLER 2005).

Finally, this chapter addresses, if available, findings from the developmental literature to emphasize that the individual differences reported by those studies were not a mere artefact of the measures used. Instead, the findings that are summarized provide evidence that subclasses of spatial abilities may either emerge during childhood or that different performances identified might be due to the involvement of conceptual frameworks that require different stages of cognitive development. This chapter does not intend to summarize the entire literature concerning the development of spatial abilities during childhood but uses findings from the developmental literature as supporting arguments for the proposed models. For a more complete account on the development of spatial abilities at both scales of space, see ELIOT (1987), LIBEN (2006), NEWCOMBE (2002), NEWCOMBE AND HUTTENLOCHER (2003), and REINHOLD (2007).

This chapter is organized as follows. Section 4.1 addresses the possible partial dissociation between two subclasses of spatial abilities. Section 4.2 summarizes the literature when it comes to different subclasses of large-scale spatial abilities. Section 4.3 reviews the literature concerning the possible relationship of spatial abilities at both scales of space.

4.1   The Dissociation of Two Subclasses of Small-Scale Spatial Abilities

Section 3.2 showed that whenever individuals make inferences from small-scale spatial information they might apply two types of mental transformations, namely object-based transformations and transformations of the imagined self. Although both types are logically equivalent when it comes to determining the appearance of an object or a configuration of objects (Figure 4.1), the analyses showed that applying these types of transformations involves qualitatively different mental operations. Consequently, Section 3.4 suggested that one possible way to model small-scale spatial abilities could be to use a model with two components (subclasses). Each subclass could represent a subset of small-scale spatial abilities enabling one of the two types of mental transformations.

Figure 4.1 Two different cognitive processes in response to the same question: (left) How does the bench look like from the side? (middle) object-based transformation (right) egocentric perspective transformation

This section summarizes the literature that addressed the question whether or not there are indeed two distinct subclasses of small-scale spatial abilities. In a first step, different models from the psychometric literature are presented and discussed. In a next step, the individual differences literature that directly compared both mental transformations is summarized. In a third and final step, findings from the developmental literature are discussed.

4.1.1   The Psychometric Literature

There are two generally accepted findings concerning spatial abilities in classical psychometric studies. First, there is general consent that spatial abilities constitute a specific domain of intelligence (e. g., ELIOT & HAUPTMAN 1981; ROST 1977). They have been identified as a separable factor in intelligence models involving a general intelligence factor (e. g., SPEARMAN 1927), multiple intelligence factors (e. g., THURSTONE 1947), and models combining both ideas (e. g., CARROLL 1993). Within intelligence models, the spatial domain has been described as the factor that "is visual or spatial in character" (THURSTONE 1938, p. 79) and being "involved in any task in which the subject manipulates an object imaginally in two or in three dimensions" (THURSTONE & THURSTONE 1941, p. 4). Second, there is a general consent that small-scale spatial abilities should be modeled as a multidimensional construct (e. g., CARROLL 1993; ELIOT 1987; HEGARTY & WALLER 2005; MCGEE 1979).

About 40 years after a spatial component of intelligence has been identified, however, there was "still vast disagreement about how best to classify standard tests of spatial ability" (MCGEE 1979, p. 893). Until today, several models, but not a unified one, have been proposed to describe this class of spatial abilities (e. g., DAVIS & SPATIAL REASONING STUDY GROUP 2015).

Thurstone's Three Factor Model
THURSTONE (1949) studied cognitive abilities that related to mechanical attitude using a test battery with 32 different spatial tests in a study with male adults.[1] He reported evidence for the existence of at least three different factors (Table 4.1, see THURSTONE & THURSTONE 1949, for description of the marker tests).[2]

THURSTONE'S (1950) third factor, spatial orientation, was statistically hard to separate from the first factor, spatial relations (PAWLIK 1968). MICHAEL, ZIMMERMAN, AND GUILFORD (1950, 1951) found empirical evidence that mark-

---

[1] THURSTONE'S (1949) study inspired a series of psychometric studies, mostly during and after World War II. Researchers produced a large, non-converting collection of studies that investigated the question of a perfect model of spatial abilities on their own (GUTTMAN, EPSTEIN, AMIR, & GUTTMAN 1990) using a broad range of tests (ELIOT 1987).

[2] Whereas the two first spatial factors, S_1 and S_2, could be interpreted from a cognitive psychological point of view, the third factor could only be defined after a careful re-analysis of the marker tests that loaded high on S_3, Lozenges A and Cubes (see THURSTONE 1950, p. 518, for further elaboration). THURSTONE AND THURSTONE (1949) further reported the evidence for a factor kinesthetic imagery k. The k-factor was interpreted to have importance in the third spatial factor and "constitutes the characteristic of the third space factor" (THURSTONE 1950, p. 519).

ers of THURSTONE'S (1950) spatial relations component loaded on the same factor as markers that were intended to represent spatial orientation.

Table 4.1 THURSTONE'S (1950) Three Factor Model

Factor	Typical marker test
S_1—*spatial relations*	
Is the ability to recognize the identity an object when it is seen from different angles.	Cards, Figures, Reversals and Rotations
S_2—*visualization*	
Is the ability to imagine the movement or internal displacement among the parts of a configuration that one is thinking about.	Surface development, Paper Puzzles
S_3—*spatial orientation*	

The Paradigm of at Least Two Separable Factors

MICHAEL, GUILFORD, FRUCHTER, AND ZIMMERMAN (1957) integrated THURSTONE'S (1950) S_1 and S_3 into a single factor, denoted as Spatial Relations and Orientation (SR-O)[3] which was defined as

> [...] an ability to comprehend the nature of the arrangement of elements within a visual stimulus pattern primarily with respect to the examinee's body as the frame of reference. In a typical test of the factor, as the entire configuration (or a principal component of it) is moved into a different position, the objects within the pattern hold essentially the same relationships to one another. (MICHAEL ET AL. 1957, p. 18)

They further proposed a second factor, visualization, which equaled S_2 in THURSTONE'S (1950) model:[4]

> Tests of this factor require mental manipulation of visual objects involving a specified sequence of movements. [...] The individual finds it necessary mentally to rotate, turn, twist, or invert one or more objects, or parts. [...] The examinee is required to recognize the new position, location, or changed appearance of objects that have been moved or modified, within a more or less complex configuration. (MICHAEL ET AL. 1957, p. 191)

---

[3] MICHAEL ET AL. (1957) related their study to factor models of other researchers by that time, for example by showing that their model was similar to the one of FRENCH (1951).

[4] Similar to THURSTONE (1950), they report the finding of a factor denoted as kinesthetic imagery (K).

Two aspects are interesting in the study of MICHAEL ET AL. (1957). On the one hand, they distinguished both factors with regard to the fact of whether a concrete sequence of movement was given within the instructions, and whether the subject had to imagine a movement during some hypothetical time interval. They emphasized that there was a specified sequence of movements to be followed in responding to test items reflecting the visualization factor, whereas the individual was not instructed regarding a sequence of movements in tests reflecting SR-O. Individuals were required to determine the direction of movement of an object occurring during the hypothetical interval of time between exposure to the stimulus component and the response component of a test item on their own. On the other hand, MICHAEL ET AL. (1957) proposed that the orientation of the subject's body might have been influential in the distinction between spatial visualization and spatial orientation:

> One hypothesized difference between the ability of spatial relations and orientation and that of visualization is that in the former ability the subject constitutes an origin, or *point of reference*, more or less at the center of the region represented by the stimulus pattern; in the latter ability the *individual is removed from the stimulus pattern* in that he appreciates spatial arrangements that to him seem to be at some distance, at least as far away as his fingertips. In the instance of visualization the examinee appears to be somewhat detached from the stimulus pattern ... (MICHAEL ET AL. 1957, pp. 190–191, emphasis added)

Although MICHAEL ET AL. (1957) admitted that their model did not show statistical independence among the identified factors, a re-analysis of the existing psychometric literature conducted by MCGEE (1979) revealed that "a plethora of factor studies since that date have provided consistent evidence for the existence of two distinct spatial abilities" (MCGEE 1979, p. 892). Consequently, MCGEE (1979) proposed that small-scale spatial abilities can be modeled as two-factor models and defined them using an amalgam of different definitions of the psychometric literature at that time (Table 4.2). He defined both factors in a much broader way than THURSTONE (1950) and emphasized, in line with MICHAEL ET AL. (1957), that visualization is about the manipulation of objects, whereas orientation involves the manipulation of the imagined body of the subject.

MCGEE'S (1979) re-analysis of the existing psychometric literature provided first evidence for the existence of two separable factors. This model, however, remained subject to intensive debates in subsequent studies (BORICH & BAUMAN 1972). CARROLL (1993), for example, failed to find evidence for such a distinction in his meta-analysis (see CARROLL 1993, p. 362).

One reason for these inconsistencies in the psychometric literature concerning the validation of a two-factor model might be that psychometric studies faced problems

Table 4.2 Summary of Studies that Show the Separability of a Spatial Orientation Factor from a Visualization Factor (all cited in GUTTMAN ET AL. 1990)

Study	Visualization	Orientation	Other factors
FRENCH (1951)	(VZ) ability to comprehend imaginary movement in 3-D-space, or to manipulate objects in imagination	(SO) ability to remain unconfused by the varying orientations in which a spatial pattern may be presented	
GUILFORD (1956)	(SV) the process of imagining movements, transformations or other changes in visual objects	(S) ability to perceive spatial patterns accurately and to compare them with each other (SO) ability to appreciate spatial relations with reference to the body of the observer; being aware that one object is higher or lower, left or right, nearer or farther than another	
MICHAEL ET AL. (1957)	(Vz) mental manipulation of a highly complex stimulus pattern	(SR-O) ability to comprehend the nature of the arrangement of elements within a visual stimulus pattern primarily with respect to the subject's body as the frame of reference	(K) one movement to vicariously move an entire configuration through turning or twisting it left or right
FRENCH, EKSTROM, AND PRICE (1963)	ability to manipulate the stimulus and alter its image	perception of the position and configuration of objects in space (observer as reference point)	
PAWLIK (1966)		observer's body orientation as an essential part of the problem	(SR) identification of an object when seen from different angles
McGEE (1979)	ability to mentally rotate, manipulate and twist two- and three-dimensional stimulus objects.	comprehension of the arrangement of elements within a visual stimulus pattern, the aptitude to remain unconfused by the changing orientations in which a spatial configuration may be presented, and the ability to determine spatial orientation with respect to one's body.	

in finding a good marker test for spatial orientation. The Guilford-Zimmerman Boat Test (GUILFORD & ZIMMERMAN 1948), for example, a commonly used marker, was found later to load on other factors involving the mental manipulation of objects (KOZHEVNIKOV & HEGARTY 2001). The lack of reliable tests that reflected the spatial orientation factor might have obscured a possible dissociation between the two subclasses of small-scale spatial abilities.

Linn and Petersen's Three-Factor-Model

LINN AND PETERSEN (1985) conducted a meta-analysis of the existing psychometric literature and computed estimates of effect sizes for sex differences in a range of spatial tasks. In a subsequent step, they classified spatial tasks into homogenous categories with respect to their magnitude of effect sizes. Those broad categories were then compared and interpreted beyond the existing cognitive literature to determine whether they shared the same underlying processes. By combining this psychometric and cognitive rationale based on data concerning sex differences and their emergence, they identified three distinct factors that were not only psychometrical homogenous groups of spatial tests but also interpretable in terms of their cognitive demands (Table 4.3).

Table 4.3 LINN AND PETERSEN'S (1985) Three Factor Model

Factor	Typical marker test
S_1—*Spatial Perception*	
Is the ability to determine spatial relationships with respect to the orientation of their own bodies, in spite of distracting information.	Rod and Frame Test, Water Level Test
S_2—*Mental Rotation*	
Is the ability to rotate a two or three dimensional figure rapidly and accurately.	Mental Rotation, Cards, Paper Form Board
S_3—*Spatial Visualization*	
Is the ability to perform complex, multistep manipulations of spatially presented information, also by using processes of spatial perception and mental rotation, and multiple solution strategies.	Embedded Figures, Paper Folding Test, Surface Development

Their meta-analysis revealed homogenous patterns of sex differences in two of the three factors. They found robust sex differences in favor of males in tests reflecting the factor spatial perception and mental rotation (see LINN & PETERSEN

1985, p. 1491). Tests that reflected spatial visualization, however, were found to be equally difficult for both groups, producing heterogenous patterns of sex differences.

Critical Reflection of the Models

This section summarized relevant psychometric studies. Some of them, for example the one of MICHAEL ET AL. (1957) and the meta-analysis of MCGEE (1979) showed that a two-component model seemed plausible both from a statistical point of view and concerning a possible psychological interpretation when mapping measures of individual differences provided by spatial tests into factors. According to them, a first subclass of abilities enables individuals to mentally perform a pre-defined sequence of objects and figures given in the task instruction while the imagined self is not involved in the spatial situation. A possible second subclass enables individuals to mentally determine a possible direction of movement of the imagined self and to make inferences on the appearance of objects from an imagined viewpoint. The first subclass of small-scale spatial abilities corresponds to the first one proposed above, namely the abilities to perform object-based mental transformations while maintaining the own perspective invariantly. The second subclass represents the second one proposed above, namely the abilities to perform a mental transformation of the imagined self with respect to a spatially invariant configuration of objects. Since both subclasses are equivalent to the ones proposed by MICHAEL ET AL. (1957) and MCGEE (1979), one might conclude that it is most likely to assume that individuals that perform well in tasks requiring the first subclass of abilities do not necessarily perform well in tasks requiring the second subclass and vice versa.

This evidence, however, has to be critically discussed with regard to at least two points. First, psychometric studies have been criticized for their exploratory method- ology that "can be used without any prior understanding of the interrelations among a set of variables" (HEGARTY & WALLER 2005, p. 133). As such, most of the models outlined above have proposed classifications of spatial tests based on statistical pro- cedures. Because those were mostly correlation-based, those classifications were highly dependent on the statistical tests used (see also LINN & PETERSEN 1985).

Furthermore, psychometric studies typically assumed homogenity of strategies. They assume that all subjects solve the psychometric test using the same intended strategy, and that an individual uses the same strategy in every test item (e. g., HEGARTY & WALLER 2005). Both assumptions have been demonstrated to be false (e. g., SCHULTZ 1991). BARRATT (1953), for example, studied ten paper-and-pencil tests, such as the Guilford-Zimmerman Boat Test (GUILFORD 1956), which was a frequently used marker for spatial orientation at that time. An analysis of verbal protocols revealed that half of the subjects did not use the intended solution strategy, spatial orientation, but a mental rotation strategy. Instead of imagining being part of the situation, the subjects rotated and moved the given stimuli in the foreground or the background (BARRATT 1953).

Empirical results have demonstrated that even children in kindergarten are able to apply a range of solution strategies, involving mentally moving the self, the object or applying analytic strategies (LÜTHJE 2010). Findings with fourth graders confirmed that children are able to adaptively apply these strategies to solve different spatial tasks, often by also successfully combining different strategies. Strategy choice was thereby found to be independent of the presentation format of the spatial tasks (see PLATH 2014, pp. 36–40, for further elaboration). Consequently, when interpreting psychometric studies as possible empirical evidence for the assumptions outlined above, one has to be aware that those determinants are neglected or that they might contribute to a possible strong relationship between both subclasses of spatial abilities.

The study of LINN AND PETERSEN (1985) overcame drawbacks of classical psychometric studies by combining cognitive with psychometric arguments. Taken together, their way of combining multiple methodical approaches in the study of spatial abilities led to the establishment of a prominent model that has further been replicated in a subsequent study (VOYER, VOYER, & BRYDEN 1995) and should therefore be considered as an alternative manner to model small-scale spatial abilities.

To discuss this issue, one has to address findings concerning sex differences in children. Indeed, sex differences were, quantified in effect sizes, the basis of LINN AND PETERSEN'S (1985) psychometric computations. There is an overall agreement in the literature, however, it is challenging to find robust sex differences in children in contrast to adults. Attempts to classify possible sex differences and to interpret them in terms of factor models did not yield strong support for LINN AND PETERSEN'S (1985) three component model (JOHNSON & MEADE 1987). Moreover, there was strong evidence that the same factor models do not hold true for boys and girls (JOHNSON & MEADE 1987). Both findings might be, among developmental aspects, also due to the fact that there are only a few age-specific tests for children of that age (e. g., NEWCOMBE & SHIPLEY 2015). Classical spatial tasks that have been developed for adults are typically too hard for children, yielding to floor effects which make the computation and interpretation of the emergence and magnitude of sex differences difficult (e. g., VOYER ET AL. 1995; S. C. LEVINE, HUTTENLOCHER, TAYLOR, & LANGROCK 1999). There is, however, a consent that, similar to findings in adults, there are robust sex differences in favor of boys in performances in mental rotation tasks emerging by the age of eight (KERNS & BERENBAUM, 1991; VOYER ET AL. 1995). In addition, sex differences in favor of boys have been found in performances in a map task involving spatial orientation demands in adolescent children (LOGAN ET AL. 2017). It remains to clarify whether possible sex differences could be identified with a set of age-specific measures, and, subsequently whether LINN AND PETERSEN'S (1985) three component model could be suitable to model children's small-scale spatial abilities. This was, however, beyond the scope of this study and consequently, the three-component model was not considered as having sufficient empirical evidence to represent a possible alternative model.

4.1.2   The Individual Differences Literature

Studies of the individual differences literature examined whether there is a difference in performances of adults in tasks requiring object-based transformation in comparison to performances in tasks requiring transformations of the egocentric perspective (HEGARTY & WALLER 2004; KOZHEVNIKOV & HEGARTY 2001). Those studies assumed that the dissociation between a set of literature-based chosen perspective taking and mental rotation paper-and-pencil tests reflected a partial dissociation between the spatial abilities involved to perform those types of transformations.

KOZHEVNIKOV AND HEGARTY (2001) examined a possible dissociation using six paper-and-pencil tests and a self-reported measure of sense of direction (SBSOD).[5] Results of confirmatory factor analyses revealed that a single-factor model significantly deviated from the data, whereas a two-factor-model fitted the data well. Although both factors were found to be highly correlated ($r = 0.7$), KOZHEVNIKOV AND HEGARTY'S (2001) findings provided first evidence that an individual's abilities to perform a mental transformation of the imagined self do not reflect his or her abilities to perform object-based-transformations. They concluded that both subclasses of spatial abilities seem not to reflect the same latent construct.

HEGARTY & WALLER (2004) showed that the findings of KOZHEVNIKOV & HEGARTY (2001) are independent of the tests chosen to reflect the egocentric perspective transformation factor. They argued that both tests reflecting this factor in KOZHEVNIKOV & HEGARTY (2001) study were very similar and differed only in the spatial array that was depicted. They questioned whether the dissociation

---

[5]The SBSOD has been proven to show a high internal reliability, to correlate with individual differences in path integration, in the systematic update of one's orientation, and to straight-line distance estimation measures and map sketching (HEGARTY, RICHARDSON, MONTELLO, LOVELACE, & SUBBIAH 2002). Moreover, the SBSOD has systematically proven to be a useful instrument to predict environmental learning (HEGARTY & WALLER 2004; HEGARTY ET AL. 2006) and to evaluate new test instruments (KOZHEVNIKOV & HEGARTY 2001). In this study, however, the SBSOD was only a weak indicator of perspective transformation abilities.

found was due to the particular answer format both tests shared. The answer format was different from the one of tests reflecting object-based transformation abilities. Consequently, in their study, subjects completed a broad range of perspective taking tests involving three different test formats. Results from confirmatory analyses indicated that a single-factor-model derived significantly from the data whereas a two-factor model explained the data sufficiently well. Despite the high latent correlation ($r = 0.8$), HEGARTY & WALLER (2004) concluded that the separability of object manipulation and perspective transformation abilities can be considered to be independent of the tests chosen to reflect each of the latent constructs.

Although the analyses of both studies revealed a tendency towards the dissociation of object manipulation abilities and the abilities to transform the egocentric perspective, the high correlations between both latent constructs which were found might explain why the psychometric literature failed to distinguish both factors. Shared variance of both factors might not only be explained due to the fact that both abilities might rely on common cognitive processes such as encoding of spatial images and the abilities to maintain them in working memory, but also due to the fact that subjects might have used different strategies than intended by the researchers (see HEGARTY & WALLER 2004, for further elaboration).

To sum up, both studies derived a hypothetical model describing spatial abilities from the cognitive literature, and subsequently tested whether or not their model could explain data stemming from psychometric testing. Although their results indicated that the latent constructs shared a considerable amount of variance, both studies, in particular when considered together with the findings of the psychometric literature, provided sufficient evidence that at least for adults, modeling small-scale spatial abilities using two partially dissociated subclasses might be appropriate.

4.1.3   The Developmental Literature

Children's spatial thought develops rapidly during primary school years (e. g., NEWCOMBE, UTTAL, & SAUTER 2013). It remains, however, unclear whether children's abilities to perform object-based transformations are also different from the ones to transform the imagined self. In a first step, this section summarizes studies that have investigated the emergence of perspective taking abilities in contrast to the abilities to perform object-based transformations. In a second step, studies are reviewed that directly compared both subclasses of spatial abilities and their emergence.

Piaget and Inhelder's studies PIAGET AND INHELDER (1948/1956) studied children's conceptual representations of space.[6] They suggested that those arise from internalized actions (operations) in perceptual space (PIAGET, INHELDER, & SZEMINSKA 1981, p. 12)[7] and allow the child to represent objects in space, even though they are no longer visually perceivable. Conceptual thought refers therefore to the abilities to generate mental representations that can be manipulated in the mind to solve a given spatial task.

PIAGET AND INHELDER (1948/1956) claimed that children undergo four developmental periods in the conception of space that similar to the organization of intelligence. These periods are hierarchically organized and denoted as sensorimotor, preoperational, concrete operational, and formal operational space. When children develop a conception of space, they construct qualitative different concepts of organizing space, starting from topological to projective to Euclidean concepts. PIAGET AND INHELDER (1948/1956) assumed that there are three types of spatial content that characterize the child's conception of space (p. 44). Each consists of a set of spatial relations that remain invariant under certain transformations. A conception of topological space relies on simple relations such as proximity, separation, order, and enclosure. Concepts of projective space involve the understanding of different viewpoints and their coordination. Finally, understanding Euclidean concepts means that the child has acquired metric, measurement (distance), and angle concepts and is able to coordinate two or more dimensions. According to PIAGET AND INHELDER (1948/1956), the child conceptualizes both the perceptual and representational space according to those levels. Perceptual conceptualization appears, however, much earlier in the development than the representational conceptualization (PIAGET & INHELDER 1948/1956, pp. 44–45).

PIAGET AND INHELDER (1948/1956) studied the emergence of perspective taking when they examined the formation of concrete operational space, stage III, that is characterized by flexible structures of spatial thought. Stage III has two substages: stage IIIA that is characterized by the emergence of concrete operations, that is, actions that have become internalized but still concrete, hence are bound to real objects (PIAGET & INHELDER 1948/1956, p. x), and IIIB that is characterized by the integration of operations into logical structures that are no longer bound to real objects but may refer to abstract concepts (see HART & MOORE 1973, for further elaboration). The differentiation of stage III is therefore closely related to

---

[6]PIAGET AND INHELDER (1948/1956) distinguished between perceptual spatial thought and conceptual or representational spatial thought (pp. 3–4). Perceptual spatial thought refers the phenomena that the child can understand from direct visual perception.

[7]Consequently, they rejected the idea that conceptual representations are due to internalized perception (PIAGET & INHELDER 1969, pp. 68–69).

the emergence of projective concepts of space and was studied by PIAGET AND INHELDER (1948/1956) using the Three Mountains Task (pp. 209–246).

Their behavioral experiments demonstrated that children in stage II (younger than 7 years) tended to fail to take a different viewpoint of their own. During stage III (7 or 8 to 11 or 12 years), they showed a progressive improvement in the coordination of perspectives as they developed these abilities in substage IIIA and IIIB. Substage IIIA (7 or 8 to 9 years) is characterized by "the child's discovery that the left-right, before-behind relations (ignoring above-below) between the mountains vary according to the position of the observer" (PIAGET & INHELDER 1948/1956, p. 233) and hence represents a transitory stage between stage II that is governed by egocentric[8] answers as well as a perfect mastery of the Three Mountains Task. Children in substage IIIB (9 or 10 to 11 or 12 years) mastered to bring all spatial relationships into accordance. Even though the task was still challenging for them, they were able to correct false relationships that occurred during the solution process. Those findings were confirmed in an experiment asking children to reconstruct a model town (PIAGET & INHELDER 1948/1956, pp. 419–446).

PIAGET AND INHELDER (1966/1971) investigated children's abilities to imagine static and dynamic objects. As a result of their experiments, they concluded:

> It comes at about 7 to 8 years and points, if not to the actual emergence then at least to the rise of the anticipatory images, that is, to the kinetic and transformation images [...] But at about 7 to 8 years a capacity for imaginal anticipation makes its first appearance, enabling the subject to reconstitute kinetic or transformation processes, and even foresee other simple sequences. (PIAGET & INHELDER 1966/1971, p. 358)

They suggested that children aged younger than seven years lack the abilities to imagine dynamic objects since they are still in the preoperational stage. Mental images in this cognitive stage were assumed to be rather static and are therefore not able to represent any kind of movement or transformation in favor of mental representations of simple elements perceived within a given configuration (see PIAGET & INHELDER 1966/1971, p. 358). When children undergo the developmental sequence, that is, when they reach the concrete operational stage of cognitive devel-

---

[8]PIAGET AND INHELDER (1948/1956) used the expression of egocentrism from a descriptive viewpoint rather than assuming that egocentrism is a typical trait of infants and younger children. It does therefore not serve to explain development but to describe a response pattern (see NEWCOMBE 1989, pp. 205; for a discussion on egocentrism in Piaget's theory and beyond).

opment by means of interaction with their environment and because of the processes of assimilation and accomodation,[9] they are also able to imagine dynamic objects:

> [...] after 7 to 8 years the child becomes capable of thinking in terms of transformations, thanks to the operational structures brought about by the equilibration of his intellectual processes, and that the retroactive and anticipatory mobility of the operations will, sooner or later, be reflected in the images themselves and will enrich the outlines of anticipation associated with them. (PIAGET & INHELDER 1966/1971, p. 359)

In summary, the developmental studies of PIAGET AND INHELDER (1948/1956, 1966/1971) predicted that children are able to solve mental rotation problems at the age of 7 to 8, whereas perspective taking abilities arise at the age of 9 to 10 years. However, they did study the emergence of both phenomena of spatial thought in different experimental settings and not in a setting that would have allowed for direct comparison.

Post-Piagetian Studies Inspired by the studies of PIAGET AND INHELDER (1948/1956, 1966/1971), researchers have systematically addressed the question of whether children might be able to coordinate perspectives before the age of 9 or 10 or to imagine dynamic objects before the age of 7 to 8.

Among the most influential studies that addressed the emergence of perspective taking were those of FLAVELL, EVERETT, CROFT, AND FLAVELL (1981) and MASANGKAY ET AL. (1974, as cited in FRICK, MÖHRING, & NEWCOMBE 2014, p. 1). They proposed that perspective taking abilities might be demonstrated to appear in a child's development quite early if one differentiates these abilities. They proposed to distinguish the child's knowledge between what another person might see (Level 1) from the child's abilities to make mental inferences on what the other person sees in detail (Level 2). A child at Level 1 thus knows that the other person sees an object in another way the child itself does, or that this person might even see an object that is invisible for him or her (MASANGKAY ET AL., 1974, as cited in FRICK ET AL. 2014, p. 1). These abilities have been shown to appear as early as in infancy (SODIAN, THOERMER, & METZ 2007). A child at Level 2 is able to compute another person's view on an object configuration by applying rules such as one observer has

---

[9] Assimilation refers to the mechanism that integrates new experiences into an already existing schema (PIAGET & INHELDER 1969, p. 5). Accomodation refers to a modification or extension of a present schema to an experienced situation that cannot be integrated into the actual scheme (PIAGET & INHELDER 1969, p. 6). Both invariant processes work hand in hand in an equilibrated way to allow cognitive progression of a mental representation that becomes more and more flexible, yet stable during maturation.

a unique view on a situation, observers at different spatial locations have different viewpoints, observers at the same spatial location have the same view. Children at Level 2 understand that the perceived size and clearness of objects varies with respect to the own viewing location (see also NEWCOMBE 1989, for further elaboration). Understanding Level 2 rules has been shown to emerge by the age of 4 to 5 (FLAVELL ET AL., 1980, as cited in NEWCOMBE 1989, p. 210) and improves rapidly between the age of 6 to 8 (SALATAS AND FLAVELL, 1976, as cited in FRICK ET AL. FRICK, MÖHRING, & NEWCOMBE 2014, p. 1).

Mental rotation studies as one example of dynamic imagery have demonstrated that precursor abilities of mental rotation might already emerge by the age of 4. At this stage, children are able to solve tasks that are equivalent to those of adults tasks MARMOR 1975. During childhood, performances increase consistently with response patterns showing answers above chance and linear response reaction time patterns starting from the age of 6 or 7 till the age of 10 (ESTES 1998; PELLEGRINO & KAIL 1982, see HARRIS, HIRSH-PASEK, & NEWCOMBE 2013; REINHOLD 2007, for further elaboration). The emergence of non-rigid transformation abilities such as paper folding is less well studied but has been shown to emerge at the age of 5.5 years with precursor abilities emerging at the age of 3 (HARRIS, HIRSH-PASEK, & NEWCOMBE 2013; HARRIS, NEWCOMBE & HIRSH-PASEK 2013).

To sum up, those studies demonstrated that children's abilities to perform mental transformations of objects and egocentric perspective transformations emerge at an earlier age than predicted by PIAGET AND INHELDER (1948/1956, 1966/1971). All these studies, however, addressed each subclass of spatial abilities for their own sake but did not compare the different subclasses directly.

HUTTENLOCHER AND PRESSON (1973) studied the cognitive processes involved in object versus perspective transformations. They investigated the children's egocentric choices when anticipating the appearance of a configuration of objects and reported that those occurred exclusively for perspective transformations, suggesting that these results are an indication that the task is too difficult for the children. They suggested that difficulties emerged because the task required children to relate the self, the (imagined) observer and the object array at two different points of time. Moreover, they found that perspective tasks become indeed more difficult if there is an incongruence between observer and the self that requires the uncoupling and recoding of the egocentric viewpoint from the one of the observer which is an additional cognitive effort. This phenomenom has also been described as the problem of conflicting frames of reference in the literature (e. g., FRICK ET AL. 2014; NEWCOMBE 1989) and might explain why the children in their study performed better in object rotation than in perspective rotation tasks.

In a subsequent study, HUTTENLOCHER AND PRESSON (1979) showed that third grader's problems in perspective tasks might be dependent on the task instruction itself. Indeed, children were facing more problems and producing more egocentric errors when solving appearance questions, asking for alternative appearances of the whole array than they were with item questions, asking for one single item in the configuration (e. g., "Which object is left of the observer?").

Two main conclusions can be drawn from the studies of HUTTENLOCHER AND PRESSON (1973, 1979). First, there is an asymmetry between object versus observer movement, that is, children produce different patterns of errors for both tasks, involving systematic egocentric errors whenever they are too much challenged too much in perspective tasks. This is not the case for array rotation tasks. This finding can be taken as an indication that children have indeed developed two distinct subclasses of spatial abilities by the age of primary school. Second, the occurrence of egocentric errors is task-dependent. In particular, identifying the position of single features seems to be easier than determining the appearance of a whole configuration (e. g., by choosing pictures). Consequently, when conceptualizing the abilities to perform transformations of the imagined self as abilities that enable individuals to solve array questions, thereby controlling for possible conflicting frames of reference, it is likely to assume that perspective taking abilities seem indeed to emerge later than the abilities to mentally rotate objects.

On the Emergence Of A Possible Dissociation   Recent studies took up those classical findings to investigate whether object manipulation abilities and perspective transformation abilities can be described by divergent developmental trajectories and understood as different subclasses of spatial abilities (CAEYENBERGHS, TSOUPAS, WILSON, & SMITS-ENGELSMAN 2009; CRESCENTINI, FABBRO, & URGESI 2014). Findings from the neuropsychological literature showed that there is no empirical relationship between the subclass of motor imagery abilities (measured by left-right judgements of hands) and the subclass of visual imagery (measured by orientation judgements of letters) of children aged from 7 to 12. Their results showed that motor imagery improved by the age of 7 or 8 when visual imagery abilities seemed to be almost fully developed (CAEYENBERGHS ET AL. 2009). These findings were completed by results of the study of CRESCENTINI ET AL. (2014) who compared reaction times and accuracy profiles of children aged from 7 to 11 years who solved both an own body and a letter transformation task. The corresponding profiles differed and an analysis based on the ratio between reaction times and accuracy revealed that there were significant developmental differences between egocentric perspective transformation abilities and object transformation abilities. The former seem to emerge by the age of 8 to 9 and therefore develop later than object transformation abilities that were found to emerge by the age of 7. CRESCENTINI ET AL. (2014) concluded that different developmental trajectories characterize both subclasses of small-scale spatial abilities which may indicate that both abilities are dissociable.

VANDER HEYDEN, HUIZINGA, KAN, AND JOLLES (2016) studied the developmental aspects between a possible dissociation of object and perspective transformation abilities. Children between 8 and 12 years old solved a paper-and-pencil mental rotation task, a paper folding task, and a cube rotation task using manipulable material. Those tasks measured object manipulation abilities. There were three tasks involving small manipulable material. Those tasks measured perspective transformation abilities. To investigate the emergence of a possible dissociation between both subclasses of small-scale spatial abilities, VANDER HEYDEN ET AL. (2016) compared four different multigroup models using confirmatory factor analyses. By doing so, they were able to demonstrate that a dissociation model explained the data of all age groups well. The best fit, however, was found with a model that described spatial abilities as a unitary construct until the age of 8, and a second model describing spatial abilities as a unitary construct until the age of 10. Despite the fact that the fit indices of both models did not differ, the authors conjectured to accept the latter model to describe the data best for reasons of model parsimony and concluded that a dissociation occurs by the age of 10.5, with object-based transformation abilities and transformation of the self abilities being highly related ($r = .75$). Although these results provide important insights concerning the development of a possible dissociation of two subclasses of spatial abilities in children, they seem to be less clear than articulated in the study. First, two of the three measures reflecting viewer transformation abilities tended to show floor effects for all age groups which might have affected the statistical estimation procedure. Second, it is less clearer which of the models outlined above should be preferred from a methodological point of view and concerning statistical significance of the difference between both models outlined.

MIX ET AL. (2016) were not able to find empirical evidence that children's small-scale spatial abilities can be modeled in a multidimensional manner. For each of the age groups under consideration (kindergarten, third graders, and sixth graders), their confirmatory analyses best explained the data with a single-factor model. One explanation for their finding might be the choice of items that was rather complex and may rely on several spatial tasks such as scaling, mental rotation and interpreting 2D representations.

Recently, FRICK (2019) investigated the latent structure of dynamic spatial abilities of kindergarten children using a set of six different paper-and-pencil tasks and exploratory factor analysis. She found empirical evidence that their spatial abilities can be described in a two component model. In particular, she found that perspective taking abilities are distinguishable from mental rotation abilities. Mental rotation, as basic spatial abilities particularly related to spatial scaling and the two of them constituted a factor that involved the manipulation of objects from a stationary point of view. Perspective taking was related to three other tasks that were strongly observer-related.

Conclusion

A review of the psychometric, individual differences, and developmental literature showed that the assumption to model small-scale spatial abilities as being a conglomerate of two closely related but conceptually distinct subclasses, namely the abilities to perform mental transformations of objects and the imagined self, is based on sufficient empirical evidence. In other words, there seem to be not only two types of mental transformations that are characterized by qualitatively different cognitive operations but they also seem to tap two distinct underlying spatial abilities. Consquently, when an individual performs well in tasks requiring object-based transformations, he or she may not necessarily perform well in tasks demanding egocentric perspective transformations. Findings from the psychometric literature and developmental studies that addressed the emergence of both subclasses provided rather indirect empirical evidence for the two dissociated subclasses. They were complemented by those individual differences and developmental studies that have shown a possible dissociation in adults and children.

4.2   The Dissociation of Two or More Subclasses of Large-Scale Spatial Abilities

Section 3.4 suggested that one way to describe large-scale spatial abilities could be to use a model with at least two components: one describing the abilities to make inferences from a survey map and another one describing the abilities to maintain the map-environment-self relating during movement in space. When staying oriented with the help of a map, individuals apply two types of transformations to make inferences, namely map-to-environment and environment-to-map transformations. Consequently, it was suggested that these different types of transformations could account for a further subdivision of the subclass of large-scale spatial abilities involved when using maps.

This section reviews the literature that has addressed the possible dissociation of large-scale spatial abilities. In comparison to small-scale spatial abilities, it is difficult to find evidence in the literature because large-scale spatial abilities have mainly been addressed in studies investigating how knowledge is acquired and stored, or what kind of knowledge is acquired by individuals. Another reason is that large-scale spatial abilities can be measured in a variety of ways and that measuring them is usually bound to some kind of specific environment, which makes it hard to compare studies and to classify them (e. g., SAS & MOHD NOOR 2009). For example, a range of measures has been developed for measuring individuals abilities to make inferences from what they experienced in space (see Table B.4 in the appendix and COLUCCIA & LOUSE 2004; WALLER 1999, for further elaboration).

In a first step, this section summarizes the few existing studies in the psychometric literature that provided evidence for a possible dissociation between the subclass of survey mapping abilities and the map-based one. In a second step, this section addresses the individual differences literature, which provided some evidence that children's map-reading abilities should not be modeled as a unitary construct. In a third and final step, developmental findings are summarized and discussed with respect to the assumed dissociations.

4.2.1   The Psychometric Literature

First attempts to identify subclasses of large-scale spatial abilities have been made in psychometric studies that focused on the relation between small-scale and large-scale spatial abilities (see Section 4.3) in adults. The factor structures of large-scale abilities has therefore been computed as a base for further statistical analyses but not for their own right. Although there was, and is, "too little empirical evidence to support unequivocal statements about the factor structure of this ability" (HEGARTY & WALLER 2005, p. 149), the following section presents results of the studies that modeled large-scale spatial abilities.

LORENZ AND NEISSER (1986, as cited in HEGARTY & WALLER 2005, p. 148) studied the factor structure of knowledge stored in cognitive maps and reported on three possible dimensions: landmark memory, route memory and sensitivity to overall geographical directions. However, a replication study failed to confirm the results in a larger experimental setup (HEGARTY & WALLER 2005, p. 148).

ALLEN ET AL. (1996) separated one major factor, denoted as topological knowledge from two manifest variables that required the use of metric knowledge (Euclidean distance and direction error). Topological knowledge was reflected by tests that relied on landmark and route knowledge such as walking an experienced route in reverse direction or picture sorting of encountered scenes. Topological knowledge was shown to be separable from Euclidean measures that drew upon survey knowledge as it is measured by indicating directions and estimating distances toward unseen landmarks.

HEGARTY ET AL. (2006) tested adults in direction and distance estimation measures in three different learning environments (direct experience, videotape and virtual environment). Participants also drew sketch maps of the environment they learned. Exploratory factor analysis revealed two factors that differed with respect to the learning medium but not the kind of measure used: "[a]n important finding of this study is that measures of spatial learning from a real environment and measures of learning from visual media are partially dissociated in that they load on different factors" (HEGARTY ET AL. 2006, p. 170).

To sum up, although both models presented by ALLEN ET AL.'S (1996) and HEGARTY ET AL. (2006) refer to knowledge encoded in cognitive maps, a tentative interpretation of those findings might be that they provide some support for the dissociations outlined above. Indeed, when interpreting survey mapping as one form of topological knowledge involving landmark and route knowledge gained from direct experience and map-based reasoning in the sense of asking subjects to draw upon metric knowledge based on information acquired from visual media, they indicate that a two-factor model describing large-scale spatial abilities could be likely to assume.

4.2.2   The Individual Differences Literature

Two studies in the individual differences literature in children addressed the question of whether there are different subclasses of the abilities to keep track of the varying map-environment-self relations from an individual differences point of view. LIBEN (1997) contributed to the literature from a methodological point of view, suggesting two conceptually distinct subclasses of map-based tasks that require map-to-environment and environment-to-map transformations. CHRISTENSEN (2011) studied individual differences in performances of children in both subclasses and related them to the children's performances in survey mapping.

Liben's Taxonomy of Map-Based Tasks in Real Space

LIBEN (1997) distinguished between different methods for investigating how individuals use maps in real space (Figure 4.2). Although she explicitly referred to studies of maps (thus studies that focus on spatial products in the sense of LIBEN 1981), her theoretical ideas are important when investigating map-based spatial thought. Her taxonomy is based on different pathways that may link the constructs representation, in particular the map (REP), the represented space (PLACE), the child, the task and the dependent variable (DV).

Production methods (Figure 4.2 (a)) "start from the child's experience with some space, and then require that the child translates some aspect of that experience to a spatial representation" (LIBEN 1997, p. 46). The crucial point in production methods is the need for referring information perceived on the environment to the map. Consequently, they can be interpreted as measures of the abilities to perform an environment-to-map transformation by means of self-location on the map (see Section 3.3.3).

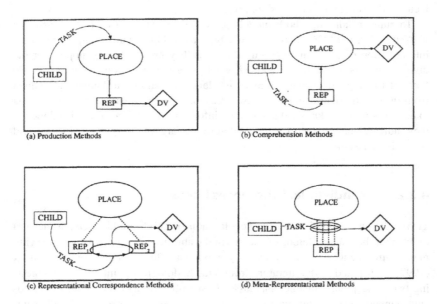

Figure 4.2 Four different pathways can be taken when studying the understanding of spatial representations such as maps (in LIBEN 1997, p. 45). ©1997 L. S. Liben, Psychological Press. Reprinted with permission

Examples of production methods are studies that involve the drawing of maps of previously experienced environments (e. g., QUAISER-POHL, LEHMANN, & EID 2004) or studies that involve placing stickers on a map that represent the experimenter (LIBEN & DOWNS 1993) or particular objects such as flags (LIBEN ET AL.,

2008; LIBEN, MYERS, CHRISTENSEN, & BOWER 2013). Those can be used when testing both adults (LIBEN, MYERS, & CHRISTENSEN 2010) and children (KASTENS & LIBEN 2010; LIBEN ET AL. 2013). Concerning the design of flag-sticker tasks, the experimenter may actively show the subject a pre-defined route with particular locations (flags) in a pre-defined order (LIBEN ET AL. 2008, 2010). The flags may also be discovered individually with the experimenter shadowing the subject or being absent (KASTENS & LIBEN 2010; LIBEN ET AL. 2013). Some flag-sticker tasks require the individuals to record their location only, whereas in other cases, the viewing perspective is additionally required.[10]

Comprehension methods (Figure 4.2 (b)) are particular interesting for the study of map-based navigation as they are "those in which the child is first given some spatial representation of place, and then asked to translate some aspect of that representation in some way to the real referent space" (LIBEN 1997, p. 51). Comprehension methods require to refer information gained from the map to the environment. Consequently, they can be interpreted as measures of the abilities to perform a map-to-environment transformation by means of place-identification (see Section 3.3.3).

Typical examples are seeking hidden toys (e. g., PRESSON 1982b) or navigation to a particular landmark (e. g., GERBER & KWAN 1994; MALINOWSKI & GILLESPIE 2001; WRENGER 2015). A simpler method not to study the wayfinding process itself but the product of it is the place-marker-test. It can be described as "a map-to-reality test, [in which] students place large, colored, numbered disks on the ground to indicate the location of similarly-colored, numbered stickers on a map" (KASTENS, KAPLAN, & CHRISTIE-BLICK 2001, p. 249).

Representational correspondence methods (Figure 4.2 (c)) investigate the abilities to establish an isomorphism between two different representations of the same environment, as the subject is "asked to relate, in some way, one spatial representation of place to another spatial representation of the same place" (LIBEN 1997, p. 54). A typical task may be one in which children would have to find flags in a small-scale model of a landscape and to put the corresponding colored stickers in a map of that landscape (LIBEN & DOWNS 1991).

Finally, meta-representational methods (Figure 4.2 (d)) are verbal in their nature as they investigate the "conscious understanding of the relations between referent spaces and their representations, as well as an understanding of place representations as symbol systems" (LIBEN 1997, p. 56). Typical studies focus on understanding the

---

[10]The second case is based upon an argumentation of LIBEN AND DOWNS (1993) who stated that it is necessary to differentiate between the two factors position and orientation to capture the problem of self-location on a map.

child's conceptions of maps and their use and the manner of how children understand advanced concepts of maps and their use.

LIBEN (1997) assumed that the abilities to solve comprehension and production tasks do not differ:

> At a somewhat deeper level of analysis, it remains an open question whether the fundamental processes that underlie the ability to perform production tasks are the same as those for comprehension tasks. As a working assumption, there is no particular reason to suppose that the kinds of basic skills and processes that underlie performance on production and comprehension tasks should differ. (LIBEN 1997, p. 53)

Christensen's Study on Children's Performances in Production and Comprehension Tasks

CHRISTENSEN (2011) questioned whether LIBEN'S (1997) assumption could be confirmed for children. He referred to first evidence concerning a possible dissociation found in a study of SIEGEL, HERMAN, ALLEN, AND KIRASIC (1979) who investigated children from kindergarten, grade 2, and grade 10 exploring a model town either in a Gymnasium or on a tabletop setting. Then, they led the children reconstruct the town either at the same scale they had experienced and learned the layout, or in the other setting. Results showed that performances of children did not differ with respect to the size of the layout initially learned but they did differ when they were asked to reconstruct the model town in the same scale as explored. Performances were best when individuals who learned the layout of the model in small-size, reconstructed it also in small-size. Similarly, individuals who learned the layout of the model in large-size, reconstructed it also better in large-size. However, significant differences were found when constructing in the other space:

> Children who were exposed to a small-scale space and then required to translate that knowledge into a large-scale space performed least accurately, while children who were exposed to a large-scale space and required to translate knowledge into a small-scale space did not perform significantly worse than children who were not required to translate their spatial knowledge. (SIEGEL ET AL. 1979, p. 584)

CHRISTENSEN (2011) interpreted the results of SIEGEL ET AL. (1979) as suggestive evidence that production (going from the large referent space to the small representation) and comprehension tasks (going from the small representation to the large referent space) may not be equivalent.

In his study, CHRISTENSEN (2011) then addressed LIBEN'S (1997) assumption by directly comparing children's performances when solving either production or comprehension tasks with maps in real space. Moreover, he studied whether solving map-based tasks resulted implicitely in a better survey knowledge than acquiring knowledge from direct experience. Children aged between 9 and 10 years old were completing one of the following two tasks: they were either placing stickers on a map to indicate their location in space after searching for flags on a campus area (production tasks) or they were placing flags on the campus according to locations indicated on a map (comprehension task). Survey knowledge was tested with a pointing task requiring the children to indicate directions to unseen landmarks and a task that required the children to determine the shortest routes to a landmark.

Results of a comparison of accuracy in sticker or flag placement showed that there were significant differences between children who completed the production task and those who completed the comprehension task (CHRISTENSEN 2011, p. 37). Children who engaged in the comprehension task were more accurate than children completing the production task. There were, however, no effects concerning the encoding of a cognitive map. Children who solved any of the two map-based tasks did not perform better in pointing to unseen landmarks or determining the shortest routes toward them than children who randomly explored the environment.

Besides those quantitative results, CHRISTENSEN (2011) found that conceptual differences in the task led to qualitatively different behaviors such as strategy use, a finding from which he concluded that

> ... it is unlikely that the two kinds of tasks are relying on the same processes. The comprehension group relies more on the map which is suggestive that they are more engaged in planning their routes through the environment to reach the desired locations. In relying significantly less on the map between flags, the production group may be taxing their visuospatial working memory. If one is not attending to the map regularly, then it would be necessary to keep one's location on the map in working memory instead of off loading information about location onto the external representation of the map. (CHRISTENSEN 2011, p. 53)

CHRISTENSEN (2011) study provides therefore empirical evidence for the assumption that there seem to be two different subclasses of large-scale spatial abilities when it comes to map use. His findings, however, do not support the assumption that survey mapping abilities are partially dissociated from the abilities to use maps in real space.

His study possibly had a drawback that might have obscured a possible dissociation. CHRISTENSEN (2011) used a between-subject research design, that is, children were either in the production-task group or in the comprehension task group. Consequently, performances in both groups were compared but children were only tested in one of the experimental conditions, solving either production or comprehension tasks. Although "[a]ssignment [to the group] was done as randomly as possible" (CHRISTENSEN 2011, p. 23), the sample size was low for the map group (n = 38) which may has led to group effects and biased findings concerning a possible relation to acquire survey knowledge.

Taken together, those studies indicate that the abilities to stay oriented and use maps during navigation should most likely not be considered as a unitary construct. Production and comprehension tasks have not only been distinguished from a conceptual point of view in LIBEN'S (1997) study but seem to involve different subclasses of abilities since it has also been shown that children who perform well in production tasks do not necessarily do so in comprehension tasks (CHRISTENSEN 2011).

4.2.3   The Developmental Literature

Although PIAGET AND INHELDER (1948/1956) did not address large-scale spatial cognition, their comprehensive research on spatial concepts inspired researchers that primarily investigated cognitive mapping (see LEHNUNG 2000, for further elaboration on how children acquire cognitive maps) and children's abilities to read maps (e. g., LIBEN ET AL. 2002).

The Emergence of Children's Cognitive Maps
SIEGEL AND WHITE (1975) described what kind of knowledge children acquire subsequently when encoding a cognitive map. They proposed that children undergo a developmental sequence leading to the formation of survey or configurational knowledge as moving from landmark knowledge over route knowledge to knowledge linking between multiple routes and landmarks. In other words, they suggested that when children move in an unknown environment, such as a new neighborhood, according to their framework, they first memorize a few salient landmarks that they recognize based on their orientation in space. As they learn important routes, such as the one to the kindergarten, to the playground, or to the park, they come to think about their neighborhood in terms of those routes that, in turn, also involve the memorized landmarks. When familiarizing with the neighborhood, those routes and landmarks become subsequently integrated into a whole, which allows them to understand the relation between the kindergarten, the playground and the park in a more nuanced manner.

SIEGEL AND WHITE'S (1975) framework is in line with PIAGET AND INHELDER'S (1948/1956) one. Indeed, knowing where a child is located with respect to memorized landmarks involves reasoning about proximity and closure. In other words, to do so, children have to apply topological concepts. When children come to reason about routes, they might think about left-right turns and how they differ with respect to the direction in which the route is walked, thereby drawing upon projective concepts. Finally, when they reason about landmarks and routes as a whole, probably to make inferences about shortcuts, directions and distances, they draw upon Euclidean concepts that allow them to grasp space in metric terms (CHRISTENSEN 2011).

Whereas SIEGEL AND WHITE (1975) addressed the kind of knowledge that children encode in their cognitive maps, HART & MOORE (1973) examined how they might represent knowledge, thereby combining PIAGET AND INHELDER'S (1948/1956) theory of the development of spatial cognition with their own conceptualization of how knowledge from direct experience may be represented in children's cognitive maps (denoted as topographical representations in their study). HART AND MOORE (1973) stated that frames of reference are essential elements of cognitive maps and that the use of those frames of reference undergoes a particular developmental sequence from egocentric to fixed and finally coordinated frames of reference.

> There is considerable evidence that in developing topographical representations of the large-scale environment, the child utilizes a framework or system of reference for interrelating different positions, routes, patterns of movements, and himself in this environment, and that this system of reference is the most important component of spatial representation. Furthermore, the ontogenetic development of systems of reference proceeds through three stages: egocentric, fixed, and coordinated, the third stage being achieved in children with the equilibrium of concrete operations. (HART & MOORE 1973, p. 283)

Egocentric orientation is characterized by an orientation system that is grounded upon actions of the child and centered around its body. The child does not have an idea of space as a whole but represents space in terms of routes. The stage of fixed frames of reference replaces the egocentric orientation system as is meant to be a "transitional stage between early preoperational egocentric orientation and concrete operational coordinated systems of reference" (HART & MOORE 1973, p. 279). The stage of fixed frames of reference is characterized by partially coordinated and interlinked landmarks that are the central organizing feature. Although the children are unable to coordinate all landmarks into one coherent representation, they are able to relate landmarks of discrete viewpoints to each other. Once the child has developed to use a coding system that is based upon abstract geometrical axes and full coordination of landmarks, it is in the final stage of the developmental sequence that is characterized by the use of coordinated frames of reference.

All three theoretical frameworks, PIAGET AND INHELDER'S (1948/1956), SIEGEL AND WHITE'S (1975), and HART AND MOORE'S (1973) might become important when concluding about the underlying concepts that children are asked to have acquired when making inferences from encoded knowledge. One of the most used measures of the abilities to make inferences from a cognitive map are pointing tasks. During a pointing task, the subject is located in the environment and has to take a pre-defined viewing direction. Then, the subject has to point toward a known but unseen landmark in the environment. The direction in which the subjects points is taken as an estimate of the direction of the landmark in the subject's cognitive or survey map. Pointing tasks have been shown to be a valid method for measuring survey knowledge in adults (MONTELLO & PICK 1993) and children (see LEHNUNG 2000, pp. 61–62).

To solve tasks such as pointing to memorized but unseen landmarks, children are required to derive metric concepts in the sense of PIAGET AND INHELDER (1948/1956), by making inferences from coordinated frames of reference in the sense of HART AND MOORE (1973), thereby relying on survey knowledge in the sense of SIEGEL AND WHITE (1975). One might conclude that children should struggle to perform well on pointing tasks until the end of the concrete operational stage at the end of middle childhood.

This assumption finds mixed support in the literature. Depending on the concrete design of the pointing task, the familiarity of the environment and the exposure time, children either demonstrate the abilities to point to memorized but unseen landmarks at the beginning of primary school (LEHNUNG, HAALAND, POHL, & LEPLOW 2001), or they tend to struggle to do so until the end of primary school. In particular when being exposed to a large, unknown environment, younger children require repeated exposure to form survey knowledge stored in fixed frames of reference, since they cannot integrate the spatial information experienced in a cognitive map as fast as older children do (HERMAN ET AL. 1987, as cited in LEHNUNG ET AL. 2001, p. 284).

To sum up, although children can be expected to be able to use pointing to externalize knowledge stored in a cognitive map by the end of primary school, the developmental theories outlined above suggest that they might remain challenged to do so until the end of primary school when it comes to solving pointing tasks in large, unknown environments.

The Emergence and Development of Map-Reading Abilities

Children aged only three years old have been shown to be able to read an aligned map during a simple comprehension task (i. e., finding a toy with the help of the map) in a controlled small-scale experimental setting (BLUESTEIN & ACREDOLO 1979; DELOACHE 1987; DELOACHE 1989). To do so, they relied on topological concepts that are particular useful whenever a single or a salient landmark was represented on the aligned map (LIBEN ET AL. 2002, p. 283).[11] In an unaligned condition, however, younger children struggle but 5-year-old children are able to read the map (PRESSON 1982a).

The emergence of the abilities to locate themselves in a simple production task has been shown for children as young as 4-year-old (BLADES & SPENCER 1987). Those abilities have been shown to develop gradually during early and middle childhood but remain sensitive to unaligned map effects (LIBEN & DOWNS 1993). In addition, younger children have been shown to struggle with maps that lack landmark information, even if the maps depict familiar environments such as the classroom (LIBEN & YEKEL 1996).

Due to the fact that younger children have not yet developed projective and Euclidean concepts, even when performing well in rather limited experimental settings as those underlying the studies outlined above, they remain challenged by tasks involving unaligned maps or ones featuring less landmark information. Even when they are in the age of middle childhood and have undergone important developmental changes, they struggle to consistently apply those acquired concepts to map tasks that are posed under realistic experimental settings and mimic the everyday use of them. LIBEN ET AL. (2013), for example, studied self-location abilities among 9 to 10 year old children as one example of production tasks. The children explored an unknown campus with a map, searched for colored flags and had to indicate the position of the flag with the help of a colored sticker. Results indicated that children at the end of primary school still struggle to do so:

> Individual data records suggest that a significant number of children were deeply confused about symbols and basic environment-map correspondence. [...] Composite maps suggest that some children have not yet mastered an understanding of vantage point, with errors commonly involving reflections around an axis of symmetry. [...] failing to understand the direction from which the map or space is being viewed. Likewise, many children appear to have not yet mastered proportion, scale, or the use of more than one axis to identify a location, all skills associated with a mature understanding of measurement: They place stickers in the correct region and on the correct kind of symbol, but are imprecise. (LIBEN ET AL. 2013, p. 2059)

---

[11] These findings do therefore not contradict PIAGET AND INHELDER'S (1948/1956) theory when interpreting the corresponding spatial behavior as being based on perceptual information (e. g., a single landmark on the map and a single object in space) rather than from representational information gained from the map. Since PIAGET AND INHELDER (1948/1956) distinguished between perceptual space and representational space and argued that spatial concepts in perceptual space are achieved much earlier than in the latter one, this might explain the findings outlined above (see BLADES 1989, for further elaboration).

KASTENS AND LIBEN (2010) reported that fourth graders preferably used topological concepts of proximity and continuity in verbal descriptions during self-location tasks. Some used projective concepts but none of the children referred to Euclidean concepts such as distance or angle when explaining how they identified their location on the map in a flag-sticker task (KASTENS & LIBEN 2010, p. 323). Consequently, tasks were easier for them when they could rely on topological information only to determine their location on a map, for example, when finding a flag next to a fountain when only one fountain symbol appeared on the map. If such pieces of information missed, children had to use projective and/or Euclidean concepts to determine their location on the map, which was still challenging for them. However, some children bypassed the use of those concepts by combining multispatial information, that is, two or more separate topological information that allowed them to determine their position on the map (KASTENS & LIBEN 2010).

Two conclusions can be drawn from what has been outlined above. First, similar to what has been found for pointing tasks, maintaining the map-environment-self relation seems challenging for children in real and probably unknown space until the end of primary school. Second, both for drawing inferences from a cognitive map and for drawing inferences from a paper map, children have to rely on topological, projective, and Euclidean concepts. In the case of pointing to unseen landmarks without a map, however, they mandatorily have to use projective and Euclidean concepts to reason about their location in space and the ones of other landmarks. When engaging in map use, however, they can bypass this cognitive demand—at least for the involvement of Euclidean concepts—by combining multiple topological cues. Consequently, children who have not yet developed those concepts entirely might be able to perform well in map-based tasks but might struggle to complete pointing-to-unseen-landmark tasks. One could therefore assume that the findings outlined above provide a tentative support for a possible dissociation between a factor representing the children's abilities to encode a cognitive map and to draw inferences from it, and another factor representing their abilities to maintain the map-environment-self relations during completion of tasks that mimic the everyday uses of maps.

Conclusion

The assumption to model large-scale spatial abilities as a conglomerate of two conceptually distinct subclasses, namely the abilities to encode a survey map and the abilities to stay oriented in real space with the help of a map, found mixed support in the literature. The only study that directly addressed this question found no evidence for a dissociation in children, a result that is in contrast with some of the psychometric findings in adults and the argumentation that both subclasses may involve different kinds of Piagetian concepts during reasoning. In contrast, the abilities to make transformations from the map to the environment during comprehension tasks and the abilities to make transformations from the environment to the map during production tasks seem not only to be conceptually different, but have also been found to be distinct from an individual differences point of view. Taken together, these findings provide some evidence for modeling large-scale spatial abilities as involving two, probably three distinct subclasses.

4.3   The Relationship Between Small-Scale and Large-Scale Spatial Abilities

Section 3.4 concluded that small-scale and large-scale spatial abilities might rely, to some extent, but not completely, on the same cognitive processes. It is safe to assume that there is at least some kind of dissociation between both classes of spatial abilities. Indeed, neuropsychological studies have shown that solving small-scale and large-scale spatial tasks rely on distinct neural structures (MONTELLO 2005). Engaging with small-scale spatial abilities is typically associated with parietal and posterior activities (ZACKS & MICHELON 2005), whereas hippocampal areas are involved whenever an individual engages with large-scale spatial information (see ARNOLD ET AL. 2013; WOLBERS & HEGARTY 2010, for further elaboration).

In line with HEGARTY ET AL. (2006), there are three possible relationships between small-scale and large-scale spatial abilities (Figure 4.3). The total dissociation model states that spatial abilities in small-scale and large-scale spaces are different, thus not sharing any cognitive processes at all. The partial dissociation model states that there are some underlying processes between both abilities, but separate, specific cognitive processes for small-scale and for large-scale spatial abilities. Whenever both abilities relate to each other via a mediating variable, their relationship is best described by the mediation model.

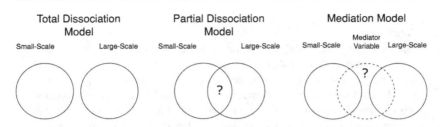

Figure 4.3 Possible relationships between spatial abilities at different scales of space (adopted from HEGARTY ET AL. (2006), p. 152). ©2006 Elsevier. Adopted with permission

This section reviews the literature that has investigated the question which of the three models might hold true for children from an individual differences perspective. In other words, this section summarizes the literature that has shown that individual differences in both classes of spatial abilities exist and to which extent performances in one class are related to performances in the other class. This section addresses findings in children and provides additional comparisons with adults when drawing conclusions from the studies outlined. In a first step, the findings of studies that have modeled children's large-scale spatial abilities as unaided navigation are summarized. In a next step, this section reviews studies that required children to navigate in real space with the help of maps.

4.3.1   Studies Without Maps

QUAISER-POHL ET AL. (2004) investigated spatial abilities of children aged 7 to 12 years at different scales of space. 439 children of German school grade two, four (primary school), and six (secondary school) completed a test involving a Water-Level Test (WLT), a Road-and-Frame Test (RFT), and a test of mental rotation (MR). Children further made a sketch map of a highly familiar environment (their neighborhood). Stages of the children's sketch maps were analyzed according to HART AND MOORE'S (1973) and SIEGEL AND WHITE'S (1975) frameworks, revealing different numbers of landmarks and route sections, different frames of reference and stages concerning knowledge representation in their cognitive maps (landmark vs. route vs. configurational). After integrating single items into latent variables, nine latent correlations (ranging from .03 to .17) were computed but did not show significancy. Results of an integrated SEM showed that small-scale and large-scale abilities were two independent constructs:

The close to zero correlations between the cognitive mapping variables and the spatial-test scores clearly show that there is high convergence between the measures belonging to the same domain (spatial ability vs. large-scale representations) but there is strong discriminant validity between the measures belonging to the two different domains. [...] the two domains of spatial ability and cognitive mapping are two uncorrelated concepts. (QUAISER-POHL ET AL. 2004, p. 104)

JANSEN (2009) investigated the effect of a short mental rotation training (< 60 min.) on performances in direction estimation tasks. 72 children aged between 9 to 10 completed a pre- and post-test on direction estimation. Between the tests, they where either trained in manual mental rotation (superposition of two Shepard-Metzler-like objects with a joystick on the computer) or played a non-spatial game. Her results showed that the intervention in mental rotation had no effects on the performance in the large-scale task. These findings contrasted, however, those from another study with adults (JANSEN, WIEDENBAUER, & HAHN 2010). She concluded from the lack of training effects that children's small-scale and large-scale spatial abilities should be modeled to be dissociated abilities.

FENNER, HEATHCOTE, AND JERRAMS-SMITH (2000) studied the impact of small-scale spatial abilities on wayfinding performance from a developmental perspective. They studied children in two age groups (5 to 6 and 9 to 10 years) who completed both a broad set of small-scale spatial items (involving items of mental rotation, visualization, mental paper folding and a Corsi Block Test) and their abilities to walk along a previously learned route on the campus in forward and reverse direction. Results showed that children who performed better in the small-scale spatial abilities tasks also showed higher performances in the wayfinding tasks. This effect, however, was only significant for the group of younger children. Older children did not differ significantly in their wayfinding performance. Although the results of FENNER ET AL. (2000) suggested that there is a certain extent of overlap between small-scale and large-scale spatial abilities in younger children on the one hand. Results for older children, on the other hand, rather suggested a dissociation between both classes of spatial abilities.

In summary, all three studies suggested that small-scale and large-scale spatial abilities are dissociated abilities when understanding large-scale spatial abilities as the abilities to perform some kind of unaided navigation or when referring to knowledge stored in the children's cognitive maps. These findings contrast some of the studies that investigated spatial abilities at different scales of space in adults. Indeed, the question whether individual performances in solving small-scale spatial tasks reflect performances in solving large-scale ones has been a consistent topic of interest in the literature (but see Table B.6 in the appendix for a summary of studies examining this question). Results of some studies have also supported the total dissociation model (e. g., SHOLL 1988), whereas other supported the partial

dissociation model (e. g., K. J. BRYANT 1982; HEGARTY ET AL. 2006), or the mediation model (ALLEN ET AL. 1996). Furthermore, age-related but not gender-related
mediation was detected in a meta-analysis (L. WANG, COHEN, & CARR 2014).

One explanation for those divergent findings between studies in children and
adults could be that developmental processes and cognitive growth[12] that children
of primary school age undergo (e. g., NEWCOMBE ET AL. 2013) allow them to use
cognitive operations that are inherent when solving spatial tasks at both scales of
space. Discrepancies between the studies in adults and those ones in children could,
however, also be a result of limitations in the research designs of the studies reported
above. All three studies, for example, measured small-scale spatial abilities through
tasks that required the children to perform mental transformation of objects only,
whereas studies in adults measured small-scale spatial abilities in a broader manner. Moreover, QUAISER-POHL ET AL. (2004) did not control for drawing abilities.
However, this variable might have contributed significantly to the sketch maps that
were made (e. g., LIBEN 1982). JANSEN ET AL. (2010), for example, did not control
for effects of the manual mental rotation training on mental rotation abilities per
se. It is therefore not clear whether there is no relationship between small-scale and
large-scale spatial abilities or whether the mental rotation training was ineffective.

4.3.2   Studies Involving Maps

LIBEN ET AL. (2013) investigated the relationship between small-scale spatial abilities and the abilities to solve production tasks in real space. Children aged between
9 and 10 years solved a water level task (WLT), a 2D mental rotation task (MR), and
a paper folding task (PFT). They further solved a hidden pictures test (HPT) and a
vantage-point test (VPT) that consisted of choosing between five photographs for a
given viewpoint on an oblique map. They further completed a flag-sticker test on an
unfamiliar campus that measured their self-location abilities. During the test, they
walked around on the campus and searched for eight colored flags, thereby marking
the correct location with a sticker on an oblique map. They were shadowed by an
experimenter who recorded the sequence of flags visited and the strategy that was
used to fulfill the tasks, in particular whether the map was turned or not.

A correlation analysis of the results revealed that WLT and VPT were significantly correlated with sticker correctness (.42 and .37, respectively). In contrast to
the results obtained in the study of adults (LIBEN ET AL. 2008), LIBEN ET AL. (2013)
did not find significant correlations between PFT and sticker correctness. Results of
subsequent hierarchical multiple regression analyses revealed that the spatial tests
significantly predicted sticker correctness at the first level of the hierarchical model,

---

[12]Defined in the sense of achievement and use of knowledge (see also BRUNER ET AL. 1966).

with WLT emerging as the only significant predictor. However, when entering strategy use in a second step, the WLT lost its predictive power and was 'replaced' by both VPT and the strategy to turn the map as significant predictors[13]. Similar results were found in CHRISTENSEN'S (2011) study.

To sum up, when understanding large-scale spatial abilities as the abilities to stay oriented with the help of a map during navigation, the findings of LIBEN ET AL. (2013) provided first evidence for the partial dissociation model, with map-aligning and perspective taking emerging as important underlying cognitive processes that may account for the relations found. Except from the fact that WLT rather than PFT was found to be an important predictor in this study, the findings are in line with a similar study in adults (LIBEN ET AL. 2013). Moreover, this partial relation between small-scale spatial abilities and the solving of production tasks has not only been shown for self-location but also for the understanding of the self-orientation on the map, another production task, and for pointing to unseen landmarks with the help of a map, a comprehension tasks (LIBEN ET AL. 2010).

All those findings were, however, based on manifest methods and are therefore dependent on the specific tasks used for the hierarchical analyses. It remains to clarify whether those results can be reproduced with other tests requiring the same cognitive processes and whether the findings could be generalized to a latent treatment of the abilities involved.

Conclusion

There are only few empirical studies on the relation between small-scale spatial abilities and large-scale spatial abilities in children. The literature has neither addressed this issue in a systematic way nor is it in line with some findings in adults. Those findings in adults point toward a partial dissociation between both classes of spatial abilities, thereby understanding large-scale spatial abilities as unaided or as map-based wayfinding. First empirical results in children indicate that both classes of spatial abilities are probably dissociated when understanding large-scale spatial abilities as unaided navigation. First evidence from a study relating individual spatial tasks to map-reading tasks suggests that spatial abilities are partially related at both scales of space. There is, however, a gap in the literature on children concerning a systematic investigation of a potential relation.

---

[13] According to the authors, this effect occured when a certain strategy was applied or the need to coordinate different frames of reference was recognized, which was measured by the WLT.

Part II
Empirical Study

Design of the Empirical Study

5

Part I presented the theoretical framework of this study. This first chapter of Part II describes the design of the empirical study. Section 5.1 specifies the general research interest of this study by presenting the research questions. Moreover, it describes the research design. Section 5.2 discusses the methodological framework and derives conditions and general methodological specifications for the design of the study. Section 5.3 presents how this study was planned to be conducted with respect to the research questions and the methodological framework.

5.1 Research Questions, Latent Constructs, and Research Design

5.1.1 Research Questions

This study investigates how spatial abilities of primary school children addressed in the context of geometry education need to be conceptualized to consider different spatial challenges that children face in their everyday life. Those involve spatial information at different scales of space, which, in turn, provides different sources of interaction with space. Whereas small-scale spatial information, either involved during manipulation of small material or when solving paper-and-pencil tasks, can be perceived from one single vantage point and require, in turn, no interaction with the spatial situations represented in the material or written tasks, large-scale spatial information require the child to move and to perceive it from multiple vantage points. This becomes meaningful during interaction with and navigation in real space, a spatial situation that represents children's day-to-day experiences with space.

© The Author(s), under exclusive license to Springer Fachmedien Wiesbaden GmbH, part of Springer Nature 2020
C. Heil, *The Impact of Scale on Children's Spatial Thought*, Studien zur theoretischen und empirischen Forschung in der Mathematikdidaktik,
https://doi.org/10.1007/978-3-658-32648-7_5

Although previous studies have acknowledged that there are different types of spatial information and that the requirement of spatial tasks might vary with respect to the relative size of information provided (SOUVIGNIER 2000), spatial abilities have been addressed in limited way only. Most studies conceptualized them as those abilities that enable children to process small-scale spatial information and argued that this class of spatial abilities reflects, in the context of geometry education, how children grasp space (e. g., BERLINGER 2015; GRÜSSING 2012; MEISSNER 2006; NIEDERMEYER 2015). They consequently implicitly assumed that scale of space, when considered not only as a spatial but also as a psychological construct, has a negligible impact on children's spatial thought. The processing of spatial information in real space, and the related construct large-scale spatial abilities, have neither been addressed at a differentiated theoretical nor empirical level. By addressing this gap in the literature, this study investigated whether there are any relations between children's small-scale and large-scale spatial abilities and if so, to which extent.

Empirical findings in the cognitive and environmental psychology literature suggest that processing spatial information at different scales of space partially reflects the same underlying abilities (Chapter 3). Those studies, however, mainly addressed adults. Concerning children there are only a few empirical findings. Moreover, previous studies have yield mixed results (Section 4.3). Depending on the chosen conceptualization of the construct, in particular the large-scale spatial abilities, spatial abilities at both scales of space have been found to be either completely dissociated (e. g., JANSEN 2009; QUAISER-POHL ET AL. 2004) or partially related (e. g., FENNER ET AL. 2000; LIBEN ET AL. 2013). Whereas findings from the neuroscientific and cognitive psychology literature have investigated different kinds of possible relationships between spatial abilities at both scales of space, this issue has not been investigated in children. Consequently, it remains an open question if, for example, there is another set of cognitive abilities that might mediate spatial thought at different scales of space, as it has been found in adults (e. g., ALLEN ET AL. 1996). Since children still undergo a fundamental cognitive development during primary school, also with respect to their spatial abilities (e. g., NEWCOMBE 1982), empirical findings from the psychology literature in adults can only be transferred in a limited way to explain a possible differentiation of the construct of children's spatial abilities to a limited extent.

This study focuses on two particular spatial settings that are inherent to geometry classes and children's everyday life: On the one hand, settings with small manipulable or written material are considered as they occur in text books, worksheets as well as individual or group work. These stimuli require children to engage in small-scale spatial abilities. On the other hand, settings are considered that mimick children's use of spatial abilities in real space, for example when they use maps to orient themselves are considered. Those require children to engage in large-scale spatial abilities.

The research question of this study can be summarized as follows:

Are primary school children's small-scale spatial abilities related to large-scale spatial abilities in the context of geometry education and, if so, to what extent do they overlap?

5.1.2   Latent Constructs

The relevant constructs for this study were specified with respect to the findings from the cognitive and psychometric literature (Sections 4.1 and 4.2). The research question involves two main latent constructs: small-scale and large-scale spatial abilities. In line with the definitions provided in Section 2.4.1, the following two definitions are considered:

Small-scale spatial abilities are a class of cognitive abilities that enable an individual to represent and make inferences from spatial information that can be perceived from one single viewpoint.

Large-scale spatial abilities are a class of cognitive abilities that enable an individual to represent and make inferences from spatial information perceived and processed over time and from more than one viewpoint due to changes resulting from movement.

This study considered spatial abilities as multidimensional constructs at both scales of space. As summarized in Section 3.4 and further discussed with respect to the empirical literature in Section 4.1, small-scale spatial abilities might be modeled by a two-factor model involving subclasses of abilities that enable an individual to perform qualitatively different types of mental transformations. In line with ZACKS AND MICHELON (2005), this study assumed that these mental transformations used during spatial reasoning imply a permanent spatial update of three different frames of reference and their interrelations. Using this representational theory, this study assumes that individuals who engage in small-scale spatial thought encode spatial relations in egocentric, environmental, and intrinsic frames of reference. The pairwise relation between those frameworks constantly change when performing a mental transformation. Consequently, a two-factor model of small-scale spatial abilities involved the following two subclasses of them:

Object-based transformation abilities (OB) are the subset of small-scale spatial abilities that enable individuals to mentally transform the relationship between the intrinsic and environmental frame of reference, that is, the abilities to mentally manipulate objects from a stationary point of view. They are also denoted as object manipulation abilities.

Egocentric perspective transformations abilities (EGO) are the subset of small-scale spatial abilities that enable individuals to mentally transform the relationship between the egocentric and environmental frame of reference, that is, to mentally transform the egocentric viewing perspective on spatially invariant objects. They are also denoted as abilities to transform the imagined self as well as perspective taking abilities.

As summarized in Section 3.3 and further discussed with respect to the empirical literature in Section 4.2, large-scale spatial abilities might be modeled to involve at least two, but probably three subclasses of abilities. Given that the encoding of spatial information perceived from multiple vantage points is an important determinant of large-scale spatial thought that is qualitatively different, but not completely dissociated from map use in real space (see also Figure 3.11, p. 100), this dissociation accounted for a distinction into two broad subclasses of large-scale spatial abilities. Map use, however, involves two qualitatively distinct types of mental transformations, map-to-environment and environment-to-map transformations that are particularly involved when individuals engage in comprehension and production tasks. Consequently, a model of large-scale spatial abilities involved the following subclasses of them:

Survey mapping abilities (SurveyMap) are the subset of large-scale spatial abilities that enable individuals to integrate the fragmented spatial information perceived from multiple vantage points in one coherent mental representation and to make inferences from it, that is, to keep track of the varying relations between the self and landmarks in the environment.

Map-environmental-self relation update abilities (MapEnvSelf) are the subset of large-scale spatial abilities that enable individuals performing mental transformation of relations between the map, the self, and the environment by means of subsequent updates of the pairwise relations, that is, to keep track of the varying relations between the self and landmarks in the environment by using a map. They might be further differentiated into the **abilities to solve production tasks (Prod)** by means of environment-to-map transformations and into the **abilities to solve comprehension tasks (Comp)** by means of map-to-environment transformations.

5.1.3 Hypotheses

An analysis and comparison of the cognitive processes involved in processing spatial information at different scales of space revealed sources of individual differences when individuals engage in spatial abilities at different scales of space. Those might be unique sources that, so the assumption, either account for a dissociation of spatial abilities at different scales of space, or those might be shared cognitive processes that account for a relation of spatial abilities at different scales of space.

The analyses revealed that a unique resource of engaging in spatial abilities at different scales of space was the encoding of spatial information. In the case of small-scale spatial abilities, children have to engage in small-scale spatial information. This is rather abstract but provides simultaneous and metric information of a situation to mentally engage in (WIMMER ET AL. 2016). In the case of large-scale spatial abilities, individuals engage in large-scale spatial information. This is specific, but since it requires the individual to move in space to apprehend it, it is sequentially perceived. Furthermore, engaging in large-scale spatial abilities either require the assimilation of different pieces of information perceived from different vantage points or the spatial relation of them to a spatial depiction, the map. In both cases, encoding is based on fragmented, non-metric spatial information, that has to be integrated over time and that is supplemented, besides visual inputs, with other multimodal sensory cues (e.g., DOWNS & STEA 1973).

The analysis of cognitive processes showed that both the maintenance of spatial information in visuospatial working memory and the abilities to perform different types of cognitive processes on spatial representation are most likely to be common sources of individual differences of spatial abilities at both scales of space. They revealed that the visuospatial working memory is inherently involved when it comes to maintaining and transforming visual information. Consequently, a possible relationship between spatial abilities at different scales of space could be explained by an individual's visuospatial capacity to the extent that both classes of spatial abilities are partially related only because both involve the same kind of processes attributed to visuospatial working memory (ALLEN ET AL. 1996; HEGARTY ET AL. 2006).

Moreover, the cognitive analyses revealed that small-scale spatial abilities involve the engagement in two types of mental transformations (Section 3.2.2) that seem to be involved in any kind of map use, at least regarding object-based mental transformations (e.,g., SHEPARD & HURWITZ 1984). In addition, transformations of the egocentric perspective require constant updates of the relations to the egocentric frame of reference (YU & ZACKS 2017), being a cognitive process that is also involved in cognitive mapping (e.g., FIELDS & SHELTON 2006) and that might resemble dimensional transformations of spatial information during map use. Consequently, object-based transformations were assumed to be particularly involved in the two subclasses of large-scale spatial abilities involving map use, whereas egocentric perspective transformations were assumed to be involved in all three subclasses of large-scale spatial abilities.

Conversely, cognitive processes involved in large-scale spatial thought such as the constant update of the body-centered movement in real space can be involved, by means of embodiment, in egocentric perspective transformation abilities (KESSLER & THOMSON 2010) but should not be involved in object-based transformation abilities. Cognitive processes that govern both the abilities to solve comprehension and production tasks involve mental alignment of the map, which can be considered as an analogue to mental rotations in small-scale space. Moreover, cognitive processes allowing for the constant dimensional transformation of spatial information related to the depicted information on the map as well as constantly updating three relations of references frames during map-use in real space (e. g., OTTOSSON 1987) could be involved in both subclasses of spatial abilities. Consequently, survey mapping abilities were assumed to be particularly involved in the subclass of egocentric perspective transformation abilities. The abilities to solve production tasks and comprehension tasks were assumed to be involved in both subclasses of small-scale spatial abilities.

All those shared and unique cognitive processes that are involved in spatial abilities at both scales of space may, so the assumption of this study, account for individual differences in performances when it comes to solving tasks that require children to use those cognitive abilities. Moreover, individual performances have been shown to be dependent on the flexible strategy use (e. g., LÜTHJE 2010) and to be sensitive to gender differences (e. g., KERNS & BERENBAUM 1991). In addition, spatial abilities have been shown to be multidimensional constructs that involve several subclasses of spatial abilities at both scales of space (e. g., HEGARTY & WALLER 2005). Those determinants of spatial thought make the construct of spatial abilities, even when considered at a single scale of space, complex and difficult to empirically grasp.

This study investigated the research question assuming strategy homogenity. Furthermore, it did not consider gender differences. This study, however, considered the multidimensional structure of spatial abilities at both scales of space by proposing models that could describe them. It intends to open a discussion in the mathematics education literature on a possible differentiation of the construct of spatial abilities. Investigating strategy use at both scales of space and investigating possible effects of gender differences are important research issues in mathematics education but they are not addressed in this study.

Based on these assumptions and the results of the cognitive process analysis, this study tests the validity of two different models (Figure 5.1). When considering spatial abilities at both scales of space independently from a possible influence of visuospatial working memory capacity, this study assumes that the partial dissociation model might hold true ($H1$). Given that spatial abilities are not a unitary construct, even when considered at one particular scale of space, this study further assumed that a possible partial relation is due to the involvement of some subclasses of spatial abilities but not all of them ($H2$). Whenever considering visuospatial working memory, this study hypothesizes that it might function as a mediator. In this case, the mediation model would hold true ($H3$).

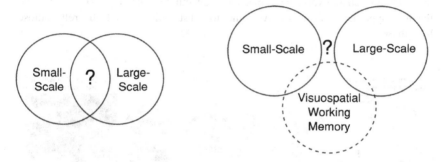

Figure 5.1 The partial dissociation ($H1$) and the mediation model ($H3$)

To sum up, the hypotheses of this study are the following:

- *H1: Children's abilities to processing spatial information in a small-scale setting partially reflect different cognitive abilities as involved in the processing of spatial information in a map-based orientation context in real space under the assumption of strategy homogenity.*
- *H2: The partial overlap between spatial abilities at both scales of space can be explained in terms of directed relations of subclasses, which are, however, not entirely related to the same extent.*
- *H3: Visuospatial working memory capacity mediates children's abilities to process small-scale and large-scale spatial information.*

5.1.4   Research Design

This study addressed the research question from an individual differences perspective, that is, a possible overlap between small-scale and large-scale spatial abilities based on quantifiable relations derived from measures that reflect both classes of spatial abilities (Figure 5.2). In other words, this study addressed the presented research question using a quantitative research perspective to empirically investigate a possible common relationship or partial dissociation between both classes of spatial abilities by means of studying homogenity and individual differences in performances in pre-defined tests (measures). Consequently, investigating the extent of overlap between small-scale and large-scale spatial abilities comes to quantifying the amount of shared variance in measures reflecting the latent constructs. Testing the three hypotheses outlined above comes to test statistical models that reflect those hypotheses.

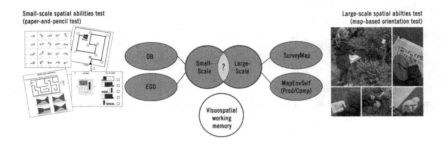

Figure 5.2 The research design of this study

The research design comprised three sets of main variables:

1. *Measures from a paper-and-pencil test*
   Those variables measured individual differences in small-scale spatial abilities by means of a paper-and-pencil test. A paper-based test was performde to test as many children as possible in order to obtain a rather large sample for subsequent statistical analyses (see also Section 5.3). This paper-based test was preferred over an individual testing method with manipulable material. In the test, both

constructs OB and EGO were operationalized in a set of tasks. Since only little standardized material for children existed, a corresponding test was developed in this study.

2. *Measures from map-based orientation test*
   Those variables measured individual differences in large-scale spatial abilities by means of tasks that involved maps as cognitive tools to foster spatial thought in real space. In the test, all three constructs, namley SurveyMap, Prod, and Comp were operationalized in a set of tasks that were conducted individually with the children. Since large-scale spatial abilities in children have not been addressed so far, no test material existed, so that corresponding material was also developed in this study.

3. *Measures of visuospatial working memory*
   Those variables measured individual differences in visuospatial working memory in the Corsi Block Test.

The research design further included the following set of extra variables:

- adoption of a perspective taking test PTSOT by KOZHEVNIKOV AND HEGARTY (2001) as a source of validation of the developed tests reflecting EGO
- means of transport to school and hobbies as simple measures for experiences in space
- experience in map-reading as simple measures to control for possible pre-existing sources of variance in the large-scale spatial abilities

The following auxiliary variables were further involved in the research design:

- *Gender:* Empirical findings have shown that males outperform females in tests of spatial abilities. The variable sex was included in the research design to understand empirical findings in greater detail and to validate whether some specific empirical findings were robust against influences of sex differences.
- *Experimental conditions:* Since the map-based orientation test was planned to take place in real space, external influences such as weather or crowdedness at the campus were expected to provide further sources of variance. To control for these, a range variables describing the experimental conditions were included.

Conclusion
This study addresses the question of whether and to what extent primary school children's small-scale spatial abilities are related to their large-scale spatial abilities in the context of geometry education. A quantitative research design is chosen to investigate this question by studying individual differences in solving spatial tasks at both scales of space. The research design is based on two test instruments, the paper-and-pencil test measuring small-scale spatial abilities and the map-based orientation test measuring large-scale spatial abilities. To address the research question, this study further investigated whether children's spatial abilities can be measured in a differentiated way using above tests. This study also addressed the question whether or not latent factors are empirically separable at both scales of space.

5.2   Methodological Framework

Section 5.1.4 described the research design of this study. It is based on the development of two tests that were intended to measure children's spatial abilities in a differentiated and reliable way. The psychometric literature has proposed different measures to investigate the quality of a test instrument. Section 5.2.1 reviews the literature and proposes benchmark values that indicate good quality of items and tests. This study also intended to quantify the extent of overlap between two latent constructs. To do so, the method of structural equation modeling was chosen. Structural equation modeling is an advanced technique of multivariate data analyses and has some requirements on the underlying data. Section 5.2.2 summarizes the relevant methodological literature concerning structural equation modeling that can be used for the design of data collection procedure of the empirical study. Since testing of children was expected to most likely involve missing values, some methods to deal with these are presented in Section 5.2.3.

5.2.1   Psychometrics

Psychometrics investigates the challenges of measurement in education and psychology, such as measurement of abilities, aptitudes, achievement, personality traits and knowledge (WILSON & GOCHYYEV 2013, p. 3). Hereby, measures can be understood as "a quantified record […], taken as an empirical analog to a construct" (EDWARDS

& BAGOZZI 2000, p. 156).[1] Among those challenges are theories of the development of measures and the development of means of evaluating the quality the measures.

According to WILSON (2005), the development of measures does not only involve the design of these, but also an reflection upon a possible outcome space, that is, "an ordered, finite, and exhaustive set of well-defined, research-based, and context-specific categories" (WILSON & GOCHYYEV 2013, p. 14) that can be interpreted with respect to the theory and the construct that is reflected by the measure.[2] Moreover, development of measures involves a conceptualization of the process of assigning numbers to the categories in such as way that they relate back to the theoretical construct (for example, correct or wrong) within a certain chosen measurement model.[3]. Among possible measurement models, classical test theory models and item response models are two prominent ones (WILSON 2005, p. 16).

Classical Test Theory (CTT) has also been denoted as the "instrument-focused approach" (WILSON & GOCHYYEV 2013, p. 15). CTT assumes that the observed score of a test consists of two components: the true score and some additional error term. Hereby, the true score is assumed as the hypothetical average score of a subject to score in endless repeated measurements under the assumption that there are no effects of learning by the test. The CTT approach focuses on the responses to the items themselves and proposes a set of quality ensuring characterizations based on descriptive statistics. There is no general assumption that the test scores reflect a single underlying latent ability (M. WU 2013). Consequently, a difference of ten in the test scores between subject A and B does not necessarily mean that the underlying ability is differing by ten.

Item Response Theory (IRT) is an "item-focused approach" (WILSON & GOCHYYEV 2013, p. 15). Other than CTT, IRT hypothesizes that a single variable, θ, is underlying the test scores (M. WU 2013). IRT therefore assumes explicitly a latent underlying ability that is not directly observable but which can predict how well a subject is answering in test items that were designed to measure the latent ability. IRT assumes that the more latent ability the subject holds, the higher he or she is likely to score. The focus herefore is not on the test scores themselves, but they are interpreted as outcomes of a probabilistic relation between the latent trait and the answer. The relationship is stated in a whole family of IRT mod-

---

[1] This study used the notions 'test' and 'task' to address the instruments that were used to collect data, understands measuring as the act of collecting data with the test and a denotes with 'measure' the actual score which is generated by applying the test to children and which can be subsequently used for further statistical analyses.

[2] Categories of outcome space are also denoted as raw variables in this study.

[3] The expression scoring is used as a synonym to measuring, that is, the process of assigning numbers. A scored raw variable is a score variable or a measure.

els, of which the RASCH (1960) model for dichotomous data is the most prominent one[4]. The Rasch model for dichotomous data assumes that the probabilistic function predicting the likelihood of solving a particular item is dependent on the item's difficulty and the subject's ability to answer correctly. Consequently, both parameters can be mapped onto a same scale by means of a transformation that maps the distance between item difficulty and subject's ability on a one-dimensional logarithmic axis (the construct continuum, denoted as logit scale).

Whenever a test, represented in a set of item responses, is proved to be Rasch conform, researchers test the central characteristics of the Rasch model using modeling software as *Conquest* (M. L. WU, ADAMS, & WILSON 2015) that allows to determine the extent a model with certain assumptions explains the given data. If diagnostic fit indices that characterize deviations between the model and the data are within a certain range, then the test is denoted as being Rasch conform. Among those, the weighted MNSQ is one fit index that is based on the weighted differences between the observed and expected scores of a person in an item (e. g., BOND & FOX 2015). In line with BOND AND FOX (2015), MNSQ values higher than 1.3 are conspicious since they point towards an underfit of the data with too much variation. *ConQuest* additionally provides estimate of the fit indices and the ninety-percent confidence-interval for the expected MNSQ values. Whenever the expected MNSQ is outside of the estimated confidence interval, the null-hypothesis that the data fits well into the model (represented by MNSQ = 1.0) is rejected and the chi-square test statistic becomes significant.[5]

IRT has gained increasing attention in the mathematics education literature since it has been used in TIMSS (Trends in International Mathematics and Science Study), and PISA (Programme for International Student Assessment) and is now accepted as a measurement model (BIKNER-AHSBAHS & VOHNS 2019) which can be used in addition to CTT analyses (e. g., MOOSBRUGGER 2012). If set of items is Rasch conform, item characteristics can be regarded independently of the subjects solving them, and differences in test scores reflect differences in the underlying latent ability. Moreover, if a set of items is Rasch conform, all items provide the same amount of information (they are all 'equally good' in measuring the latent construct) and differ

---

[4] Whenever referring to IRT, this study mostly refers to the Rasch model for dichotomous data. In the case of polytomously scored data, an extension of the Rasch model for polytomous data exists (MOOSBRUGGER 2012).

[5] One makes therefore inferences from a non-significant result which is generally problematic in the issue of statistical hypothesis testing. Other programs, such as *R eRm* (MAIR & HATZINGER 2007) bypass this problem by providing a number of tests that explicitly test the characteristics of the Rasch model. This approach is more stringent and more rigorous from a psychometric point of view but results in the elimination of a lot of items.

only with respect to their item difficulty (KOLLER, ALEXANDROWICZ, & HATZINGER 2012).

Item and Test Analysis

CTT as well as IRT provide a range of concepts that can be applied to analyze the test as well as single items. CTT allows to analyze the items on the basis of descriptive statistics which is sufficient to determine problematic items that are too easy, too hard or that do not distinguish sufficiently between subjects of different levels of ability. CTT does not, however, provide any justified characteristics at the level construct validity of the items and provides item statistics in dependence of the subjects that solved them. In particular, CTT does not characterize whether a set of items that is scored together does indeed reflect the same underlying latent ability or not. At this point, IRT approaches are a useful supplement. Whenever a Rasch analysis shows that the model with its assumptions explains the data sufficiently well, one might conclude that items reflect the same underlying construct and item characteristics might be determined independently of the sample.

CTT provides the following three item characterizations (e. g., BORTZ & DÖRING 2016):

- **item difficulty**
  The item difficulty (facility) p_i is the percentage of the total number of correct answers divided by the total number of subjects solving the item. The item difficulty is therefore the higher, the more subjects are solving the item. Higher values therefore characterize easier items.
- **item variance**
  The item variance $var(x_i)$ characterizes the differentiation ability of an item. The item variance is related to the item difficulty by $var(x_i) = p_i \cdot (1 - p_i)$, thus implying that the highest variance occurs for items with medium (0.5) item difficulties.
- **item discrimination**
  The item discrimination $r_i t$ characterizes how well the differentiation between the subjects on the basis of one item reflects the overall differentiation in all items. It is defined as the correlation of the subject's scores in the item and the sum score in the test (hereby, the item for which the value is computed is excluded). However, the discrimination value is dependent on the item's facility; CTT discrimination values therefore tend to be lower for very easy and very hard items which should be considered when interpreting them.

To judge whether or not an item has good CTT statistics—while keeping the interaction between item facility and discrimination in mind—the following cutoff values are considered in this study: items with a difficulty between 0.2 and 0.8 are preferable (BORTZ & DÖRING 2016, p. 477). Items with a discrimination value $\leq$ 0.3 are denoted as having a low discrimination value and need further revision. Values between 0.3 and 0.5 are denoted as acceptable and all values above 0.5 are denoted as high discrimination values (BORTZ & DÖRING 2016, p. 478).

IRT provides estimates of item difficulties. They range from low to high values (of logits) with an average of zero. An item difficulty in IRT is defined as the amount of ability required by a subject to have a chance of 50% to answer the item correctly. A pure consideration of those values therefore does not allow to judge whether the item is easy or difficult, but it gives an indication of how the item behaves with respect to other items. This study intended to interpret item difficulties of several items included in a test to gain insight to what extent the tests and the single items were able to differentiate among individuals with high and low scores. The item-person-map (Figure 5.3) consists of two panels that are separated by a vertical axis that corresponds to the latent person ability and item difficulty for the

Figure 5.3 Example of a map of latent abilities and item difficulties as computed by *ConQuest*

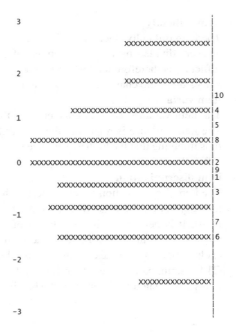

test (the logit-scale). Logit values under zero correspond to low latent ability scores, zero corresponds to medium ability scores and higher values of latent ability are expressed by high logit-values, respectively. The distribution of the scores for a test is represented on the left panel of each item map being scaled on the difference scale. It represents a transformed distribution of the relative frequencies that are observed when computing summary statistics, each 'X' constitutes a certain number of cases of the whole sample. On the right panel of each map, the items inserted into the Rasch analysis are arranged according to their difficulty level. Items that are clustered below zero on the vertical axis correspond to easy items. Their difficulty can be interpreted to be harder, the higher the items are mapped on the axes.

Objectivity, Validity, and Reliability

To evaluate the quality of a test, it is important to consider its objectivity, reliability, and validity (RAMMSTEDT 2010).

Objectivity refers to the degree in which measuring with the test is independent of external influences. A test is objective if the measure is only dependent on the subject that completed the test but not on external factors such as experimenters or subjective evaluation of the raw variables. Objectivity can be achieved by using few experimenters who are all trained and test according to a test manual. In addition, using items of closed answer format and a standardized coding manual result in a higher objectivity of the test (RAMMSTEDT 2010).

Reliability refers to the consistency of a test and specifies the vulnerability of the corresponding measure against random measurement errors. It further refers to a test's stability over time (P. KLINE 2013, p. 7). It is an "index of how consistently a test measures whatever it is supposed to measure (i. e., the construct)" (WILSON & GOCHYYEV 2013, p. 21). In other words, the reliability of a measure is the ratio of variance that can be explained by individual differences measured but not by measurement errors. Whenever a measure is highly reliable, it is free of influences from random measurement errors and is likely to produce similar results when applied several times under the same test conditions. In a reliable test, the observed scores are close to CTT true scores.

Measuring internal consistency of a test involving several items is one way to control for reliability (RAMMSTEDT 2010). Cronbach's α (CRONBACH 1951) and McDonald's ω (MCDONALD 1970, 1999) are means to do so. Cronbach's α is an indicator to determine the internal consistency of a set of tau-equivalent items that contribute to a measure (i. e., items that do NOT differ in their discrimination values such as items that are Rasch conform) (EID & SCHMIDT 2014, pp. 273–279). McDonald's ω is the reliability indicator of choice whenever the items fit a congeneric model (see EID & SCHMIDT 2014, pp. 312–339) that is, whenever items are allowed to differ according to their discrimination values. Using the Cronbach's α would lead to an underestimation of the reliability. There are no clear cutoff criteria for Cronbach's α. P. KLINE (2013) emphasized that Cronbach's α should not be $\leq .70$ (p. 13). Whether or not the internal consistency can be considered high or low remains subject to careful interpretation during test development.

A test is valid if "it measures what it claims to measure" (P. KLINE 2013, p. 17). There are different methods for showing validity of a test (P. KLINE 2013, pp. 17–31):

- face validity: the test appears to measure what it claims to measure
- concurrent validity: the test correlates highly with tests that claim to measure the same construct
- predictive validity: the test predicts some criterion that should be related to what the test claims to measure
- content validity: the test is accepted to measure what it claims by asking experts
- construct validity: the test is demonstrated to reflect the way the construct that it claims to measure, that is, the psychological nature of the construct that is measured (e. g., being multidimensional)

5.2.2 Structural Equation Modeling

Structural Equation Modeling (SEM) refers to a "family of related procedures" (R. B. KLINE 2015, p. 9) that "takes a confirmatory (i.e., hypothesis-testing) approach to the analysis of a structural theory bearing on some phenomenon." (BYRNE 2013, p. 3). Consequently, SEM is not an exploratory tool but requires theoretical assumptions that are tested in SEM analyses to make it plausible or to quantify a theoretical assumption. SEM is therefore a priori a family of confirmatory methods but not in a strictly sense (R. B. KLINE 2015, p. 11). Best practice approaches include testing alternative models that include the same set of variables but assume another interaction among them. Those alternative models are typically derived from the literature as well and tested against the model under consideration. Testing a single-factor model against a two-factor model is an example for exploratory approaches in SEM.

General principles of SEM

One key characteristic of SEM is the involvement of two types variables, namely manifest (observable) ones which represent the data collected and latent (non-observable) ones that represent the theoretical construct (R. B. KLINE 2015, p. 12). Manifest variables reflect the construct in an indirect way. Their statistical realization are denoted as indicators, and the concrete statistical realization of a construct based on indicators is denoted as factor (R. B. KLINE 2015, p. 13). Consequently, SEM allows for studying both, latent and manifest variables in one analysis.

The relation between manifest variables and a latent variable are specified in the corresponding measurement model, a hypothesis of how the manifest variables relate to the latent ones. They might, for example, be assumed to be tau-equivalent (as it is the case in Rasch models) or tau-congenetic (EID & SCHMIDT 2014). Relations between latent variables are specified in the structural model, which can be interpreted to be a conglomerate of measurement models, that specify the overall number of factors and to what extent various indicators relate to the factors as well as a structural model that describes the interrelationships between the latent variables.

SEM further includes estimates of error terms which are modeled as latent variables and represent, for indicators, the amount of variance that cannot be explained by the underlying factor. The explicit handling of measurement errors is both characteristic and advantage of SEM and distinguishes SEM from other multivariate analysis methods (BYRNE 2013, p. 3). Multiple regression, for example, assumes that all variables are free of errors, which is practically not the case when dealing with empirical data from quasi-experimental setups.

SEM relies on the empirical covariance matrix (R. B. KLINE 2015, p. 13). The covariance between two continuous random variables X and Y is defined as $cov_{XY} = r_{XY} \cdot SD_X \cdot SD_Y$[6]. Hereby, r_{XY} denotes the Pearson correlation that is a measure of linear association between two variables. Hence, the covariance matrix, which is a collection of pairwise covariances, contains pairwise quantifications of linear association between two variables in a single number. The central goal of SEM analysis is to specify patterns of linear associations (factors) for the given data set in a theoretical model by trying to explain as much variance as possible for all observed variables with a given model. When fitting a hypothetical model to the empirical data, the researcher investigates therefore whether or not the covariance matrix implied by the theoretical model is equal to the empirical covariance matrix. Deviations between both matrices may be quantified which leads to indicators that characterize the goodness of fit (R. B. KLINE 2015, p. 13).

Confirmatory Factor Analysis (CFA)
Confirmatory factor analysis (CFA) is a special member of the class of SEMs that involves only measurement models. Consequently, CFA is only concerned with relationships between indicators and factors. Similar to SEM, the goal of CFA is to establish a model that accounts for as much variance as possible for the indicators and

---

[6] $cov_{XX} = 1 \cdot SD_X^2$, so the covariance of a variable with itself is its variance.

that explains patterns of linearity among them. Indicators are therefore assumed to be intercorrelated because they share a common cause (the factor) (MOOSBRUGGER & SCHERMELLEH-ENGEL 2012, p. 334). During CFA, similar to SEM, the researcher specifies a theoretical model that includes the number of factors and patterns of factor-relation in advance on the basis of a literature review. Although testing an alternative model is a common practice, as the name suggests, CFA is confirmatory in nature and distinguishes from exploratory factor analysis (MOOSBRUGGER & SCHERMELLEH-ENGEL 2012, p. 334). CFA is therefore indispensable for construct validation of test instruments that operationalize a theoretical construct.

Since CFA deals with measurement models, CFA is a precursor to SEM analysis. Remember that SEM is a conglomerate of measurement models that specify how latent variables relate to their indicators and regressions among latent variables captured in the structural model. Establishing a reasonable measurement model by running CFA analysis is therefore always a precursor step before estimating and interpreting structural relations among latent variables.

Sample Size
Using SEM requires a reflection on the necessary sample size since SEM requires large samples (R. B. KLINE 2015, p. 14). The question about the minimum sample size to estimate a model robustly and with adequate statistical precision remains an issue of debate. In practice, the following factors affect the required sample size: (1) Complexity of the model (i. e., number of parameters to be estimated); (2) the nature of the indicators (i. e., number per factors, reliability, distribution and metric); (3) choice of a certain estimator. Consequently, there is no general guideline on the required sample size (R. B. KLINE 2015, p. 15), and sample size must be reflected case-specifically against the background of the research design.

Although those have been criticized for not being model-sensitive (e. g., WOLF, HARRINGTON, CLARK, & MILLER 2013), a number of rules of thumb have been proposed in order to give a guideline for determining optimal sample sizes. BENTLER AND CHOU (1987), for example, proposed that 5 or better 10 cases per parameter to be estimated are sufficient (see R. B. KLINE 2015, p. 16, for further elaboration on those $N : p$ ratios). The sample size should be, however, at least 200 (BARRETT 2007; R. B. KLINE 2015).

Goodness of Fit
When estimating the parameters of a specified SEM, a hypothetical model is fitted to empirical data which is an arbitrary numerical computation procedure. A converging computation does not indicate that the model explains the data well. Fit statistics describe how well the model explains the data. Software performing SEM analyses,

such as *lavaan* (ROSSEEL 2012), compute a broad range of fit indices with different characteristics. Whereas some are particular sensitive to model complexity, other compare the results of the fit to alternative models. Consequently, SEM requires researchers to report several fit indices. According to R. B. KLINE (2015), a minimal set that needs to be reported in SEM analyses is:

1. The χ^2-fit statistic that describes the magnitude of difference between the covariance matrix of the data and the fitted model with its degree of freedom and the corresponding p-value. A non-significant χ^2-fit statistic indicates that the model explains the data well enough. A rule of thumb further proposes that the χ^2-value that results from the fit testing should be smaller than two times the number of freedoms, $\chi^2 \leq 2 \cdot df$ (e. g., MOOSBRUGGER & SCHERMELLEH-ENGEL 2012, p. 337).
2. The Steiger–Lind Root Mean Square Error (RMSEA) is an absolute measure of fit and a badness-of-fit statistic where a value of zero indicates that the fitted model explains the data perfectly.
3. The Bentler Comparative Fit Index (CFI) quantifies the deviation of the fitted model to an null model in which all structural paths are hypothesized to be zero. The CFI indicates how much percent of the fitted model is better than the null model, with a CFI of 1.0 indicating perfect fit (R. B. KLINE 2015, p. 276).
4. Bentler's Root Mean Squared Residual (SRMR) is a badness-of-fit statistic which measures the mean absolute covariance residual, that is, a measure of taking into account the differences between the empirical and model predicted correlations (R. B. KLINE 2015, pp. 277–278).

This study further refers to the Tucker-Lewis Index (TLI) that measures fit similar to the CFI but controls for the degrees of freedom in the fitted and baseline model (R. B. KLINE 2015, p. 276). It therefore represents a conservative statistic that rejects more models than the CFI and that penalizes complex models (HU & BENTLER 1999)

BYRNE (2013) and HU AND BENTLER (1999) proposed ranges of acceptance for these fit indices. HU AND BENTLER's (1999) simulation study provided more restrictive cutoff values for those indices indicating very good fit of the model to the data. Those cutoff values are used in this study (Table 5.1).

Although SEM is a confirmatory method, testing alternatives, that is, nested models to a preliminary model[7], is a common practice in the analysis process. In addition

---

[7]A model A is nested in model B if you can always obtain model A by constraining some of the parameters in model B.

Table 5.1 Cutoff-Values for a Very Good Model Fit in SEM as Proposed by HU AND
BENTLER (1999)

Index	Criteria for good fit
TLI	.95 − 1.00
CFI	.95 − 1.00
SRMR	.00 − .08
RMSEA	.00 − .06

to fit indices that are reported for alternative models, there exist model difference
tests that allow to analyze whether or not the alternative model is better than the preliminary model (MOOSBRUGGER & SCHERMELLEH-ENGEL 2012, p. 337). Another
way of evaluating models, even for the case of being not nested, is a comparison of
the information criteria AIC and the sample size sensitive BIC. Hereby, the model
with the smaller AIC and BIC can be interpreted as the one that explains the data
better (R. B. KLINE 2015, p. 286–287).

Indicators
To estimate the model parameters, the model must be identified, that is, there should
be enough empirical data for fitting a model of a certain complexity. A model is
identified if the number of empirical data (variances and covariances of variables)
is larger than the number of parameters to be estimated (factor loadings, factor variances[8], factor covariances, measurement errors) (MOOSBRUGGER & SCHERMELLEH-
ENGEL 2012, p. 336).

The number of indicators per factor has been discussed with regard to model
identification and robust parameter estimation. Too many parameters may lead to
improper solutions. In contrast, using more indicators results in a better operational-
ization of the construct. Rules of thumb propose to use at least three indicators per
factor that load significantly (R. B. KLINE 2015, p. 303). Using only two indicators
per factor is critical and may lead to non-positive parameter matrices.

It is recommendable to use indicators with good psychometrics such as factors
loadings of $\geq .71$ (R. B. KLINE 2015, p. 303). This can be interpreted that 50% of
the variance in the indicator is accounted for by the latent factor (if the indicator
loads on one factor only). Evaluating factor loading in SEM is critical because it
might give an idea of how much a single indicator reflected the latent construct or
not. According to COMREY AND LEE (1992), factor loading has been described as

---

[8]During the estimation process, factor variances is obtained by either fixing it to 1.0 or by
fixing the loading of the first factor to 1.0.

excellent ($\geq$ .71), very good (.63 to .71), good (.55 to .63), fair (.45 to .55), and poor (.32 to .45).

Choice of the Estimator for CFA and SEM

There are different numerical estimators for parameter estimation CFA and SEM. Among those, maximum likelihood (ML) estimation is the most commonly used. ML assumes that the given data are continuous and follow a multivariate normal distribution of the joint population distribution (R. B. KLINE 2015, p. 235). Under these conditions, ML estimations neither underestimate nor overestimate the true parameters (asymptotic unbiasedness), are consistent (with increasing population size, the estimated parameter converges towards the true parameter) and are asymptotically efficient (the estimator has the lowest asymptotic variance in comparison to other estimators) (FINNEY & DISTEFANO 2013, p. 441).

Whenever ML estimator is applied to non-normal data, biases occur in standard errors and in χ^2 statistics (FINNEY & DISTEFANO 2013, p. 443). To estimate in a way that is robust to violations of the conditions outlined above, there are robust ML-estimators that estimate parameters by the normal ML, but correct the standard errors and chi-square test statistics afterwards (FINNEY & DISTEFANO 2013; LI 2016). *R lavaan*, for example, uses similar to *MPlus* (L. K. MUTHÉN & MUTHÉN 2004) the MLR-estimator that involves the Yuan-Bentler correction for the chi-square-statistic and robust standard errors (LI 2016; ROSSEEL 2012). Robust ML should be used whenever observing that the data is skewed or widely distributed (kurtosis). Best practice approaches propose to check each variable (indicator) whether or not it does deviate substantially from a normal distribution and to apply MLR[9]. Substantial deviation from normality is, for example, in line with WEST, FINCH, AND CURRAN (1995) given if the skew of the distribution of an indicator is larger than two and the kurtosis larger than 7.

The normality assumption about the indicators and the assumption of being continuous becomes critical whenever the sum scores only have a few response categories (see FINNEY & DISTEFANO 2013, pp. 450–451). Although sum scores are inherently metric, a distribution with only a small number of categories can hardly be interpreted as normally distributed and is better characterized being discrete realizations of a continuous variable rather than being continuous. One way to deal with this kind of indicators is to use weighted least squares estimation with adjusted variance (WLSMV) (B. O. MUTHÉN 1984; ROSSEEL 2012). In addition, WLSMV has been shown to be unaffected by distributional characteristics (e. g., HOOGLAND

---

[9]Personal correspondence with Prof. Dr. Tobias Koch, Leuphana University, 11th October 2017.

& BOOMSMA 1998) but is affected by the sample size (i. e., bad behavior for small samples such as N = 200) (FLORA & CURRAN 2004; LI 2016).

WLSMV estimations in comparison to MLR estimations have widely been discussed in regard to ordered categorical data such as for Likert-scales (e. g., FLORA & CURRAN 2004; HOLGADO-TELLO, CHACÓN-MOSCOSO, BARBERO-GARCÍA, & VILA-ABAD 2010; LI 2016; LIONETTI, KEIJSERS, DELLAGIULIA, & PASTORE 2016). Whenever applying MLR estimation to Likert-like scales with only a few number of categories, it has been shown that this leads to effects of biases in parameter estimates, standard error estimates and fit indices (e. g., FLORA & CURRAN 2004) . Those effects decrease with an increasing number of categories. Whenever the categorical variable has two to four categories, results of simulation studies revealed that MLR estimation consequently underestimates factor loadings and standard errors of the parameters; other estimators such as WLSMV should be used instead. For categorical variables with five to seven categories that are roughly symmetric, MLR produces estimates in an acceptable range. In case there are seven categories, MLR outperformed estimators such as WLSMV RHEMTULLA, BROSSEAU- LIARD, AND SAVALEI (2012).

5.2.3   Treatment of Missing Data

Missing data might occur whenever the subjects skip an item in a psychometric tests, when they do not finish the test due to time problems or when they are not provided the items. The latter case occurs in case of problems with the test booklet in case of paper-and-pencil testing, test procedure errors by the experimenter in individual testing (for example, if the experimenter forgets to ask a particular item or procedures the item other than stated in the standardized manual), organizational errors in individual testing (general time problems that do not correlate with the subjects abilities in the test, bur occur due to external time limitations), or is part of the research design (e. g., rotational designs). Treatment of missing values is challenging, in particular with regard to SEM analyses, because it might decrease the sample size dramatically. To avoid this, missing data requires handling by means of particular estimators or imputation strategies.

Patterns of Missing Data and Their Implication in Statistical Inference

RUBIN (1976) classified missing data within a taxonomy that relied on the process that caused missing data and described implications for making inferences from data sets with missing values in those different categories. RUBIN (1976) assumed that each subject features a latent probability of having a missing in variable Y.

Dependent of whether or not this probability is correlated to other variables in the data set, he differentiates between data missing completely at random (MCAR), data missing at random (MAR), and data not missing at random (NMAR). In practice, the character of the missing data determines the specific handling in SEM (ENDERS 2013, p. 496).

Data have a missing completely at random mechanism (MCAR) whenever the probability that there is a missing in variable Y does not depend on the observed scores in all other variables as well as the observed scores in Y itself. In psychometrical testing, this implies that a subject's likelihood to skip a certain item is not related to all other items and the subject's hypothetical answer in case of not skipping the item (would-be-value). In particular, the item is not skipped due to a lack of the underlying abilities are intended to be measured by the test (ENDERS 2013, p. 497). Whenever the missing data mechanism is MCAR, the data set with missing values can therefore be interpreted as a random sample from the full data. LITTLE (1988) proposed a test to investigate whether or not the data are MCAR. An analysis with Little's test can give further evidence whenever there is evidence from the study design that the missing data mechanism could be MCAR.

Data have a missing at random mechanism (MAR) whenever the probability that there is a missing in variable Y can be explained by one or more observed values in other variables but not to the would-be-score in Y itself. In psychometrical testing, a subject's likelihood to skip an item might depend on whether the item is reading-intensive, for example. Thus, skipping the item might be explained by a variable that measures reading competence. However, the item is not related to a lack of the underlying abilities being measured by the item. The non-response is fully explained by a construct-independent variable (reading competence, for example) (ENDERS 2013, p. 497).

Data have a not missing a random mechanism (NMAR) whenever the probability that there is a missing in variable Y does depend upon observed scores in other variables and the would-be value of Y itself. Once again, in psychometrical testing, a subject's likelihood to skip an item depends therefore on a deficit in the underlying abilities that is intended to be measured by the item that was skipped. Subjects skip items because they know that they would score low, for example. Therefore, NMAR-data features systematic patterns of missings that constraint their utility in further analytics (ENDERS 2013, p. 498).

Ad-Hoc Handling Techniques
A common way to deal with missing data is to use ad-hoc-approaches. Ad-hoc techniques either refer to the removal of incomplete cases, denoted as listwise deletion, or to techniques such as pairwise available analysis, denoted as pairwise deletion,

and to single imputation methods. In the case of listwise deletion, further analysis is based upon a set of complete cases which might result in a dramatic increase in the sample size. Small sample size, however, might lead to biased parameter estimates in the SEM model (ENDERS 2013, p. 499). Pairwise deletion, which maximized the use of available data, is preferable in terms of the conserved sample size. However, it implies that the data are MCAR. Single imputation methods rely on replacing missings by single values such as the arithmetic mean of the observed scores. However, even applied under MCAR-condition, single imputation methods generally produce biases in the subsequent SEM analysis (ENDERS 2013, p. 499).

Full Information Maximum Likelihood
Full information maximum likelihood (FIML) treatment of missing data is an easy way to apply an advanced approach that is implemented in the SEM software such as R lavaan or MPlus as an add-on to the use of (robust) maximum-likelihood estimation. FIML does not impute or replace data with missings but uses the data that are available. It is an integrated missing-value approach that manages SEM parameter estimation and missing value treatment simultaneously. FIML estimates population parameters that most likely produce the estimates from the available sample data that is analyzed (see ENDERS 2013, p. 499).

FIML becomes problematic whenever the data are not continuous but categorical (discrete) since the FIML estimation assumes that all data stem from a multivariate normal joint distribution. Whenever the researcher wishes to use an adequate estimation method for categorical data (e. g., WLSMV), data have to be completed (imputed) prior to further analytical steps. Moreover, since FIML estimates missing values at the level of the sum scores, it denies existing information at the level of single items that contribute to the sum score. If including all single items into the SEM would demand for very large sample size, multiple imputation techniques can be an option to deal with missing data.

Multiple Imputation
Multiple imputation (MI) approaches to missing data are highly flexible since they do not rely on model assumptions such as multinormal distribution of all endogenous variables.[10] MI handles MAR data that consist of categorical as well as continuous variables with missings and produces a number of filled data sets with plausible

---

[10]I would like to express my sincere gratitude toward Dr. Alexander Robitzsch (IPN Kiel) who helped me not only with the methodological choice for treatment of missing data but also with the implementation in R.

values before analyzing them (ENDERS 2013, pp. 507–508). MI is a conglomerate term for the following three main steps (Figure 5.4):

1. **Imputation phase**
   The algorithm creates a set of copies of the original data set which is respectively filled with plausible values when missings occurred. Historically, an imputation of 5 data sets has been proposed but state of the art is to generate 10 to 20 imputations. In addition, the more imputations are generated, the more power is to be expected in the subsequent statistical test power (ROBITZSCH & PHAM, GIANG, YANAGIDA, TAKUYA 2016).

2. **Analysis phase**
   For each of the generated data sets, a separate SEM is estimated. If one imputes 20 data sets, one estimates the SEM model on the basis of 20 different data sets which produces 20 different parameter estimates and standard errors.

3. **Pooling phase**
   The collection of parameter estimates and standard errors is combined according to RUBIN's (1976) rules in a single set of output values.

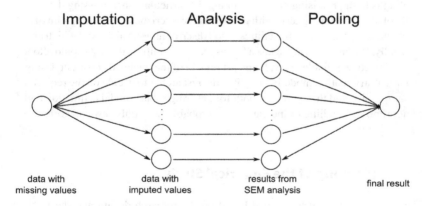

Figure 5.4 Three main steps of the multiple imputation method

There is a range of methods for the imputation phase that deal with a variety of data characteristics (see LITTLE & RUBIN 2002, p. 200, for further elaboration). SI AND REITER (2013) proposed a non-parametric Bayesian model approach for multiple information of high-dimensional categorical data. The approach is based on Dirichlet process mixtures of a multinomial distribution model derived from contingency tables of all categorical variables. The realization of the theoretically derived process is numerically approximated by a Markov Chain Monte Carlo (MCMC) simulation that has the desired probability distribution as a steady state. This approach has successfully been applied in the context of the Trends in International Mathematics and Science Study (TIMSS) to estimate, among others, missings in background variables (ROBITZSCH & PHAM, GIANG, YANAGIDA, TAKUYA 2016).

Multiple imputation approaches rely on an imputation model that contains at least the variables under consideration of the proceeding SEM-analysis. As a general rule, the complexity of the statistical inferences that are drawn from the data has to be represented in the imputation model (ROBITZSCH & PHAM, GIANG, YANAGIDA, TAKUYA 2016). Whenever the complexity of the model is high, this might result in a computation intensive imputation procedure which is a possible drawback of MI.

Conclusion
Examining individual differences in performances in solving spatial tests using multivariate statistical approaches such as structural equation modeling is both promising and challenging, in particular when missing data are involved. Developing tests with good test characteristics, that is, a test involving items with good item facilities, good discrimination values and which are, ideally, Rasch conform are just a few aspects. In addition, the test should allow for objective, reliable and valid measurement of the intended construct. Using structural equation modeling requires the researcher to decide on the required sample size and the estimator, both for ensuring robust model estimation and appropriate handling of indicators that probably have only few categories.

5.3   Planning of the Empirical Study

The empirical study was planned based on the research design and the methodological background that imposed articular requirements on the sample size and constitution of the data. Ideal conditions were limited due to personal and temporal resources available. This section describes the organizational framework, preliminary consideration concerning the sample and the development of tests.

5.3.1 Location and Organizational Framework

The whole study was planned to take place in the last weeks of the school year in summer 2016. The testing period was chosen to maximize the chance of good weather because a positive impact on the childrens' motivation for outdoor testing was expected.

Since tests of large-scale spatial abilities are sensitive to the test environment (e. g., SAS & MOHD NOOR 2009), the empirical study was planned to take place at a single location. Although it reduced the potential amount of students to be enrolled in the study and diversity among them, this allowed for controlling for environmental impacts during testing and for designing a standardized testing procedure for both small-scale and large-scale spatial abilities.

The study was planned to take place at the campus of Leuphana university, Lüneburg. This test environment was favorable for three reasons. First, this study intended to measure childrens' large scale spatial abilities in an unknown environment in order to be in line with previous studies with children (e. g., CHRISTENSEN 2011; LIBEN ET AL. 2013). It was assumed that the campus was an unknown environment for the children and that this would stimulate map use. An additional variable was included in the research design controlling for this assumption of whether or not the environment was unknown.

Second, the architectural characteristics of the campus were favorable to study how children make inferences from the map on geometrical correspondences. According to WEISMAN's (1981) characteristics of an environment, the campus is complex since it had been transformed from military buildings. The buildings are uniformly designed with three floors, an equal facade and roof design. The general layout of the campus is ordered in a perpendicular way, which, in turn, could be used for item design involving difficulty-generating features such as the number of road intersections taken. Important landmarks (restaurant, library,...) are visible from one single viewpoint and are not signposted which makes it difficult for inexperienced visitors of the campus to recognize them. By the time of the study, the most important landmark, the central building, was still a construction site that could not be seen from ground view without approaching.

Third, a testing room was available during the whole testing period. This allowed for standardized and groupwise testing of the small scale spatial abilities under a controlled experimental setup.

5.3.2   Requirements Regarding the Sample and Testing Material

Aspects of Sample Choice

This study investigated children at the end of primary school in Germany (fourth graders) who were aged about 9 to 10 years. There were two major reasons for choosing this age group. First, the transition from primary to secondary education is an important challenge that children face during their development. To accompany this transition, it is important to have empirical insights into the spatial knowledge that children have acquired at the end of primary school and that they bring to secondary school. Knowing in a differentiated way which aspects of spatial thought they have developed well, knowing about the relation between different subclasses of spatial abilities and understanding how well children can reason spatially with maps is important for geometric learning in secondary school. Vice versa, having differentiated insights into children's abilities at the end of primary school may help to design pedagogical intervention in primary school. This refers too to their state of knowledge according to the national curricula, to put it simply, to know what they don't know but should be knowing.

Second, from a methodological and developmental point of view studying children at the end of primary school was interesting. With respect to the chosen research design, the sample for the empirical study need to demonstrate sufficient variance in performances in all small-scale and large-scale tasks. In line with classical findings by PIAGET AND INHELDER (1948/1956) and HART AND MOORE (1973) children at the end of primary school should be in the concrete operational period in which cognitive abilities are developed to a considerable, but not complete extent. According to the theory of PIAGET AND INHELDER (1948/1956) and PIAGET AND INHELDER (1966/1971) children's abilities to solve tasks of mental object and perspective manipulation should be sufficiently but not fully developed. Applying projective and Euclidean concepts might still be challenging, which, in turn, could result in large individual differences when solving spatial tasks. PIAGET AND INHELDER's (1948/1956) strict developmental sequences and some of their findings were questioned in subsequent studies (see NEWCOMBE 1989, for further elaboration). Results from those studies and empirical results of the mathematics education literature (e. g., GRÜSSING 2012) indicated that spatial abilities are not fully developed at the end of primary school and this, consequently, should yield large individual differences in performances.

In addition, the cognitive and developmental literature has shown that children are able to understand representational correspondences at the beginning of primary school. Reasoning about geometric correspondences, however, remains subject to difficulties until the end of childhood (e. g., LIBEN ET AL. 2013).

Sample Size

As discussed in Section 5.2.2, a reasonable sample size is important for applying SEM to guarantee robust parameter estimation. Although no detailed a-priori sample size analysis was conducted, approximate values for sample size were computed using the 1:5-rule of thumb outlined above. Table 5.2 summarizes the estimations for a different number of possible indicators and latent factors (the constructs of small-scale and large-scale spatial abilities with an uncertainty how to model the large-scale ones). Numbers of parameters to estimate were computed under the assumption that latent variables will be standardized during model estimation, that is, no variance need to be estimated.

In line with the recommendation of R. B. KLINE (2015), this study planned a sample size of at least 200 children. In addition, to include a reasonable number of indicators into the model, up to 250 children were required for the sample. This corresponded to a testing of more than 10, better 15 different classes. Given the limited temporal framework, this required groupwise testing of small-scale spatial abilities in a paper-and-pencil test. Since the children did not know the campus, they had to be supervised during completing of the map-based orientation test, a limitations that required the involvement of further experimenters for large-scale spatial abilities testing.

Table 5.2 Approximate Sample Sizes for Varying Indicators and Numbers of Latent Constructs in a SEM

Nb. of indicators	Nb. of factors	Nb. of parameters to estimate	Df	Sample size according to 1 : 5-rule
10	4	26	29	130
15	4	36	84	180
20	4	46	164	230
10	5	30	25	150
15	5	40	80	200
20	5	50	160	250

Requirements Regarding the Testing Material

Based on the methodological framework, the following two preliminary requirements on the test material were taken into consideration during test development.
First, from a psychometric perspective, this study intended to measure spatial abilities in a differentiated way. For this reason, the development of individual tests
did not only involve the task instructions and the item design but it also considered
possible outcome spaces and a scorings. Moreover, the items and tests used should
have appropriate item facilities and discrimination values and should not yield floor
and ceiling effects. In addition, each test should be objective, reliable and construct
valid. If a test involved several items, they should be conform with the Rasch model.
A maximum completion time of 45 minutes was set for each of the two tests. The
test length was therefore long and demanded full concentration.

Second, with regard to the methodological framework of SEM, the following
requirements on the testing material were necessary. Each latent construct needed
to be reflected by at least to different measures to ensure the identifiability of the
model. To apply MLR, measures (indicators) possibly should involve at least five
categories, that is, at least four different items.

Testing Material for Small-Scale Spatial Abilities

One contribution of this study was to develop an age-specific paper-and-pencil test involving a set of tasks that measured the construct of small-scale spatial abilities with the two subclasses object-based transformation abilities (OB) and egocentric perspective transformation abilities (EGO).[1] To do so, potential tasks were adopted from those for adults. In addition, new material was developed. A preliminary version of the test material was piloted and subsequently revised for the main study.

Section 6.1 presents the development of the paper-and-pencil test. It specifies possible sources for the development, briefly addresses gaps that required the development of a few tasks from the scratch, and presents the tests that were developed to measure OB and EGO. [2] Section 6.2 presents results of the pilot study and describes which revisions had to be made for the main study.

---

[1] The term test will be used to address the collection of individual tasks that were developed to reflect the latent constructs. From a psychometric point of view, they are tests, but they will be referred to as tasks when the emphasis is on the children's perspective and what children need to do. Since the analyses of the pilot test resulted in data, they are referred as measures when reporting statistical results.

[2] A short description of the tasks can be found in the ESM, p. 1

Electronic supplementary material The online version of this chapter (https://doi.org/10.1007/978-3-658-32648-7_6) contains supplementary material, which is available to authorized users.

C. Heil, *The Impact of Scale on Children's Spatial Thought*, Studien zur theoretischen und empirischen Forschung in der Mathematikdidaktik, https://doi.org/10.1007/978-3-658-32648-7_6

6.1 Development of a Set of Paper-and-Pencil Tasks

6.1.1 Preliminary Considerations

The development of the paper-and-pencil test was based on the following preliminary considerations:

- There is an imbalance between the existence of a rich set of psychometric tests that were mainly developed for adults (e. g., ELIOT 1987) and only a few existing and psychometric validated tests for children (e. g., NEWCOMBE & SHIPLEY 2015; VEDERHUS & KREKLING 1996).
- Although it has been shown that tests for adults may be adopted for children, adopted tests might suffer from poorer reliabilities (e. g., GRÜSSING 2012; JOHNSON & MEADE 1987). Consequently, a pilot study was needed.
- There is a range of adult test material that was adopted for children involving tasks of mental rotation, mental folding, and object shifting (e. g., BERLINGER 2015; GRÜSSING 2002, 2012; MEISSNER 2006). These materials could be used for developing tasks that were intended to reflect OB.
- There are only a few tests for adults measuring EGO. Up to now, the PTSOT is the most valid test (KOZHEVNIKOV & HEGARTY 2001) but no adoption for children exists. EGO abilities have mainly been addressed in experimental one-by-one settings (e. g., FRICK ET AL. 2014; NIEDERMEYER 2015). Paper-and-pencil tests for children have been reported to be either too difficult, too easy, or to feature bad psychometrical characteristics (e. g., GRÜSSING 2012; MEISSNER 2006). Consequently, measures that were intended to reflect EGO needed to be developed from scratch.
- The whole test was intended to measure individual differences in terms of the complexity of processes involved rather than by speeded tests. Although particular abilities, such as mental rotation, have been conceptualized and measured as speed-dependent ones, this study did neither intend to use different kinds of tasks within the test neither limit the time for individual tests. It was assumed that a large number of tasks that required the children to cognitively engage with spatial stimuli for a long time (45 minutes) and that were intended to be scored rather strictly using complex multiple choice scorings[3] would be sufficient to observe large differences in individual performances.

---

[3]Complex Multiple Choice scores an item only with one point if a certain subset of 'miniitems' is correctly solved.

– For both OB and EGO, a set of tests involving various task instructions was intended to be developed to avoid that potential construct validity could be found exclusively due to the effect of similar task instructions and answering formats (see HEGARTY & WALLER 2004).

Since this study intended to contribute to research in mathematics education, aspects of the concrete requirements from the competences *Space and Shape* sub-content *to orient oneself in space* of the German education standards in mathematics education (KMK 2004) were included:

1. The demand that children need to recognize, describe, and use spatial relations in various representations such as spatial configurations, maps, routes, and perspective drawings has been given particular attention.
2. Moreover, the demand that children are able to relate two to three-dimensional representations of situations was incorporated into the tasks.

A first subset of tasks involved either a representation of a spatial object or a configuration of objects. In addition, tasks were developed that demanded the children to cognitively engage with maps or plans.

Based on these preliminary considerations and the ones subsequently outlined for OB and EGO, a preliminary version of the test material was developed and tested in a pre-pilot study. This study investigated the comprehension of the test material, the encoding manual and the temporal framework with 60 fourth graders (see ESM C.1, pp. 7). The following two sections present the revised set of those tasks.[4]

6.1.2   Tasks Measuring OB

The abilities needed to perform object-based transformations were not only assumed to involve rigid transformations such as mental rotations but also semirigid ones such as mental folding (see also Section 3.2.2). Grouping mental rotation with mental folding into one subclass of small-scale spatial abilities was justified because both kinds of tasks were not only found to be similar with respect to the underlying cognitive processes but also with respect to their neural basis, their malleability, and predictive validity for STEM (HARRIS, HIRSH-PASEK, & NEWCOMBE 2013). Since

---

[4]The tasks are presented in German. A translation of the tasks instructions is provided in the ESM, p. 1.

non-rigid transformations are non-reversible changes of the object structure, these types of transformations were not included in the test.

It was difficult to develop tasks that involved representations such as plans or maps. The mental rotation measure Maps is the only one which required the children to reason about maps, at least to some extent. Table 6.1 shows the tasks that were developed to reflect OB.

Table 6.1 Classification of Tasks Reflecting OB

Type of transformation	Specified mental transformation	Task/measure
Rigid	mental rotation	2DMR (2D), Maps (2D), 3DMR (3D)
	reflection	2DMR
Semirigid	folding	PFT
	(dis)assembly	–
Nonrigid	compression, melting	–

The following difficulty-generating features where considered during development:

1. *Complexity of the object or the configuration*
   Tasks with an increasing number of pieces involved in one object (e. g., blocks) could be considered as a difficulty-generating feature, since the object might become more difficult to be encoded in the working memory. However, this has not always been considered to be a unique determinant of complexity in mental rotation (see REINHOLD 2007, pp. 102–104). In addition, it was assumed that objects having a single distinct feature are easier to mentally manipulate since the subject may concentrate on this feature while solving a task (HEGARTY 2018).
2. *Angle of a possible rotation to perform*
   The studies of SHEPARD AND METZLER (1971) have shown that the reaction time of a mental rotation process increases with the given angular deviation. When interpreting reaction times as a means of difficulty, the degree of rotation of an object could be a difficulty-generating feature.
3. *Type and number of wrong answers (distractors)*
   The difficulty of a task can be changed by introducing distractors, that is, wrong but plausible answers. These can avoid one-to-one assignments in answers and force the individual to perform the intended mental transformation for every given stimulus. In addition, they decrease the guessing probability. It was assumed that the more distractors were included in a task, the more difficult it

became. Besides the number of distractors, their type itself can be interpreted as a difficulty-generating feature. Distractors might display objects that are structurally different to the original objects, or they might display objects that are reflections of correct solutions. The latter can be considered as making tasks more difficult.

Mental Rotation Tasks
Definition: Mental rotation tasks require individuals to decide whether or not two depicted figures or objects are identical after performing a pre-defined rotation of the visual stimulus.

Intended Cognitive Processes: To solve a mental rotation task, the individual needs to perceive and encode a visual stimulus in, probably, depictive format. Then, the subject performs a single mental rotation and compares the mentally generated image with a depiction in the task instruction. This cognitive operation is similar to a perceptual process of a physical rotation of the object. The cognitive processes of mental rotation are those described by SHEPARD AND METZLER (1971).

Tests: Three different mental rotation tests were developed: two 2D and one 3D mental rotation tasks. Although mental rotation has been presented as a speeded-component of spatial abilities (HEGARTY & WALLER 2005), all tests were developed to be solved without time limit, thereby involving rather few single distinct features in the visual stimuli as well as a complex multiple choice scoring.

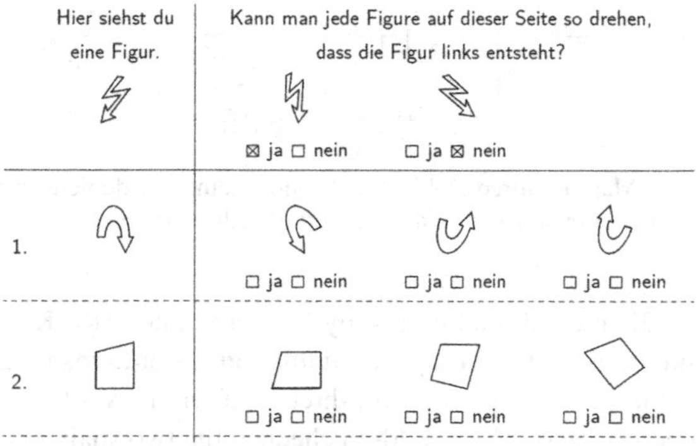

Figure 6.1 The task 2DMR required children to decide whether the depicted figures on the right were rotated versions of the original figure to the left or reflections of it

This study adopted the 2D mental rotation task by GRÜSSING (2012) in the measure 2DMR by using the same general setting of items and by choosing similar figures as in her study (Figure 6.1). The 2D mental rotation test corresponded to an adoption of the classical Cards, Flags, and Figures Test (THURSTONE & THURSTONE 1949) as well as the Card Rotations Test (EKSTROM 1976). These tests have been classified tests involving mental rotation procedures (e. g., CARROLL 1993 p. 329). The outcome space of the measure 2DMR are one, two, or three crosses per item that were intended to be scored dichotomously.

Since GRÜSSING (2012) reported low discrimination values for some of the items in her 2D mental rotation test (p. 217), an additional task of 2D mental rotation, Maps, was developed in this study (Figure 6.2). Both tests were similar in terms of their instruction, outcome space and scoring, but the latter one depicted small maps instead of concrete figures.

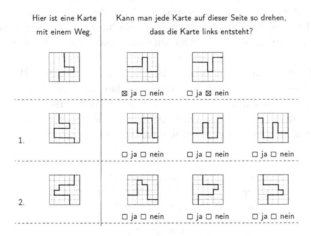

Figure 6.2 The task Maps required children to decide whether the depicted maps on the right were rotated versions of the original map to the left or reflections of it

Moreover, the 3D mental rotation test by VANDENBERG AND KUSE (1978) was adopted for children using less complex stimuli and fewer items (Figure 6.3). The outcome space of an item were one, two, three, or four crosses that were intended to be scored dichotomously when children checked the two snakes correctly.

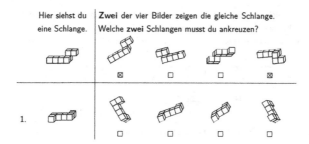

Figure 6.3 The task 3DMR required children to decide which of the two cube-objects (denoted as snakes) to the right were rotated ones of one on the left and which ones were distractors

Paper Folding Task

Definition: Paper folding tasks require individuals to identify the pattern of holes after a sheet of paper has been folded several times and punched with an object or cut out.

Intended Cognitive Processes: To solve paper folding tasks, individuals have to engage in multistep operations that ask them to imagine the folding and punching/cutting process of a piece of paper while keeping the perceived and imagined stimulus in working memory, probably using analytical strategies (LINN & PETERSEN 1985). Reasoning about folding operations has been shown to be isomorphic to perceiving the physical folding of paper. The cognitive processes of paper folding are those described by SHEPARD AND FENG (1972).

Tests: This study adopted the PFT for children (Figure 6.4) as proposed by GRÜSSING (2012). Although several adoptions for children exist, among them the ones of the national mathematics challenges in Germany (*Mathematikolympiade*), for example in task 550425, or the one proposed by BERLINGER (2015), this study used the same item setting as proposed by GRÜSSING (2012) to control the quality of the resulting data with her findings reported for a large sample size. The outcome space of each item is a set of crosses that were intended to be scored dichotomously according to a Complex Multiple Choice mechanism.

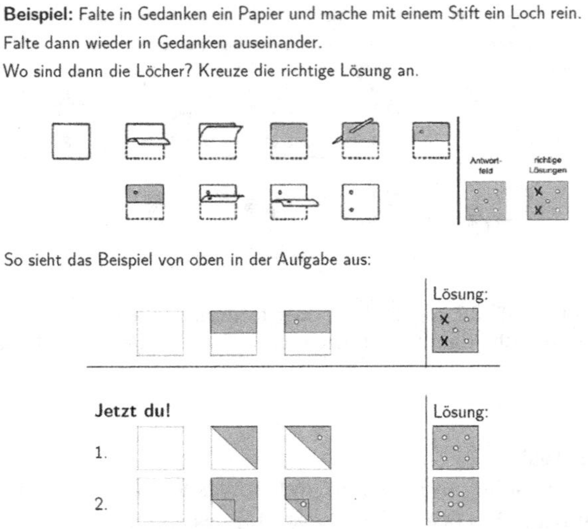
Beispiel: Falte in Gedanken ein Papier und mache mit einem Stift ein Loch rein.
Falte dann wieder in Gedanken auseinander.
Wo sind dann die Löcher? Kreuze die richtige Lösung an.
Antwort-feld
richtige Lösungen
So sieht das Beispiel von oben in der Aufgabe aus:
Lösung:
Jetzt du!
1.
2.
Lösung:

Figure 6.4 The task PFT required children to imagine how a paper was folded and punched. Then, they had to identify where the resulting holes were located

6.1.3 Tasks Measuring EGO

This study intended to measure EGO within a broad spectrum of possible tests, in particular with respect to those requiring the child to master conflicting frames of reference, that is, Level-2 tasks according to Flavell et al. (1981). Their development was based on Niedermeyer (2015) classification of possible perspective taking tasks (Figure 6.5), which was developed in line with Huttenlocher and Presson (1973, 1979) findings (see also Section 4.1.3).[5] This study intended to propose one task for each of the three dimensions of Niedermeyer (2015) classification:

- Does the test rely on a single feature (object) or on a configuration of objects?
- Is the projected viewpoint depicted or is the subject asked to imagine it?
- Is the position of the projected viewpoint specified or does is need to be identified?

[5]Tasks that demand for model-rotation, turning and reconstruction tasks were omitted since they rely explicitly on concrete operations with material which was beyond the score of this study.

Moreover, this study proposed an additional fourth dimension for the classification based on whether children had to reason about routes, for example when sorting pictures, or discrete positions in space. This distinction was based on the findings that previous studies used adoptions of the same task but found important differences in children's performances while testing the same age group. When children were required to sort pictures, they struggled to solve the task (GRÜSSING 2012), but when they were presented a similar stimulus and were asked to assign pictures to discrete positions, they performed much better (BERLINGER 2015).

Consequently, one might assume that a qualitative difference of the task may be achieved by either asking for selecting pictures at discrete positions or by asking for sorting pictures, given a route (continuous enchained positions). In addition, tasks involving simulated mental movement of the own body have been argued to be more likely to be solved by children using the intended strategy, transformations of the imagined self, rather than using alternative strategies such as mental rotation (VANDER HEYDEN ET AL. 2017).

Therefore, the following fourth dimension was important for the development of EGO tasks (Figure 6.5):

- Does the task require either the individual to transform the imagined self to adopt discrete positions and viewpoints, or to imagine perspective shifts along routes?

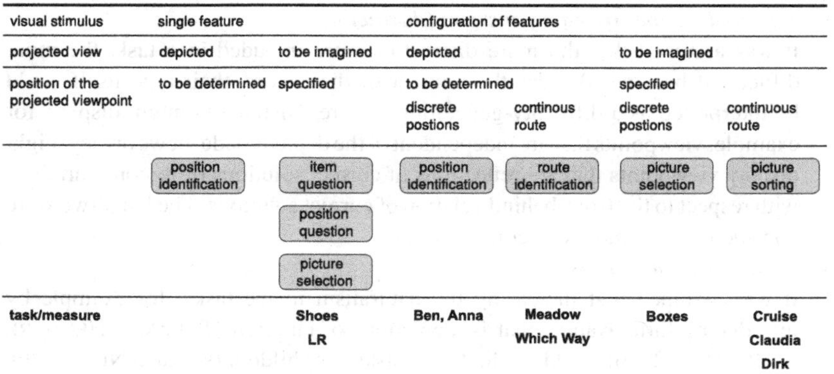

visual stimulus	single feature		configuration of features			
projected view	depicted	to be imagined	depicted		to be imagined	
position of the projected viewpoint	to be determined	specified	to be determined		specified	
			discrete postions	continous route	discrete positions	continuous route
	position identification	item question	position identification	route identification	picture selection	picture sorting
		position question				
		picture selection				
task/measure		Shoes LR	Ben, Anna	Meadow Which Way	Boxes	Cruise Claudia Dirk

Figure 6.5 Classification of tasks measuring EGO. (Adopted and extended from NIEDERMEYER 2015, p. 17)

During development of the test, the following difficulty-generating features were considered:

1. *Complexity of the depicted configuration*
Some features were considered that might generate complexity in perspective taking tasks. First, the number of objects that are in the configuration. Configurations with few objects were assumed to be easier to perceive, to encode in working memory and to be mentally transformed (see NIEDERMEYER 2015, p. 49). Second, the characteristics of the configuration. Although the symmetry of a configuration is not a difficulty-generating feature per se (NIEDERMEYER 2015), it was assumed that configurations that feature a number of directed objects are easier to process since they may contain single, distinct features that help to solve a task. Third, a task might become harder by choosing a number of abstract, undirected objects in the configuration (e. g., boxes) since they demand subjects for identifying their orientation and their interrelations (see NIEDERMEYER 2015, p. 51, for further elaboration). Moreover, it has been shown that directional cues that represent familiar objects with which the individual frequently interacts and which can be imagined during the solution process, facilitate performance in perspective tasks (GUNALP, MOOSSAIAN, & HEGARTY 2019).

2. *Type of the mental transformation*
Mental rotations around the horizontal axes may be considered to be more difficult with an increasing angle of rotation, rotations by 90° being therefore easier than a rotation of 180° (KEEHNER ET AL. 2006; MICHELON & ZACKS 2006).

3. *Type and number of wrong answers (distractors)*
It was assumed that the more distractors were included in a task, the more difficult it became. Besides the number of distractors, their type itself could be interpreted as a difficulty-generating feature. Distractors might display, for example, viewpoints that are independent of the depicted side views or they might display viewpoints that are reflections of correct solutions or re-configurations with respect to the front-behind-relation of correct solutions. The latter two were considered to be more difficult.

4. *Involvement of avatars*
It was assumed that increasing the naturalism in the tasks, for example by introducing little avatars, that is, depictions of children (TARAMPI, HEYDARI, & HEGARTY 2016) could facilitate the tasks for children (see also NEWCOMBE 1989). In addition, it was expected that these little avatars should also minimize the chance that children used alternative strategies, in particular mental rotations (e. g., BARRATT 1953).

Position Questions

Definition: Position questions require individuals to determine the spatial relation of a specific feature with respect to the point of view in a given position (NIEDERMEYER 2015, p. 41).[6]

Intended Cognitive Processes: To solve position questions tasks, the individual has to perceive and encode the spatial stimulus. Then, the subject focuses on a specific feature (i. e., left/right hand, a certain object, etc.) and performs a single transformation of the imagined self to answer the question. The cognitive processes are thus equivalent to those outlined in experiments on embodied rotations by PARSONS (1987b) and SAYEKI ET AL. (1997). To solve these, the child has to become aware and correct his or her own egocentric perspective which is ad odds with the one to take during an adjustment process as described in EPLEY, KEYSAR, VAN BOVEN, AND GILOVICH (2004).

Tests: The tests Shoes and LR required the children to solve position questions. The task Shoes (Figure 6.6) was similar to the one proposed by WIEGAND (2006, p. 128). It was assumed that low spatial achievers, in particular children who have not sufficiently developed their projective concepts, could struggle in this task. The outcome space of each item is a set of colored shoes and each shoe was intended to be scored dichotomously.

In addition, the map-based task LR (Figure 6.7) was developed as a child's adoption of the Money Road Test proposed by MONEY AND ALEXANDER (1966, as cited in FIELDS & SHELTON 2006, p. 509). The test required children to perform single transformations (rotations) of the egocentric viewpoint to determine whether a turning is to the left or to the right. The outcome space of this task is the abbreviation of a direction (e. g., 'L'). Scoring was based on a Complex Multiple Choice mechanism on item pairs representing equal cognitive demands to control for guessing probabilities. (Table 6.2).

---

[6]Item question ask vice versa for the specification of a feature, given the spatial relation (left/right) with respect to the point of view in a given position.

Hier siehst du 4 Kinder auf einem Spielplatz. Male jeweils **aus der Sicht der Kinder** alle linken Schuhe blau und alle rechten Schuhe gelb aus!

Figure 6.6 The task Shoes required the children to color the shoes of the depicted LEGO figures

Position Identification Tasks
Definition: Position identification tasks require individuals to identify single, spatially discrete positions that correspond to depicted projected viewpoints (NIEDERMEYER 2015, p. 43).

Intended Cognitive Processes: To solve position identification tasks, the individual has to perceive and encode the visual stimulus (depictions of projected viewpoints). Then, the individual has to perform multiple transformations of the imagined self toward discrete spatial positions, thereby probably establishing a spatial framework that allows the individual to reason about locations of objects with respect to imagined body axes. The imagined viewpoint needs to be retained in working mem-

Stelle dir vor, dass du vom Start in Richtung Schatz gehst und schreibe an
jede Ecke in den Kreis, ob du dabei nach **rechts (R)** oder **links (L)** abbiegst.

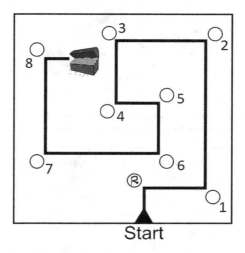

Figure 6.7 The task LR required the children to make left-right judgements on a map

Table 6.2 Pairs of the Same Cognitive Demand in LR

Item pair	Item/ turn	Cognitive demand
StepLR1	2 and 8	right-left discrimination from egocentric perspective
StepLR2	3 and 7	right-left discrimination with imagined rotation by $+90°$
StepLR3	1 and 5	right-left discrimination with imagined rotation by $-90°$
StepLR4	4 and 6	right-left discrimination with imagined rotation by $180°$

ory and has to be compared to the given picture. Therefore, the cognitive processes
of position identification tasks equal the ones described by YU AND ZACKS (2017).

Tests: The tasks Anna (Figure 6.8) and Ben (Figure 6.9) were developed in analogy to a set of other tasks that have been proposed in the literature, such as DE
LANGE's (1984) Picture Show that was used in a modified way by BERLINGER

Anna sieht sich einen Spielplatz an. Der Spielplatz sieht von oben so aus:

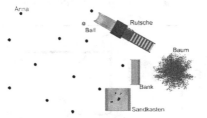

Anna hat ein Foto an dem Punkt gemacht, wo ihr Name steht (Foto 1). Danach hat sie an 5 weiteren Punkten auch ein Foto gemacht (Foto 2 bis Foto 6).

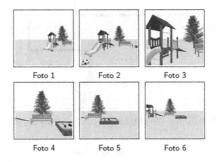

Wie ist Anna gelaufen um ihre Fotos zu machen?
Verbinde die richtigen 6 Punkte oben in der Karte!

Figure 6.8 The task Anna required the children to identify five different locations from which a picture had been taken using shifts of the imagined self

(2015), GRÜSSING (2002), and MEISSNER (2006).[7] However, ceiling effects have been reported among fourth graders in this task (e. g., BERLINGER 2015, p. 235). To develop a task that allows for a differentiation between individual performances, a range of further locations from which the picture could have been taken were added to decrease the guessing probability and to increase complexity of the task. The outcome space of the task consists of a set of connected locations that can each be dichotomously scored.

---

[7]The tasks Anna, Ben, Meadow, Which Way, Claudia, and Dirk were designed using Sweet Home 3D, copyright (c) 2005–2017 E. Puybaret. The Software includes 3D models and textures distributed under a free license.

Ben sieht sich einen anderen Spielplatz an. Der sieht von oben so aus:

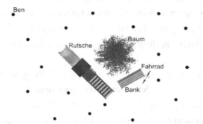

Ben hat ein Foto an dem Punkt gemacht, wo sein Name steht (Foto 1). Danach
hat er an 5 weiteren Punkten auch ein Foto gemacht (Foto 2 bis Foto 6).

Wie ist Ben gelaufen um seine Fotos zu machen?
Verbinde die richtigen 6 Punkte oben in der Karte!

Figure 6.9   The task Ben required the children to identify five different locations from which
a picture had been taken performing mental shifts and rotations of the imagined self

Route Identification Tasks

Definition: Route identification tasks require individuals to identify a route that
corresponds to a set of depicted projected viewpoints.

Intended Cognitive Processes: To solve a route identification task, individuals
have to perceive and encode the visual stimulus. They have to understand that the
depicted viewpoints (two or more) correspond to different time points and that there
was movement in between. Then, the subject has to integrate the given stimulus
into a fixed frame of reference by means of multiple egocentric transformations,

probably drawing on embodied experiences made in real space. In a last step, the subject retains the imagined route in working memory and makes inferences from it. Cognitive processes of route identification tasks are perspective taking processes as described by KESSLER AND THOMSON (2010) and involve spatial frameworks that are constantly updated as described in YU AND ZACKS (2017).

Tests: The tests Meadow (Figure 6.10) and Which Way were developed as tasks requiring route identification. Meadow, for example, was an adoption of the Guilford-Zimmerman-Boat Test (GUILFORD 1956). It involved less intriguing transformations of the viewpoint than the original test and rather focused on the imagined self-movement in real space along canonical axes than to imagine sitting in a moving object. The outcome space of each item in the task is a set of crosses that are scored dichotomously according to a Complex Multiple Choice mechanism.

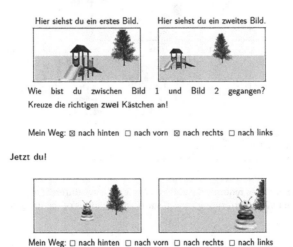

Figure 6.10 The task Meadow required the children to imagine the movement that was performed in between the given two viewpoints

Picture Selection Tasks

Definition: Picture selection tasks require individuals to select the picture of a viewing direction that corresponds to a given position among a set of possible depictions (NIEDERMEYER 2015, p. 35).

Intended Cognitive Processes: To solve a picture selection task, the subject has to perceive and encode the visual stimulus and to adjust the egocentric perspective using a single mental transformation of the imagined self at the given position. Then, the individual imagines what can be seen using a spatial framework, retains the imagined viewpoint in working memory and compares it with the set of possible solutions to select the correct picture. The cognitive processes of picture selection tasks equal those described by EPLEY, KEYSAR, ET AL. (2004) and YU AND ZACKS (2017).

Tests: The task Boxes (Figure 6.11) was developed similar to the test 'Von welcher Seite?' by GRÜSSING (2002). In her study, the empirical results of a qualitative interview showed that children solved this tasks, besides using a flexible analytical strategy, by performing mental transformations of the egocentric viewpoint. The task Boxes consists of three items that present three different configurations of the two or three boxes. The outcome space of each item consisted of single letters that were written into the five boxes. Each box was scored dichotomously.

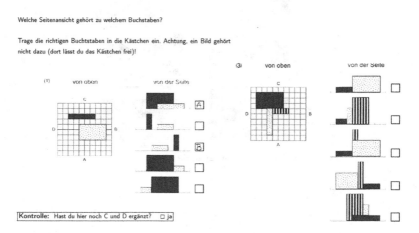

Figure 6.11 The task Boxes required the children to select the correct four out of five possible side views to a given bird's eye perspective

Picture Sorting Tasks

Definition: Picture sorting tasks require the subject to sort a set of depictions of projected viewpoints that correspond to a pre-defined spatial route (e. g., a given path).

Intended Cognitive Processes: To solve a picture sorting task, the individual has to encode the bird's eye view of the labyrinth. By mentally simulating a movement of the imagined self along the pre-defined route in the labyrinth, the individual has

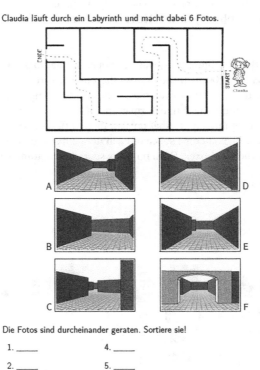

Figure 6.12 The task Claudia required the children to sort pictures that Claudia had taken along the pre-defined route in a labyrinth

to subsequently reason what could be seen along the route, thereby probably using the spatial framework, adjusting and controlling for possible conflicts with the own perspective on the tasks, and to emulate embodied experiences from own movement in space. The imagined scenes have to be maintained in working memory and subsequently be compared to the depicted viewpoints in the tasks. The cognitive processes are therefore those described by EPLEY, KEYSAR, ET AL. (2004), FRANKLIN AND TVERSKY (1990), KESSLER AND THOMSON (2010), and YU AND ZACKS (2017).

The outcome space of picture sorting tasks (Figure 6.12, 6.13) is a sequence of up to six letters, probably involving double letters or empty spaces. The scoring of

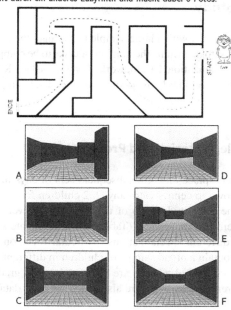

Dirk läuft durch ein anderes Labyrinth und macht dabei 6 Fotos.

Die Fotos von Dirk sind durcheinander geraten. Sortiere sie.

1. _____   4. _____

2. _____   5. _____

3. _____   6. _____

Figure 6.13 The task Dirk, another labyrinth tasks

picture sorting tasks is rather complex since all sorting entries are interdependent. Different scorings have been provided such as dichotomous scoring of the whole sequence (e. g., GRÜSSING 2012). This implied, however, that the measure became difficult since stepwise solutions were not accepted. Another possible scoring is a ranking-wise scoring (e. g., BERLINGER 2015), implying that the measure would be low for all those children who fail in sorting one of the foremost pictures. This study proposed to score the length of a correctly sorted sequence behind a picture and added whether the first picture was recognized or not. Children who failed to sort in one of the foremost pictures were therefore able to score high if all other pictures were sorted correctly.

6.2   Pilot Study

The material was piloted with a large sample to evaluate the preliminary quality of the test instruments, to determine the number of tasks to be completed in 45 minutes, and to pilot the organizational framework. For the tasks LR and for the sorting tasks Claudia, Cruise, and Dirk, the pilot study intended to empirically evaluate the proposed scoring.

6.2.1   Sample, Material, and Procedure

The pilot study took place on eight subsequent school days in mid February 2016 in Jena, a small town in central Germany. 222 children from five different schools participated in the pilot study. Two of the five schools were public schools and three of them were private schools (Table 6.3). The private schools were chosen because of their different specializations (particular focus on language, on social learning, etc.) to obtain a broad sample of children in different educational settings in class four. In one school, classes are organized across grades. For this reason, some older children in class five were also tested but their data was not considered during analyses.

Table 6.3 Sample of the Pilot Study

Type	Classes	Grade 4	Grade 5	Sum
public schools	7	125	–	125
private schools	5	66	31	97
sum	12	191	31	222

The children were given exactly 45 minutes to complete the test after the general introduction. They had to complete the tasks in the order of presentation in the test booklet. During the introduction, the tasks 2DMR and Boxes were explained in front of the class with manipulable material.

The estimated maximum completing time per measure was half a minute per item and one minute for reading the task instruction. Some extra completing time for all measures that involved complex spatial stimuli (e. g., Claudia, PFT) was added. The measures were then arranged in the test booklet so that the first ten measures had a maximum completing time of 45 minutes and that there were three additional measures which had a maximum completion time of 10 minutes (Table 6.4).

Table 6.4 Test Booklet for the Pilot Study

Item	Estimated time	Item	Estimated time
1. Shoes	3 min	6. Meadow	4 min
2. LR	3 min	7. PFT	5 min
3. 2DMR	3.5 min	8. Ben, Anna	7 min
4. Claudia	5 min	9. Cruise	4 min
5. 3DMIR	3.5 min	10. Boxes	7 min
Which Way (3 min) Maps (3 min) Dirk (4 min)			

Five children had to leave after 25 minutes for attending their swimming course. Their data were deleted from subsequent analyses. Moreover, data from children completing the test booklet by randomly choosing answers in under 10 minutes were also excluded. The resulting data stemmed from a sample of 185 fourth graders. The sample consisted of 98 boys and 87 girls aged between 8 years 6 months and 11 years 5 months ($M = 10.05$, $SD = 0.47$).

The test booklets were encoded using SPSS (CORP 2017) on the basis of a preliminary version of the encoding manual. Within the framework of the pilot study it was accepted that the quality of the data would be sufficiently high by encoding the test booklets only one time since the codes of the encoding manual were objectively evaluable.

6.2.2   Preliminary Analysis

Table 6.5 shows the number of missing values per task. For the last three additional measures, the percentage of missing values was higher than 10%, a result that was interpreted to indicate that a considerable number of children were not able to solve those tasks within 45 minutes. The subsequent analyses of those tasks were based on the sample of children who completed them to compare their characteristics to their parallel tasks, that is, tasks with the same cognitive demand. This was done to decide whether or not they can be included in the test booklet for the main study.

Table 6.5 Number of Missing Values to Solve a Task in 45 Minutes

Range	Name of the task	Missing values	Percent
1–7	Shoes-PFT	0	0%
8	Ben	3	1,6%
9	Cruise	7	3,7%
10	Boxes	16	8,6%
11	Which Way	30	16,2%
12	Maps	37	20,0%
14	Dirk	60	32,4%

6.2.3   Detailed Analysis of All Measures

To decide on possible revisions that had to be made for the main study and to detect inappropriate task instructions, bad item designs, or problems with item scoring, the measures were analyzed in detail. The same item analysis routine was applied for all the measures.

The following sections present results from analyses of three measures. They exemplarily demonstrate CTT and IRT item analyses for three different cases (Section 5.2.1). First, for the measure LR, the analytical framework helped to empirically verify the scoring which was initially proposed based on a theoretical analysis of cognitive processes. Second, for the measure 3DMR, the analytical framework helped to empirically detect items with bad psychometric properties that required further revision. Third, an analysis of a task requiring picture sorting, which was the task Claudia, is shown to present how these tasks were analyzed for possible revisions.

Detailed analyses for all other measures that were conducted in a similar manner as the examples outlined in this section can be found in the ESM C.3 (pp. 33). The results of the detailed analyses for all measures are also summarized at the end of this Section (Table 6.13, p. 210) since the results were the basis of the latent construct analysis in Section 6.2.4 as well as the revisions for the main study that are described in Section 6.2.5.

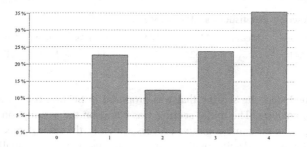

Figure 6.14 Distribution of sum scores for measure LR (n = 185)

LR

An analysis of the relative frequencies of the sum scores for the measure LR (Figure 6.14) revealed that

(1) almost all subjects solved at least one pair of directions,
(2) the distribution showed slight effects of bimodality,
(3) the measure was subject to ceiling effects, that is, the measure was not able to differentiate among subjects with high achievements.

Given that one pair of items required the children to distinguish between left and right without performing an additional mental transformation of the imagined self, this might have yielded the result that almost all children scored with at least one point. In line with observation three, the bimodal distribution that was observed may has been due to the fact that there were not enough difficult items to distinguish between high performers. It might also has been due to sex differences, but those were not considered in the pilot study.

The descriptive statistics represented what had been observed for the distribution of the sum score (Table 6.6). The mean value indicated towards a clear shift towards higher scores that were achieved more often than lower scores. The distribution showed medium skewness which was not an indication for a slight deviation from the normal distribution but might be explained by the bimodality. This was supported by a high kurtosis value of the distribution which indicated toward a wide distribution of achieved scores.

Table 6.6 Descriptive Statistics for LR (n = 185)

	Range	Median	Mean	SD	Skewness	Kurtosis	Cronbach's α
LR	[0,4]	3	2.62	1.32	−.44	−1.19	.70

In a second step, the scoring that was proposed in Section 6.1.3 was verified using Rasch analysis. The initial scoring was based on the assumption that pairs of items requiring the same cognitive processes, that is, the same degree of mental rotation of the imagined self, could be parceled because they were equally difficult (Table 6.2). To test this assumption, the item-person map was computed using a trial version of ConQuest 3 (M. L. WU ET AL. 2015).

Figure 6.15 shows the item-person map with the corresponding parameter estimates and the latent ability distribution for all eight unpaired items of the measure (left) and for the paired items (right). The figure on the left demonstrates that there are no items to differentiate between subjects that have a high level of latent ability; all items are of low to medium difficulty. This is in line with the ceiling effects that were observed. Moreover, four pairs of items could be detected and interpreted by means of the cognitive processes that were necessary to solve them. Item pair 1 that consisted of item 2 and 8 represents items that do not need a transformation of the egocentric frame of reference at all. Item pairs 2 and 3 that represented imagined rotations of 90 degrees each approximately had the same difficulty. Item pair 4 that represented items that demanded for an imagined rotation by 180 degrees of the egocentric viewpoint, represented the most difficult item pair. Relations of the item difficulties remain unchanged in the item pairs as confirmed by a Rasch analysis for the four item pairs (right side of Figure 6.15).

The results indicated that the approach to build item parcels was reasonable. Table 6.7 presents the corresponding summary statistics. Similar to results shown in Figure 6.15, the item pairs show different item facilities and all item pairs show reasonable discrimination values. Item 1 has a low discrimination value which should be due to its high facility value. All weighted MNSQ-values were within the acceptable interval ($MNSQ < 1.3$). The four item pairs can therefore be considered to be Rasch conform.

LR was kept for the main study. It was expected to differentiate among low and medium scorers and it had acceptable reliability (.70). Item 1 was kept since it measured relevant abilities, that is, left-right discrimination from the egocentric viewpoint. Resolving the answer of how to control for ceiling effects in this measure was difficult since all possible rotations of the imagined self were already considered during task development.

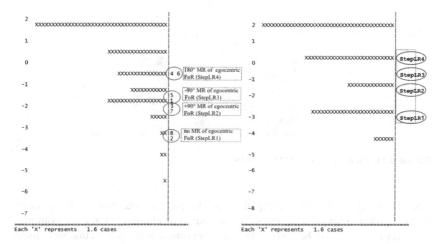

Figure 6.15 Distribution of item difficulties in LR (left) and their mapping into pairs with the same cognitive demand (right)

Table 6.7 Summary of Items Statistics for LR (n = 185)

Item no.	Item label	Facility (CTT)	Discrimination (CTT)	Weighted MNSQ+CI (IRT)	Difficulty (IRT)
1	StepLR1	.86	.32	1.14 (0.70, 1.30)	−3.02
2	StepLR2	.70	.60	0.85 (0.79, 1.21)	−1.47
3	StepLR3	.58	.56	1.00 (0.80, 1.20)	−0.61
4	StepLR4	.48	.47	1.04 (0.80, 1.20)	0.16

3DMR

Summary statistics for the task 3DMR were computed. An analysis of the relative frequencies of the sum scores (Figure 6.16) revealed that

(1) almost all students scored with one point or higher,
(2) the distribution was approximately normal,[8]
(3) only a few children achieved the full score.

---

[8] As outlined in Section 5.2.2, variables with more than 5 categories can, technically spoken, be considered as continuous, therefore being normally distributed. This study uses this expression, but understands the different categories as sampling from a normal distribution.

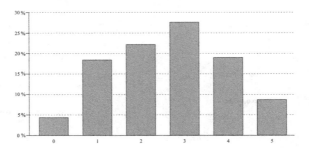

Figure 6.16 Sum scores for 3DMR (n = 185)

The first observation could be explained by the fact that one item of the measure probably was too easy since it was solved by almost all children. Item 1, for example, seemed to be very easy because of the different forms of the snakes that were used in the construction of the item (Figure 6.17). If a child identified snake one and three to be of another shape, no mental rotations were additionally needed to detect the correct solutions.

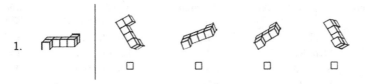

Figure 6.17 Item 1 of 3DMR was too easy

The second and third observation indicated that the measure featured items of different difficulties that differentiated among individuals with low and high mental rotation abilities. One item seemed to be particularly difficult which could explain the third observation. A re-analysis of the items revealed that the number of cubes that build a snake seemed to be a difficulty-generating characteristic. Item 5, for example, featured complex snakes with up to 8 cubes (Figure 6.18).

A re-analysis of the measure showed that all items contained distractor snakes that differed substantially from the snake on the left side (e. g., snake two for item 5) which allowed children to solve the items not only by mental rotation but also by using analytical strategies and procedures of exclusion. Moreover, the single items were not constructed in a homogenous way in a sense that distractors were not always reflections of the snake on the left.

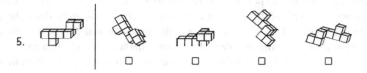

Figure 6.18 Item 5 of 3DMR was complex, but involved easy distractors (second snake)

The descriptive values of the measure (Table 6.8), in particular the low skewness, high kurtosis values and the low reliability, confirmed what has been observed before.

Table 6.8 Descriptive Statistics for 3DMR (n = 185)

	Range	Median	Mean	SD	Skewness	Kurtosis	Cronbach's α
3DMR	[0,5]	3	2.62	1.34	−.02	−.79	.50

In a second step, the summary of item statistics was computed (Table 6.9). Facility values for item 1 and 5, for example, demonstrated that item 1 was very easy and item 5 was very difficult for the children. Consequently, both items had low discrimination values. The weighted MSNQ indices were within the acceptable interval ($MNSQ < 1.3$) and all items could therefore be assumed to be described sufficiently well by the Rasch model.

Table 6.9 Summary of Items Statistics for 3DMR (n = 185)

Item no.	Item label	Facility (CTT)	Discrimination (CTT)	weighted MNSQ + CI (IRT)	Difficulty (IRT)
1	3DMR1	.77	.03*	1.12 (0.81, 1.19)	−1.46
2	3DMR2	.72	.42	0.92 (0.84, 1.16)	−1.12
3	3DMR3	.42	.34	0.99 (0.87, 1.13)	0.42
4	3DMR4	.51	.32	0.97 (0.88, 1.12)	−0.07
5	3DMR5	.23	.27*	1.02 (0.80, 1.20)	1.46

*denotes discrimination values < 0.3

An analysis of the corresponding person-item-map (Figure 6.19) confirmed the observations outlined above. Item 1 was the easiest one and item 5 the hardest one. The single items had different item difficulties and seemed therefore to be useful for measuring the abilities to perform 3D mental rotations in a differentiated manner.

Although the items of the measure showed a variety of item difficulties, the measure needed revision. The different shapes of the distractor snakes in items with even complex snakes made it impossible to determine whether the items measured mental rotation abilities or analytical abilities, an observation that was probably confirmed

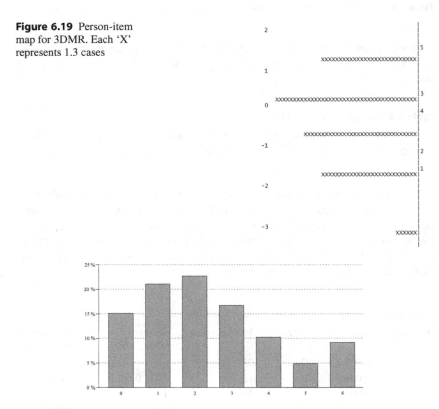

Figure 6.19 Person-item map for 3DMR. Each 'X' represents 1.3 cases

Figure 6.20 Sum scores for Claudia (n = 185)

by the low test reliability (.50) and low item discrimination values. For subsequent analyses within the pilot study, item 1 was withdrawn but item 5 was maintained despite the item's low discrimination value, since it allowed to differentiate among high achievers.

Claudia

An analysis of the relative frequencies of the sum scores for the measure Claudia (Figure 6.20) revealed that

(1) a considerable number of children did not achieve one point at all in the measure,
(2) about three quarters of the children achieved 3 points at maximum,
(3) the distribution of the sum scores revealed a slight effect of bimodality.

The first two observations indicated that the measure Claudia was difficult for the children. This might be explained by the format of the measure that was new and unknown to them. Identifying all locations of the pictures correctly and sorting them in a photo sequence is a task of high cognitive demand involving visuospatial working memory capacity. Whereas 15% of the children failed completely to do so, 44% of the children achieved one or two points, it was harder for them to achieve higher scores. However, almost 10% of the children achieved the full score. Scores of 4 seemed to be a threshold between high-achievers and low-achievers. This might explain, besides possible effects of sex differences not examined in the pilot study, the bimodality. Descriptive statistics confirmed the observations (Table 6.10).

Table 6.10 Descriptive Statistics for Claudia (n = 185)

	Range	Median	Mean	SD	Skewness	Kurtosis
Claudia	[0,6]	2	2.37	1.78	.57	−.53

In a second step, typical errors that occurred in the first pictures to sort were analyzed. 137 of 185 children chose the correct picture F as a starting point and 132 decided for a second and third picture.[9] In Table 6.11, the corresponding solutions for picture 2 and 3 are quantified. Apparently, 67 children of the 132 chose the correct picture B as the second picture, but only 33 of those 67 children managed to identify A as the third picture. Instead, 29 children of the 67 chose C or E by making

---

[9] 5 children stopped sorting after the first picture

egocentric errors at that point. Interestingly, 33 children identified the picture D to be the second one. This result could be interpreted as an error in estimating distances given by the pictures in the measure. They misinterpreted the length of the first corridor and immediately attributed picture D. 18 of those children continued to sort correctly in picture three.

Table 6.11 Choices for Pictures Two and Three if Picture F Is Chosen First (n = 132)

Second picture	Third picture					
	A	B	C	D	E	total
A	0	5	1	0	1	7
B	33	0	16	5	13	67
C	1	3	0	1	6	11
D	4	18	4	0	5	33
E	4	7	2	3	0	16
	42	33	23	9	25	132

The wrong estimation of the length of the first corridor was a problem in the measure. 25 of the 185 children ignored picture F to be the first one and started their sorting with picture D. 11 of them continued sorting correctly with picture B. According to the scoring that was initially proposed, those 25 children were scored with low scores because the scoring function was not flexible enough to capture those solutions. Consequently, the measure had to be revised for the main study. It was not objectively scorable and reflected abilities that were not intended to be measured. The format of a labyrinth, however, was decided to be kept. It was intended to be modified and a second labyrinth task with scorable solutions (Section 6.2.5) was added.

Scoring of the Sorting Task Cruise
An analysis of the scoring proposed for Cruise by means of the same approach (see ESM C.3) revealed that certain solutions were underscored. An adopted scoring is proposed in Table 6.12.

Table 6.12 New Scoring for Cruise

Score	Description	Example
0	random guessing	BADCFE
	incomplete wrong solution	BF…
1	one pair of correctly sorted pictures	BEADFC
	incomplete solution with two correct elements	BD…
2	two pairs of correctly sorted pictures	BAEDFC
	triple of correctly sorted pictures	BAECFD
	incomplete solution with three correct elements	BDF…
3	one picture is mixed up	BDFEAC
	almost correct solution with missing end	BDFEA.
4	correct solution	BDFAEC

Additional Observations

The children were also observed during the introduction of the test and while completing it. Three questions were risen regularly by shouting them out loud into the class whenever the test experimenter did not react to a hand gesture:

(1) "I cannot see the way to the treasure! Do I have to walk along the black line?" (LR)
(2) "I do not understand the description. What does the picture mean?" (PFT)
(3) "Do I have to check the 'yes' in the control questions?"

The task instruction using 'you walk along' in LR seemed to be abstract, a problem that could probably be resolved using an avatar. The depiction of the paper piercing demanded for a high level of abstraction to be understood, a problem that could be resolved by presenting the example during the general introduction of the test. Finally, the control questions were read but checking them was found to be confusing, a problem that could be resolved by omitting the checking box while keep asking the question (Table 6.13).

Table 6.13 Summary of Item Statistics of the Pilot Study and Observations for Revision

Item	Range	Mean	SD	Item facility (CTT)	Item discrimination (CTT)	Cronbach's α	summary of observations
Shoes	–	–	–	–	–	–	measure is not scorable objectively, keep as starter
LR	[0,4]	2.62	1.32	[.48, .86]	[.32, .69]	.70	slight ceiling effects, bimodality
2DMR	[0,5]	2.25	1.56	[.29, .60]	[.26, .48]	.65	withdraw item 2, sufficient differentiation between subjects with high and low abilities
Claudia	[0,6]	2.37	1.78	–	–	–	revision of the measure necessary due to picture D
3DMR	[0,5]	2.62	1.34	[.23, .77]	[.03,.42]	.50	withdraw item 1 and 5, revision of the whole measure
Meadow	[0,6]	3.57	2.21	[.54, .63]	[.51, .67]	.85	ceiling effects, all items have the same difficulty
PFT	[0,6]	3.37	1.53	[.21, .93]	[.23, .54]	.68	replace item 1 (low discr.), good measure
Anna	–	–	–	–	–	–	withdraw (non-human movement)
Ben	[0,5]	3.08	1.83	[.49, .69]	[.55, .68]	.82	ceiling effects, bimodality
Cruise*	[0,4]	1.91	2.5	–	–	–	scoring requires adoption, bimodality, floor effects
Boxes	[0,10]	6.70	2.81	[.61,.74]	[.34,.60]	.65	strong ceiling effects, item 1 has a cutoff in the threshold values, the distractor in item 3 depicts only two boxes
Which Way	–	–	–	–	–	–	withdraw, high guessing probability
Dirk	–	–	–	–	–	–	withdraw, no unique solution
Maps	[0,4]	1.92	1.40	[.26, .62]	[.43,.54]	.70	no items with low difficulty

*adopted scoring

6.2.4   Confirmatory Factor Analyses

Confirmatory factor analyses were conducted to investigate which latent factors are empirically separable and whether or not the test instrument was construct valid or not. Based on the analysis of missing values and the choice between the two measures that required children to perform mental rotations of two-dimensional figures, the data of the measures Which Way, Dirk, and Maps were withdrawn from further analyses. In addition, data of the task Shoes were withdrawn for not being objectively scorable.

Nine measures were kept for further analyses. The sum score of the items was computed for seven of them and for the two sorting tasks Claudia and Cruise, the scoring was in line with the theoretical sum-score outlined above[10]. Because the CTT and IRT analyses revealed that items 2DMR2, 3DMR1, and PFT1 had low discrimination values or did not fit into the Rasch model, those items were withdrawn for subsequent analyses and the sum score of these measures was computed using the remaining items.

Table 6.14 shows the Pearson correlations, based on pairwise deletion of missing values.[11] Measures of OB tended to correlate moderately with each other (.43 to .46). Correlations between measures of EGO were lower (.13 to .38). An unexpected result was the high correlation between LR reflecting EGO and PFT reflecting OB (.41).

Since the data was collected class-wise, the intraclass coefficients (ICs) for every task were computed in a preliminary step. They determined possible biases that may have occurred in the subsequent analyses due to an underlying multilevel structure in the data (B. O. MUTHÉN 2002, p. 90). Table 6.15 shows the results of a *Mplus* (L. K. MUTHÉN & MUTHÉN 2004) analysis.

---

[10]Claudia and Boxes were entered into the analyses with the initially proposed scoring (since both measures underwent revisions in the main study), Cruise with the revised scoring.

[11]Pearson correlations are slightly smaller than polychoric correlations that have been computed when treating all the variables as categorical. Because subsequent analyses were conducted with MLR-estimations that rely on continuous variables, Pearson-correlations are reported here.

Table 6.14 Pairwise Pearson Correlations in the Pilot Study (n = 185)

Measure	(1)	(2)	(3)	(4)	(5)	(6)	(7)	(8)	(9)
(1) LR	1								
(2) 2DMR	.26	1							
(3) Claudia	.18	.17	1						
(4) 3DMR	.25	.45	.18	1					
(5) Meadow	.19	.23	.13	.20	1				
(6) PFT	.41	.46	.32	.43	.24	1			
(7) Ben	.17	.24	.20	.24	.35	.32	1		
(8) Cruise	.27	.10	.25	.18	.21	.25	.22	1	
(9) Boxes	.24	.32	.27	.19	.37	.33	.25	.38	1

Table 6.15 Intraclass Coefficients for the Small-Scale Spatial Abilities Measures in the Pilot Study

LR	2DMR	Claudia	3DMR	Meadow	PFT	Ben	Cruise	Boxes
.07	.04	.03	.05	.01	.13	.06	.03	.02

Six out of the nine tasks had an $\rho_{IC} \leq .05$, a result that showed that variance in those measures was mainly due to individual differences but not due to a considerable effect of belonging to a certain class (GEISER 2012). The tasks LR, PFT, and Ben showed a higher effect of class-clustering. As to PFT, for example, 13% of the variance measured could be explained by class effects and only 87% by interindividual differences. As a consequence, for these three measures, effects of belonging to a certain class may have had an influence on thei corresponding variances. In subsequent analyses, intraclass effects were neglected since they were not of interest for this study, but parameters were estimated using robust corrections to standard errors to avoid possible dramatic increases of α-errors, a commonly reported effect when high ICs are not considered (GEISER 2012).

To examine whether a set of measures reflecting OB was empirically separable from a set of measures reflecting EGO, confirmatory factor analyses (CFA) were conducted in R lavaan (ROSSEEL 2012) using the complete data sets only (n = 169).

All measures were treated as continuous variables in the analyses since all sum scores had at least five categories. Robust maximum-likelihood estimation (MLR) was used. This estimator was expected to adjust standardized errors to adjust estimation and significance testing to possible biases occurring due to non-normal distributions (Section 5.2.2).

In a first step, a two-component model was fitted to the data, thereby assuming that the measures 2DMR, 3DMR, PFT would load on a factor OB and that the measures LR, Claudia, Meadow, Ben, Cruise, and Boxes would load on a second factor EGO. Due to the analysis of the correlations (Table 6.14), there was a doubt whether LR reflected EGO or OB from an empirical point of view. To investigate this question, a preliminary model was computed in which LR was allowed to load on both factors OB and EGO. Results indicated that LR loaded significantly on the factor OB but not on EGO. In other words, only OB was able to explain variance in LR above what has already been explained by EGO, which was an unexpected finding.

In subsequent analyses, LR was therefore considered to be a task reflecting OB. Table 6.16 (p. 213) shows the robust, adjusted fit indices for the two-factor model solution. The p-value of the χ-square test of model fit indicated that the model did not significantly deviate from the data. The Root Mean Square Error of Approximation, RMSEA, was well below .06 (.03), the Bentler Comparative Fit Index, CFI was larger than .95 (.98), the Tucker-Lewis Index, TLI was larger than .95 (.97), and the Standardized Root Mean Square Residual, SRMR was smaller than .08 (.04). All fit indices therefore met the criteria for good fit (see Table 5.1 and HU & BENTLER 1999). Furthermore, an inspection of the residuals showed that they were all low ($\leq |0.1|$) (R. B. KLINE 2015, p. 278). Only the residual of Meadow correlated slightly higher with the residual of 3DMR (-0.113). Those slight deviations between model and data could be neglected.

Table 6.16 Comparison of Fit Indices for the Single- and Two-Factor Model in the Pilot Study

Model	χ^2	df	RMSEA[CI]	CFI	TLI	SRMR	AIC	BIC
1-factor-model	48.80**	27	.07 [.03, .10][†]	.91	.88	.056	5673	5720
2-factor-model	30.32	26	.03 [.00, .07][†]	.98	.98	.043	5656	5715

**indicates that the χ^2-Test of model fit is significant at .01-level.
The RMSEA is shown with its 90% confidence interval (CI). [†]indicates if $p_{RMSEA \leq .05}$ is non-significant.

Figure 6.21 shows the completely standardized solution of the two-factor model.[12] The standardized factor loadings can be interpreted as standardized regression coefficients in a multiple regression (R. B. KLINE 2015, p. 301). In line with

---

[12]Standardized estimates are based on both the variances of latent variables and indicators being rescaled to 1.0.

COMREY AND LEE (1992), the factor loadings were poor (Claudia), fair (LR, Meadow, Ben, Cruise), good (2DMR, 3DMR), very good (Boxes), and excellent (PFT).

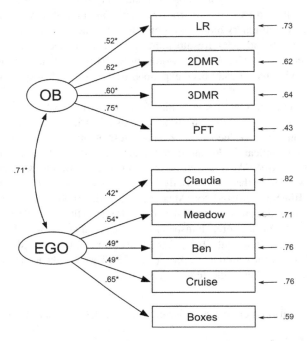

Figure 6.21 Completely standardized solution of the confirmatory factor analysis in the pilot study revealing two distinct subclasses of small-scale spatial abilities. *indicate significant path coefficients ($p < .001$)

Since all indicators were assumed to load only on one factor, their path coefficient estimates can be interpreted as correlations between the indicator and the latent variable. In addition, the squared factor loadings represent the proportion of variance in the indicator that is explained by the factor only and not by measurement errors. The factor OB, for example, accounted for $.75^2 = .56$ (56%) of variance in the indicator PFT but only for 27% of variance of LR. The factor EGO accounted for 40% of variance in the indicator Boxes, but only for 18% of variance of Claudia which was very low. This result indicated that a variance of most of the tasks could not be sufficiently explained by the underlying result but involved high measurement errors. In particular the EGO measures had low factor loadings. These results were not surprising because most of the measures for testing the abilities to perform egocentric transformations have been developed from the sketch. They indicated that those measures should, if possible, undergo revision for the main study, a finding that was in line with the results of the descriptive analyses performed before.

An analysis of MCDONALD (1999) ω coefficient for reliability that acknowledges heterogeneous indicator-factor relations showed that both the subscales and the total test featured acceptable reliabilities. OB had a McDonald's $\omega_{OB} = .73$, EGO an $\omega_{EGO} = .66$ and the total omega for the test of nine indicators was $\omega_{tot} = .77$.

In the two-factor model, both factors OB and EGO were highly correlated ($r = .71$) and shared a considerable amount of variance (50%). A single-factor model in which all indicators loaded on one factor was computed to test whether these factors were identical. The fit indices of this model are also shown in Table 6.16 (p. 213). The overall Chi-square-test for this model was significant which indicated that the model deviated from the data. Furthermore, the RMSEA, CFI and TLI do not meet the criteria for good fit (see Table 5.1). The single-factor model did therefore not explain the data sufficiently well. Consequently, it was more likely that the two-factor solution explained the data better, a result that was underlined when comparing the AIC and BIC-values that were smaller for the two-factor model than for the single-factor model. Significance of this result was confirmed by the chi-square difference test comparing the fit of both models ($\chi^2(1) = 15.69$, $p \leq .001$).

In addition to the factor analysis, a scatterplot mapping the latent estimates performances in all measures of OB against those in EGO for each child (Figure 6.22) demonstrated that a two-factor model could explain the data well.

In conclusion, confirmatory analyses demonstrated that the set of measures that reflected OB did not represent the same set of measures as those ones reflecting EGO. In other words, the results indicated that two latent factors, OB and EGO, were empirically separable. This empirical finding was in line with the initial assumption that was considered during test development. Consequently, the results indicated that the test instrument of the pilot study was construct valid. An analysis of measures at the level of items and at the level of their discrimination values, reliability coefficients, and factor loadings, however, indicated that the test required revision for the main study.

6.2.5 Implications for the Main Study

The empirical findings of the pilot study (Sections 6.2.3 and 6.2.4) suggested that the paper-and-pencil test which was developed in this study was suitable for the main study. It was shown to be construct valid, most items featured good psychometric

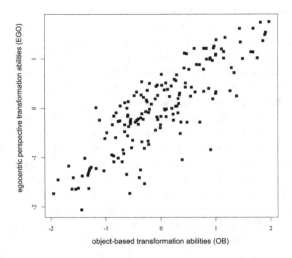

Figure 6.22 Comparison of estimates of performances in the latent standardized constructs OB and EGO in the pilot study

item characteristics and overall, the tests measured children's small-scale spatial abilities in a differentiated manner.

Revision of Single Items and Tests

To correct for psychometric weaknesses found in the statistical analyses (see ESM C.3) and the additional observations of the children, several items and measures were revised for the main study. Instructions of individual measures, changes in items or modifications of the whole measure was done due to reasons of poor item characteristics (high or low item facility, low discrimination, no fitting into the Rasch model), poor test design (e. g., kind of distractors, scoring of the tests), and difficult task instructions. Some of the measures were deleted from the test booklet. Table A.1 (p. 429) in the ESM shows all results from the pilot study and the intended revisions for the main study. The following sections present revisions for the three examples outlined above.

An avatar was included in the revised version to stimulate the children's processes to perform egocentric transformations during imagined navigation along the path since the measure LR did not reflect EGO as intended during the development but was empirically found to reflect OB.

In a revision of the test 3DMR, for example, items were based on mental rotations of five basic snakes that consisted of a line of four cubes matched together and additional joint cubes (Figure 6.23). For each item, the snakes were rotated around the x-, y-, and sometimes z-axis. The same hold true for the distractors. Those were all designed as reflections of the basis snake. The revised version of the measure 3DMR was added to the booklet of the main study without further piloting.

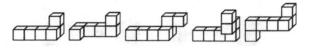

Figure 6.23 Snakes for the revised 3DMR

The analysis of the measure Claudia revealed that 33 children sorted picture D to be the second one. Those children had over-estimated the length of the first corridor. Since the measure did not intend to test visual distance estimation but perspective taking processes, it was revised by adding concrete locations at which the pictures were taken (Figure 6.24). The revised version of the measure Claudia was a picture-selection-measure that considered the differences of possible additional requirements of length estimation by suggesting positions at which the photos were taken.

The task Emil was developed as an alternative picture-sorting task. Similar to the original task Claudia, Emil required the children to imagine being the protagonist Emil who walks through a labyrinth while taking pictures. The route through the labyrinth required the children to manage conflicting frames of reference (Figure 6.25). The scoring was designed similar to the one of Cruise (Table 6.12). The following scoring was proposed for Emil

- number of correctly sorted pictures + score for the first picture,
- score of 5 for a misplaced picture C in a correctly sorted sequence,
- score of 4 for a misplaced picture other than A, E in a correctly sorted sequence,
- score of 3 for 'BFFACE' (problems with A or E).

The new measure Claudia and Emil were piloted with 31 fourth graders. For this sample, the Claudia task was slightly too easy and the corresponding sum score revealed ceiling effects (almost 50% of the children scored with the full score). The

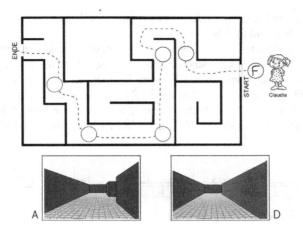

Figure 6.24 The revised task Claudia involving fixed positions

Table 6.17 Descriptive Statistics for Claudia and Emil in a Small Pilot Study (n = 31)

	Range	Median	Mean	SD	Skewness	Kurtosis
Claudia	[0,5]	3	3.48	1.71	−.75	−.57
Emil	[0,5]	4	3.45	1.59	−.65	−.78

measure Emil was also slightly too easy for those children, a result that could be explained by the fact that those children completed only those two tasks. Table 6.17 shows the statistics of the sum score of all single items of the measure Claudia and the computed sum score of the measure Emil. Both were kept for the main study.

Final Testing Material
The pilot study confirmed that small-scale spatial abilities can be measured in a group test in paper-and-pencil format with a general introduction of about 15 minutes. On the basis of these results, a set of tests that can be completed in 45 minutes was chosen for the main study.

The general introduction of the test was slightly modified. Since the instructions for the task PFT were difficult to understand, the example of the booklet was included into the introduction using a piece of paper in front of the class. The explanations for 2DMR and Boxes were kept.

NOCH EIN LABYRINTH

Emil läuft durch ein Labyrinth und macht dabei 6 Fotos.

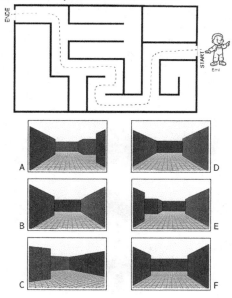

Die Fotos von Emil sind durcheinander geraten. Sortiere sie.

1. _____   4. _____
2. _____   5. _____
3. _____   6. _____

Figure 6.25 The task Emil required the children to sort pictures taken in a labyrinth

The final paper-and-pencil test for measuring small-scale spatial abilities is provided in the appendix A. The order of the tasks, their characteristics, and estimated maximum completion times are shown in Table B.2 in the appendix.

Figure 25 ...

The ...

Testing Material for Large-Scale Spatial Abilities

7

Another important contribution of this study was the development of an age-specific map-based orientation test involving a set of tasks that measured the construct of large-scale spatial abilities with the two broad subclasses which are survey mapping abilities (SurveyMap) and the abilities to reason about the map-environment-self relation (MapEnvSelf) during map-based navigation in space. The latter subclass was assumed to consist of two distinct subclasses, map-to-environment transformation abilities applied while solving comprehension tasks (Comp) and environment-to-map transformation abilities while solving production tasks (Prod).

Section 7.1 describes the development of the map-based orientation test. Because the test involved real space, a range of methodical choices had to be made. Those are also reported in the first section. Section 7.2 presents the individual tests of the main study and their integration into a treasure hunt. [1]

---

[1] A short description of the tasks can be found the ESM, p. 2.

Electronic supplementary material The online version of this chapter (https://doi.org/10.1007/978-3-658-32648-7_7) contains supplementary material, which is available to authorized users.

C. Heil, *The Impact of Scale on Children's Spatial Thought*, Studien zur theoretischen und empirischen Forschung in der Mathematikdidaktik, https://doi.org/10.1007/978-3-658-32648-7_7

7.1   Development of a Set of Map-Based Orientation Tasks

The map-based orientation tasks were planned to take place at the university campus of Leuphana University (Figure 7.1).

Figure 7.1 The university campus (Chr. Gatzert, Leuphana Videoredaktion)

7.1.1   Preliminary Considerations

The development of the map-based orientation test was based on the following preliminary considerations:

- There are no tests in the mathematics education literature to measure large-scale spatial abilities. The development of a new test instrument was required.
- A range of different measures has been used in the psychological and geographical literature, but those are all not standardized due to the involvement of real space (e. g., SAS & MOHD NOOR 2009). Consequently, each test will be influenced by the surrounding environment, the chosen map, the chosen routes, and finally the external parameters during data collection.
- Pointing tasks represent prominent non-graphical measures for adults and children when it comes to the estimation of directions of fixed locations in space. Pointing tasks can be used for externalizing knowledge stored in a cognitive or survey map (e. g., LEHNUNG, LEPLOW, HAALAND, MEHDORN, & FERSTL 2003). In addition, they were used for measuring map-based spatial thought.

- Flag-sticker tasks are important production methods (Section 4.2.2). The conceptual analogue comprehension tasks are disk/flag-placing tasks (e. g., CHRISTENSEN 2011).
- Concerning the use of maps to study spatial thought in children, there might be an influence of the viewing angle which is chosen for the map (opaque vs. oblique). The impact of viewing angle has been shown to decrease with the developmental changes. Children can be considered to be able to read opaque maps by the end of primary school (LIBEN 2003).

7.1.2 Conceptual Choices

Measuring Large-Scale Spatial Abilities in a Broad Way

This study developed tasks to reflect the three subclasses of large-scale spatial abilities outlined above. A conceptual distinction was included concerning the movement of the individual. This study distinguished between tasks in which the individual was not required to move in space and those in which the individual needed to move to complete the task. The following Table 7.1 summarizes the tasks that were developed and included in the map-based orientation test. Different kinds of tasks are shown in italics, the names of the tests/measures are not emphasized.

Table 7.1 Classification of Map-Based Orientation Tasks

	MapEnvSelf		SurveyMap
	Environment-to-map transformations	**Map-to-environment transformations**	
	Production tasks	*Comprehension tasks*	
Static	*symbol identification**	*pointing tasks*	
		MapRot	
		ActiveMapUse	
Dynamic	*flag-sticker tasks*	*disk-placement task*	*pointing tasks*
	Dots	Disks	Homepointing
	Dir		MentalMap

*not for testing, introduction to the map-based orientation test

Integration of Individual Tasks Into A Treasure Hunt

This study intended to integrate the different tasks outlined in Table 7.1 into a treasure hunt involving fixed routes and standardized locations for the tasks on the campus. For a pilot study with four third graders who were recruited from a summer holiday activity, possible tasks of the map-based orientation test were integrated into a preliminary treasure hunt. The full design of this preliminary version is presented in the Electronic Supplementary Material D. Important conclusions from the pilot study for the design of the main study were:

- Presenting the tasks as a treasure hunt was considered to be a reasonable means of raising motivation.
- The children should not perform more than six pointing tasks at more than one location on the campus for reasons of loss of motivation and concentration.
- Walking time between locations involved in the treasure hunt should not exceed 5 minutes for reasons of motivation and overall testing time.
- Walking straight along a route, thereby passing road junctions, did not increase the difficulty of items in flag-sticker and disk-placement tasks.
- Multiple left-right turns along a short route made those tasks more difficult for the children.
- Using global directions for task design and standardization (N, S, E, and W) turned out to be useful for test standardization.
- When integrating several tasks into the treasure hunt, children needed to be corrected when they were misaligned to guarantee task independence.

7.1.3   Methodical Choices

The Map
This study used an opaque map of the university campus (see ESM D.2, p. 61). Besides five landmarks that were provided on the map (Mensa, the restaurant; Wiese, the football meadow; Bibliothek, the library; Garten, the garden; Baustelle, construction site), all other labels were omitted. In other words, although the buildings on the campus were numbered, the children were not given this information. Instead, they had to establish configurational relationships for staying oriented during navigation.

To measure the children's spatial abilities to engage in spatial thought when using a map in an unknown environment, the following main methical decision was chosen:

> During the test, the children were not allowed to physically turn the map but were asked to hold the map in such a way that the geographic north pointed up.

Pointing Tasks

Pointing tasks have been proven to be good measures for externalize children's survey knowledge (LEHNUNG 2000, pp. 61–62). When children point to pre-defined landmarks, it has been shown that the accuracy of testing can be increased using the projective convergence approach that requires them to give an estimate of the direction to an unseen landmark from three distinct locations (HARDWICK ET AL. 1976; SIEGEL 1981; WALLER 1999). Although projective approaches to measuring in pointing tasks have been shown to be more accurate, this study relied on single-direction estimations when children made inferences on directions. By doing so, several landmarks could be included.

In line with what has been found by LEHNUNG ET AL. (2003), this study assumed that children are able to use circle-and-arrow devices, a circular cardboard or plastic with a mobile arrow attached to the center of the circle, to indicate directions (Figure 7.2) by the end of primary school. For purpose of easy and standardized data collection, this study used an arrow-and-circle device with 12 equal sections.

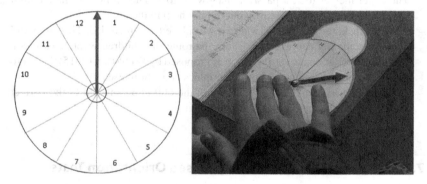

Figure 7.2 The arrow-and-circle device used as answering instrument during pointing tasks

Flag-Sticker-Tasks

It was assumed that navigating an unknown environment with few landmarks and no signage with the help of an unaligned map would shift the cognitive demands from land-mark-based orientation towards the use of spatial thought to perform map-to-environment and environment-to-map transformations.

In addition, to ensure the most standardized manner of possible testing, a range of methodical decisions were made. In line with what has been summarized in Section 4.2.2 concerning possible ways of conceptualizing production and comprehension methods, this study opted for the following methodical decisions during flag-sticker tasks:

- The children walked along a pre-defined route. That allowed for the development of a standardized scoring of the items and for controlling for the difficulty of the single paths.
- The experimenter took an active role. They led the children towards the single locations without pronouncing the directions.
- To guarantee independence of each item, the children were corrected after each item of the flag-sticker task.

Disk-Placement Task
This study took the following methodical decisions during disk-placement tasks:

- The children walked to a pre-defined location. This allowed for the development of a standardized scoring and to control for the difficulty of the single locations.
- The experimenter took a passive, shadowing role. They followed the children and guaranteed for their security but did not help in the item.
- For each disk, the location of where it was placed was recorded. This study did neither decide to keep track nor score the route the children went. This was a methodical reduction to settings such as proposed by WRENGER (2015) that were designed under the viewpoint of geographics education research.
- To guarantee independence of each item, the children were led to the correct location after placing their disk on their own.

7.2   Presentation of the Map-Based Orientation Tests

Based on the preliminary methodical choices and first observations of the pilot study, all map-based orientation tests were integrated into a treasure hunt. Figure 7.3 presents important landmarks, the starting point, routes to the flags, and locations of the disks. The following sections describe the tasks that were developed to measure children's large-scale spatial abilities. The design of the treasure hunt, that is, how hints were collected by the children and provided by the experimenters, is outlined in the Manual of the test (ESM D.2, p. 58). In short, the location of the treasure was the intersection of the line between the yellow and the red flag and the line between the start and the blue flag (little pink cross in Figure 7.3). The tasks and their characteristics are also shown in Table B.3 in the appendix (p. 433).

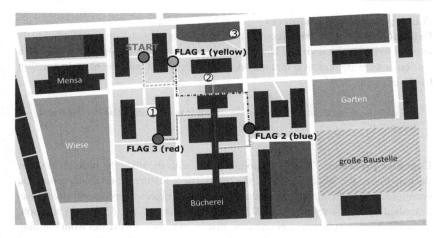

Figure 7.3 Campus map with routes of the treasure hunt in the main study

7.2.1   Tasks Measuring MapEnvSelf

In line with what has been outlined in the geography education literature (e.g., WRENGER 2015), the following difficulty-generating features were considered during development:

– It was assumed that items are the more difficult the more decision points (i.e., road junctions) they contained in a route. In addition, the involvement of possible turnings at road junctions were considered as another difficulty-generating feature. During production tasks, it was assumed that this would increase the number of pieces of information to maintain in working memory. During comprehension tasks, this feature was assumed to increase the cognitive demand of keeping the map constantly aligned while extracting information from it.
– It was assumed that items are the more difficult the more invisible crossings they contain (walking through buildings).
– It was assumed that visibility of landmarks might influence the difficulty of an item both in a favourable (orientation help) as well as in an undesirable (misinterpretation of landmarks) way.
– It was assumed that items were more difficult when they required the children to mentally align the map since this would involve extra cognitive operations.

Pointing to Unseen Landmarks

Definition: Pointing in the direction of unseen landmarks tasks require individuals to point to a visually invisible landmark that is depicted on the map while taking a particular viewing direction that may not be in alignment with the map. Instruction by the experimenter:

> *"Could you please show me on the (answering) disk in which direction the [landmark] is located?"*

Cognitive processes: To solve a pointing task with the map, the individual scans the map to find the landmark and to identify possible features to establish geometric correspondences. Scanning can be understood as an object-based transformation as described by KOSSLYN ET AL. (1978) and WIMMER ET AL. (2016). In addition, to establish geometric correspondences, that is, to refer information from the map to the environment, the individual has to become aware of conflicting frames of reference that occur due to a possible map misalignment. To align the map with the environment, the individual performs a mental rotation as described by SHEPARD AND HURWITZ (1984) and maintains this information in working memory. In a last step, the individual identifies the direction of the landmark on the map and develops a spatial hypothesis concerning the direction of the landmark with respect to the egocentric frame of reference. Those processes correspond to the ones denoted as visualizing (LIBEN 1999) and hypothesis formulation according to OTTOSSON (1986).

Tests: Two tests requiring children to point to unseen landmarks were developed, MapRot and ActiveMapUse. The tasks were inspired by pointing tasks in real space, primarily used in studies to measure survey knowledge (e. g., HEGARTY ET AL. 2006). Both tasks involved the same landmarks to be pointed to, but varied with respect to the location where children had to indicate directions. MapRot required the children to point to landmarks at a single location, the Start Point, whereas ActiveMapUse required the children to point to unseen landmarks at different locations (Table 7.2). The outcome space of both tasks is a number, representing a section (direction) on the answering disk that is scorable in a dichotomous way.

Table 7.2 The Tasks MapRot and ActiveMapUse Required the Children to Indicate to Unseen Landmarks with the Help of Their Map

Location	Landmark	Viewing direction	Misalignment map-viewing direction	Pointing direction
MapS1	restaurant	South	180°	right
MapS2	library	South	180°	front left
MapS3	garden	South	180°	front left
MapW1	restaurant	West	90°	front left
MapW2	library	West	90°	back left
MapW3	garden	West	90°	back left
MapUseYellow	restaurant	North	0°	back left
MapUseRed	library	South	180°	front right
MapUseBlue	garden	West	90°	back left

Flag-Sticker-Tasks

Definition: Flag-sticker tasks require individuals to determine the self position and viewing direction on the map after moving in real space. Instruction by the experimenter:

> *"We are now arriving at the [yellow/blue/red] flag. Could you please stick the [yellow/blue/red] round sticker on the map in order to indicate where we are? [Correct position if necessary] Can you please stick the [yellow/blue/red] arrow on the map in order to indicate our viewing direction?"*

Cognitive processes: To solve a flag-sticker task requires the individual to perceive and process the constantly changing spatial information in the environment and maintain this information in working memory. To find the correct position and viewing direction, the individual engages in a self-location process involving subsequent hypothesis formulation as described in Figure 3.11 (p. 100) and by OTTOSSON (1986), thereby keeping the map aligned using a mental rotation as described by SHEPARD AND HURWITZ (1984).

Tests: The tasks Dots and Dir required children to find their position and direction on a map. The tasks were developed based on the adults and children's literature on map use (e. g., KASTENS & LIBEN 2010; CHRISTENSEN 2011). Each of the tasks consisted of three items that differed in their difficulty according to the featured number of decision points, number of turns and degrees of misalignment of the map at the final location (Table 7.3).

Table 7.3 The Tasks Dots and Dir Required Children to Identify Their Position and Viewing Direction on the Map After Walking to the Yellow, the Blue, and the Red Flag

Location	Path	Nb. decision points	Nb. turns	Viewing direction	Degrees of misalignment
yellow flag	start-yellow path	2	2	North	0°
blue flag	yellow flag-blue path	4	2	South	180°
red flag	blue flag-red route	5	5	West	90°

To ensure independence of the items, the experimenter corrected both the location (mark correct location with a dot in map) and the heading (mark arrow next to the location) during item formulation (between formulating the items self-location and self-heading) and before starting the next item in the test.

This study used colored round stickers with a diameter of 8 mm for indication of the position and small arrow stickers of about 1.5 cm length for indication of viewing direction. The outcome space of Dots and Dir therefore consisted of two sets of stickers. Location sticker placement was polytomously scored based on the Euclidean distance between the sticker and the correct location on the map. Viewing direction sticker placement was scored dichotomously.

Disk-Placement Task

Definition: Place recognition tasks require individual to place a disk in the environment that corresponds to a position depicted on the map. Instruction by the experimenter:

> *"I have three disks for you. [Show Disk 1,2,3]. On the map, you can see the locations where you have to drop the disk. We'll start with Disk 1, then proceed to Disk 2 and finally Disk 3. Can you please place the disk correctly here on campus?"*

Cognitive processes: To solve a disk-placement task, the individual needs to scan the map to identify the location of the symbol representing the place where the disk has to be placed, thereby performing mental operations as described by KOSSLYN ET AL. (1978) and WIMMER ET AL. (2016). Then, the subject has to plan a route to the location and decide for a sequence of landmarks that should be encountered along the route (involving objects and topological landmarks). To place the disk, the individual then engages in a series of place identification processes as described in Figure 3.11 (p. 100) and by OTTOSSON (1986), thereby keeping the map aligned using a mental rotation as described by SHEPARD AND HURWITZ (1984). This involves subsequent hypothesis development of how the two-dimensional information on the map could look like in the environment, thereby relying on visualizing processes and hypothesis testing against the spatial information perceived in the environment.

Tests: The task Disks was developed in line with a similar task proposed by CHRISTENSEN (2011). The measure consisted of three items that differed in their difficulty according to the featured number of decision points and number of turns to manage to reach the final location in the environment (Table 7.4).

Table 7.4 The Task Disks Required the Children Two Place Three Disks in the Environment According to the Locations Provided on the Map

Disk	Shortest path	Nb. decision points	Nb. turns
disk 1	straight on	1	0
disk 2	straight-right-straight-left	3	2
disk 3	(left)-straight-left-straight	4	1–2

To ensure independence of the items, the experimenter corrected the disk location after each item. The experimenter walked to the correct location and explained the viewing direction before the child started in their own to the next disk. The outcome space of each item consisted of a set of locations where the disks have been placed by the children. Each item could be polytomously scored according to a certain tolerance range in which the disks have to be placed.

7.2.2   Tasks Measuring SurveyMap

When individuals navigate real space with the help of maps, they engage in processes of survey mapping (Section 3.3.2). In addition, they engage in processes of path integration (Section 3.3.3) that allows them to have a sense of direction of the starting point. Besides completion of other tasks along the route on the campus, this test intended to measure their abilities to indicate the direction of the starting point. To measure spatial thought that was map-mediated, but not directly related to maps, this study proposed pointing tasks to memorized landmarks to measure the abilities of making inferences from a survey map.

Pointing to the Start Tasks

Definition: Pointing to start tasks require individuals to indicate the direction of the starting point without using a map after having moved. Instruction by the experimenter:

> *"I'll cover your map now. Can you please indicate the direction of our starting point on the answering disk?"*

Cognitive processes: To complete a pointing to the start task, the individual has to effectively monitor the self-movement by subsequent updates of the relation between the egocentric frame of reference and the ones of landmarks and the environment. To do so, individuals engage in path integration as described by MONTELLO (2005) or LOOMIS ET AL. (1999).

Table 7.5 The Task Homepointing (HP) Required the Children to Indicate the Direction of the Starting Point

Location	Landmark	Viewing direction	Misalignment map-viewing direction	Pointing direction
Start1	restaurant	North	0°	back left
Start2	library	South	180°	front right
Start3	garden	West	90°	back left

Tests: In line with similar tasks proposed in the literature (e. g., NEIDHARDT & POPP 2010), but using another device for measuring, the task Homepointing was developed (Table 7.5). The measure consisted of three items, performed at the yellow, the blue, and the red flag. The items differed with respect to the length of the path walked until reaching the flag. The outcome space of the items is a number, representing a section (direction) on the answering disk that is scorable in a dichotomous way.

Pointing to Memorized Landmarks

Definition: Pointing to memorized landmarks tasks require the subject to point to a range of landmarks after active exposure to the real space without using a map. Instruction by the experimenter:

> *"I'll take your map now. We have learned about nine different landmarks today. I'd like to know in which direction they are located. Can you show me on the answering disk in which direction the [...] is located?"*

Cognitive processes: To indicate the direction to unseen but memorized land-marks, individuals have to make inferences from their cognitive map. This study assumed the representational theory of MEILINGER (2007), that is, the idea that a survey map is represented in form of a network of vista spaces in memory (Section 3.3.2 and 3.3.3). According to MEILINGER (2007), pointing to unseen landmarks can be understood to be enabled by the process of imagining to integrate represented vista spaces step-by-step into an environmental one, thereby retrieving knowledge stored on the vista spaces and perspective shifts (p. 349).

Tests: The measure MentalMap (Table 7.6) was developed in line with similar tasks proposed in the literature (e. g., LEHNUNG ET AL. 2001). It consisted of three items each, representing three distinct sets of landmarks to which the individuals were

Table 7.6 The Tasks MentalMap Required the Children to Indicate to Unseen Landmarks Without the Map

Location	Landmark	Viewing direction	Misalignment map-viewing direction	Pointing direction
MMap1	restaurant	South	180°	right
MMap2	library	South	180°	front left
MMap3	garden	South	180°	front left
MFlag1	yellow flag	South	180°	left
MFlag2	blue	South	180°	front left
MFlag3	red	South	180°	front left
MDisk1	disk 1	South	180°	front left
MDisk2	disk 2	South	180°	front left
MDisk3	disk 3	South	180°	back left

asked to point to. The first set of landmarks were those that were encountered only formally on the map (restaurant, library, and garden), the second set of landmarks were the three flags, and the third one were the three disks. The outcome space was a section (number) on the answering disk that is scorable in a dichotomous way.

Additional Testing Material

This chapter presents the extra materials that were used in the main study. Section 8.1 describes a tabletop version of the Corsi Block Test that was administered to the children to measure individual differences in visuospatial working memory capacity. Section 8.2 presents a perspective taking test involving manipulable material, the PTSOT-C, that was used to further measure children's abilities to perform imagined transformations of the self.

8.1 The Corsi Block Test

The Corsi Block Test (CBT) was used to measure visuospatial working memory. In its original version, the test requires the subject to correctly imitate the experimenter who taps a sequence of wooden blocks (forward tapping task). In an alternative version, the test requires the subject to tap the sequence of wooden cubes in reverse order than it had been tapped by the experimenter.

The CBT is a typical test for investigating the visuospatial working memory capacity (KESSELS, VAN ZANDVOORT, POSTMA, KAPPELLE, & DE HAAN 2000) since it had been developed to provide a spatial alternative to assessing memory span by verbal sequences (BERCH, KRIKORIAN, & HUHA 1998; KAUFMAN 2007; KESSELS ET AL. 2000). Although it is still a matter of debate what constitutes the underlying information processing operations, it had been shown that the CBT measures a conglomerate of active spatial attention performed by the central executive

Electronic supplementary material The online version of this chapter (https://doi.org/10.1007/978-3-658-32648-7_8) contains supplementary material, which is available to authorized users.

235
C. Heil, *The Impact of Scale on Children's Spatial Thought*, Studien zur theoretischen und empirischen Forschung in der Mathematikdidaktik, https://doi.org/10.1007/978-3-658-32648-7_8

(BERCH ET AL. 1998; EYSENCK, PAYNE, & DERAKSHAN 2005) and performance of the visuospatial sketchpad (EYSENCK ET AL. 2005).

Although the CBT is widely used, there exists no standardized version of the test (e. g., BERCH ET AL. 1998; KESSELS ET AL. 2000). CBT display characteristics have been varied with respect to the color, the number and size of the blocks and their placement on the display (BERCH ET AL. 1998). Furthermore, the CBT has been varied with respect to test administration, namely in the pointing procedure, the block-tapping rate, trails per error and the item sequences. Other variances in the CBT have been found in the task itself (asking for forward or backward tapping) and in the scoring (BERCH ET AL. 1998).

8.1.1 Materials

In this study, the CBT was designed in alignment with the experimental setup similar to the one described and tested with children at the end of kindergarten in the USA by the Peabody Research Institute at Vanderbilt University (see their appendix FARRAN, LIPSEY, & WILSON 2011, for detailed information on the experimental setup)[1]. For their setup, results showed that 443 first graders scored with a mean longest forward sequence of 4.6 (SD = 1.1) out of 9 and a mean longest backward sequence of 3.7 (SD = 1.3) out of 6 (FARRAN ET AL. 2011). The technical report of the project suggested that the described testing material suits to test VSWM of younger children. The reported results for the forward span are in line with normative studies that have been proposed (e. g., FARRELL PAGULAYAN, BUSCH, MEDINA, BARTOK, & KRIKORIAN 2006) which found that the average forward span of first graders is about 5 and increases to 6 by the end of primary school. Although there are no normative studies concerning the backward span, it was assumed that the findings of FARRAN ET AL. (2011) were also within a reasonable developmental scope. Since VSWM gradually improves during childhood (e. g., VANDER HEYDEN ET AL. 2017), the children in the sample of the present study were expected to score higher but not with full score in both forward and backward span. For these reasons, the experimental setup proposed by FARRAN ET AL. (2011) was chosen to be used. It was not subject to an extra pilot testing.

The CBT design in this study consisted of nine $20\,mm \times 20\,mm \times 20\,mm$ wooden cubes mounted on a black $206\,mm \times 210\,mm$ piece of paper that was fixed upon a blue clipboard. The cubes were arranged in the classical arrangement of Corsi's

---

[1] https://my.vanderbilt.edu/toolsofthemindevaluation/files/2012/01/Corsi2.pdf (last access: 23/11/2017)

apparatus (see also BERCH ET AL. 1998). From the experimenter's point of view, each cube was labeled with a number from one to nine (Figure 8.1).[2] Those numbers were not visible from the child's opposite point of view. Other than in the original setup, the cubes were labeled in their order of arrangement for a better orientation, starting from the top left to the bottom right.

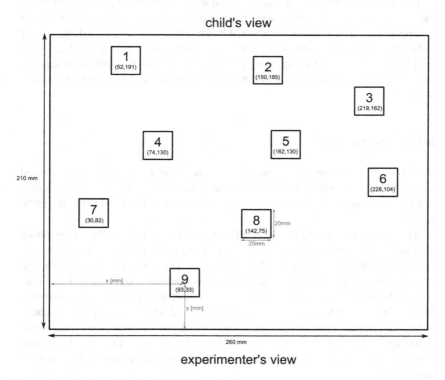

Figure 8.1 Design of the Corsi Block Test display

For forward and backward tapping, two separate testing blocks were introduced with sequences whose length increased by one. For each sequence, there were two different trails of the same length but not with the same sequences. In line with the argumentation in the literature, items were assumed to be more difficult whenever

---

[2]The coordinates were measured from the bottom-left corner of the board to the middle of each cube.

the length of the block sequence to tap increased since the immediate memory load becomes larger (BERCH ET AL. 1998). The block tapping sequences were the same as in the study of FARRAN ET AL. (2011). They were originally pulled from number sequences from the WAIS digit span task[3] which has been reported to be a typical way to generate test sequences (BERCH ET AL. 1998).

During test construction, it was not taken into consideration that item difficulty has also been shown to depend on the number of criss-crosses in the path that has to be tapped (see ORSINI, PASQUADIBISCEGLIE, PICONE, & TORTORA 2001, for further elaboration). Since the path configuration itself is related on the general layout of the block display and the sequences that are tapped, it is therefore influenced by two independent variables. These variables in the item design were neglected in this study.

8.1.2   Procedure

The CBT in this study used test administration characteristics that have been reported in a high number of other studies: index finger pointing, a tapping rate of one second, and two trails per level (in comparison to studies named in BERCH ET AL. 1998, p. 325).

The CBT was mounted on a table. The experimenter and the child were sitting face-to-face. The experimental setup was standing in the middle so that the experimenter was able to read the numbers. After a short introduction into the test, the experimenter started with two testing blocks of forward sequences and two testing blocks of backwards sequences. The experimenter tapped the corresponding blocks with the index finger according to the manual without naming the numbers. In between tapping there was a break of one second. After tapping, the experimenter encouraged the child to repeat or reverse the sequence by tapping slowly the corresponding cubes. The child's tapping was scored of being correct or false but not commented.

The CBT started with the two testing blocks of forward sequences which were introduced in a short example. The first sequence consisted of two blocks and length was subsequently increased by one block whenever the child was correct. If the child did not repeat a sequence correctly, the experimenter tapped a second sequence of the same length. If the child was correct in the second trail, the experimenter proceeded

---

[3]E-mail correspondence with Dr. Kimberly Nesbitt [19th August 2016], former Vanderbilt University, now University of New Hampshire.

to the first trail of the next length. If the child failed again, the testing block was finished.

After a short introduction to backward tapping, the CBT continued with the two testing blocks of backward sequences. The testing procedure was the same as for forward tapping, but the maximum length of backward sequences was six blocks. During the CBT, the experimenter documented whether or not the child completed a sequence correctly and in which trail. The documentation of the experimenter was integrated into the manual so that the experimenter was able to read aloud the manual and testing at the same time (see ESM E.2, pp. 69).

8.1.3   Scoring

Different scorings of the CBT have been proposed in the literature. A classical scoring is the maximal length of the block sequence that was repeated (forward span) or reversed (backward span) correctly in one of the trails. Although this scoring is straight forward, it has proved to be a less sensitive measure of individual differences since the range of possible scores is very limited. In contrast, the total score that takes into account the number of trails used to complete a certain sequence has been reported to be more sensitive (KESSELS ET AL. 2000). In this study, the children completed two testing blocks for forward and backward tapping. Table 8.1 shows four different scorings that were applied to both forward and backward span.

Table 8.1 Proposed Scorings for the CBT

Name	Description	Measure
MFS/MBS	maximum forward and backward span	compute maximum over the two test blocks
AFS/ABS	average forward and backward span	averaged span over the two test blocks
MTFS/MTBS	maximum forward and backward total score	sum over the maximum completes sequences, weighted with a score of how many trails were needed
ATFS/ATBS	average forward and backward total score	average over the total score, computed for both testing blocks

High correlations were expected to be found among all the forward and the backward spans and total scores. Maximum span measures (MFS, MBS) were expected to be the least sensitive measures when it comes to distinguish between individuals with low or high VSWM capacity. The average measures (AFS, BFS) and the total scores introduced (MTFS, MTBS) were expected to measure more sensitively. Finally, the averaged total scores (ATFS, ATBS) were expected to be the most sensitive measures since they included all the raw data from the experiment (correctness in all sequences and the number of trails used to complete the sequences). Within the analytical framework of this study, it was planned to include one forward and one backward measure with good psychometrical characteristics into the analyses.

8.2   PTSOT-C

In this study, the Perspective Taking Spatial Orientation Test (PTSOT) by KOZHEVNIKOV AND HEGARTY (2001) was adopted for the use with primary children. The PTSOT is a frequently used test instrument for investigating the abilities to perform egocentric transformations in adults and has been proven to be construct valid (KOZHEVNIKOV & HEGARTY 2001). In the present study, the PTSOT was considered to be a potentially extra measure based on small three-dimensional rather than purely paper-based spatial stimuli as proposed in the small-scale spatial abilities test. This section describes the adaption and testing procedure. Moreover, it presents a formal framework for item analysis and item construction and specifies the set of items used in this study.

8.2.1   Design of the Adopted Instrument

The adopted version of the PTSOT for children, denoted as PTSOT-C, was designed with respect to a set of 12 parameters (see NEWCOMBE 1989, for a summary of theoretical arguments used to justify the adaption). Figure 8.2 compares the original PTSOT for adults with an adaption for children with respect to the chosen design parameters.

As shown on the right side, the adopted instrument consisted of a three-dimensional, small-scale field of animals that were stuck onto a green sheet of paper, denoted the 'meadow' (see ESM E.1, pp. 66 for a precise outline of the animals' position on the display). The setting further included the pointing device of the large-scale spatial abilities test (compare Figure 7.2, p. 225) with 12 numbered

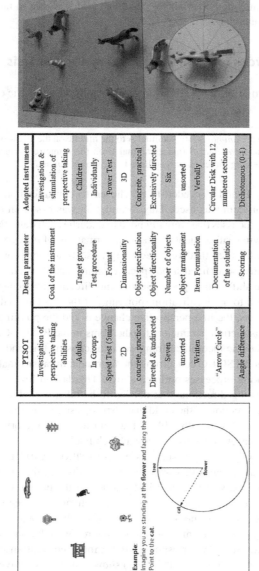

PTSOT	Design parameter	Adopted instrument
Investigation of perspective taking abilities	Goal of the instrument	Investigation & stimulation of perspective taking
Adults	Target group	Children
In Groups	Test procedure	Individually
Speed Test (5min)	Format	Power Test
2D	Dimensionality	3D
concrete, practical	Object specification	Concrete, practical
Directed & undirected	Object directionality	Exclusively directed
Seven	Number of objects	Six
unsorted	Object arrangement	unsorted
Written	Item Formulation	Verbally
"Arrow Circle"	Documentation of the solution	Circular Disk with 12 numbered sections
Angle difference	Scoring	Dichotomous (0-1)

Figure 8.2 Adaption process of the PTSOT into the PTSOT-C (in HEIL 2017, p. 3)

sections and a mobile arrow on it that was placed in front of the object field. The whole setting was mounted on a table so that the child was able to face the whole scene from an oblique view.

8.2.2   Framework for Item Construction and Analysis

A standardized item of the PTSOT-C was verbally introduced by the experimenter as follows:

> "Imagine that you are *animal A* [stick animal A to the middle on the arrow on the disk] on the meadow facing *animal B* [places animal B on the semicircle attached to the disk]. In which viewing direction do you have to turn to see *animal C*?"

Sets of animals A, B and C needed to be chosen to construct items. To do so, a item formalization item was used. This formulation was based on two intended perspective taking processes that were intended to underlie each item. For each item, a set of four underlying, subsequent mental transformations of the egocentric frame of reference were identified that were necessary to solve an item correctly. Those transformations pairwise defined a perspective taking process. The first process corresponded to the shift of the egocentric viewing direction by an angle α and the second corresponded to the localization (pointing) of the third animal by an angle β. Out of this analysis, the first perspective change α and the pointing direction β were chosen as construction parameters for the items in the PTSOT-C.

Figure 8.3 depicts the theoretical framework for constructing and analyzing an arbitrary item of the PTSOT-C concerning the four mental transformations and the two perspective taking processes outlined above. The dashed lines correspond to the two mental viewing directions that needed to be taken to solve an item. The egocentric viewing direction of the child that also corresponds to the orientation of the answering disk is shown with a black line. Furthermore, the original viewing direction of the animal A is illustrated with a black line.

In a qualitative study with eight subjects, KOZHEVNIKOV AND HEGARTY (2001) found three solution strategies that were applied by adults in the original PTSOT: angle identification (analytical-geometrical solution of the items), mental rotation (rotation of the object array) and perspective taking (change of the egocentric viewpoint). The authors classified the first two strategies as being object manipulation strategies since they involve the manipulation of interobject relations or the objects' orientation. The latter strategy corresponded to the intended solution strategy of the items. KOZHEVNIKOV AND HEGARTY (2001) showed that perspective changes $|\alpha|$

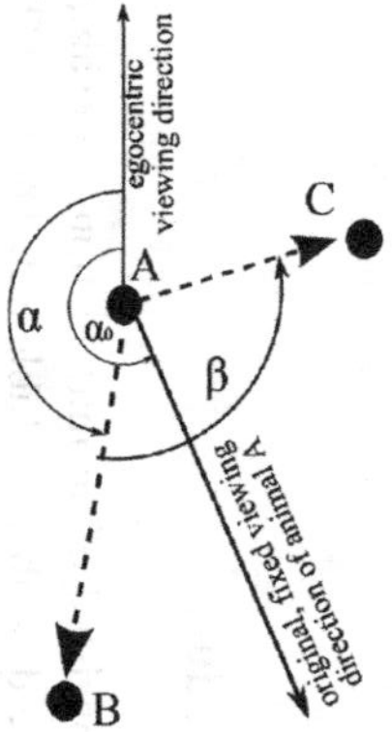

Process	Item construction parameters	Defining mental transformations
1	**Shift of egocentric viewing direction - angle α** *"Imagine that you are animal A and you are facing animal B ...."*	T_1 transformation of the egocentric frame of reference from an opaque view into the field view of animal A (adoption of viewing angle α_0) T_2 rotational transformation of the egocentric frame of reference by $\alpha-\alpha_0$ towards animal B (adoption of viewing angle α)
2	**Localization angle β** *"...in which direction do you have to turn in order to see animal C?"*	T_3 rotational transformation of the egocentric frame of reference by β towards animal C (adoption of viewing angle $\alpha+\beta$) T_4 transformation of the egocentric frame of reference from the resulting field view to an egocentric opaque view

Figure 8.3 Analytical framework for one item of the PTSOT-C (in HEIL 2017, p. 4)

with less than $|90°|$ were solved mainly by object manipulation strategies, especially when the pointing direction was frontal to the imagined viewing direction. Perspective changes with more than $|90°|$, independently of the pointing direction were found to be solved by a perspective taking strategy.

These findings contributed to the design of the items in the present study. The first item served as a motivating starter. It did not involve any perspective changes (item 1). Then, there were three items that demanded for a perspective change $|\alpha|$ of less than $|90°|$ but with pointing directions that were in the back of the imagined viewing direction (items 2–4). Four subsequent items had an increasing perspective change $|\alpha|$ and varied the pointing direction with respect to the four possible canonical directions left/right and back/front (item 4–8). For all items, the animals were carefully chosen in such a way that the exact pointing direction of the target animal C was in the middle of the arrow-and-circle device.

Table 8.2 summarizes the set of items that were designed for the main study. All items were intended to be scored dichotomously. All angles were measured from head to head of the corresponding animals using the animal display and Geogebra (RUPPERT, BAUER, HOHENWARTER, & MEYER 2013).

Table 8.2 Item Set Used in the PTSOT-C in the Main Study

Item	Animal A	Animal B	Animal C	Perspective change α	Pointing direction β	Pointing orientation
1	pig	cow	dog	2°	48°	right front
2	dog	chicken	pig	−4°	132°	left back
3	horse	dog	pig	−16°	100°	left back
4	rabbit	chicken	pig	−45°	159°	right back
5	rabbit	dog	cow	−98°	156°	left back
6	cow	dog	chicken	−110°	33°	left front
7	rabbit	horse	cow	−167°	135°	right back
8	chicken	horse	cow	170°	68°	right front

8.2.3   Testing Procedure

The testing procedure consisted of an introduction, an example item and the set of eight test items that are described in Table 8.2. The PTSOT-C was mounted on a table so that the child was sitting in front of the 'meadow' and the answering instrument. Both were stuck on the table. The experimenter held the documentation paper and a second set of animals that ought to be stuck to the disk with glue dots. The experimenter was sitting on the side of the table, left to the child.

The experimenter introduced the test as follows:

> "There are six animals on this green sheet of paper, the meadow. In front of you, there is a disk that equals a clock [show]. The black arrow represents your viewing direction. In the following game, your task will be to pretend to be the animal that I stick to the middle of the arrow [show]. Then, you imagine heading towards a second animal that I stick here straight forward [show]. If you are imagining the situation, I'll name you a third animal. Then, you have to show me on the disk where the third animal is located. Please orientate always towards the head of the animal."

After the introduction, the experimenter presented the example:

> "Imagine that you are the horse [stick horse in the middle on the arrow on the disk] on the meadow and you are facing the rabbit [place rabbit on the semicircle attached to the disk]. Can you imagine that? [Wait for answer of the child.] I've stuck the animals on the disk as a reminder. In which direction do you have to turn to face the pig? Please focus on the heads of the animals."

The child subsequently chose the answer on the disk. When the child chose the correct answer (segment 10, 9 was also considered to be correct), the experimenter continued with the test. If the child chose the egocentric solution (segment 11 or 12), the experimenter explained the example again and repeated it until the child understood the tasks of the PTSOT-C.

During the example and the whole test, the child was allowed to gesture but not to turn the array (the meadow), the answering disk or themselves. During the testing procedure, the experimenter documented the number of the segment on the disk that was chosen by the child for each item. The experimenter did not comment on the solutions of the child, but motivated them. Every two items, the experimenter reminded the child to imagine being animal A and facing animal B when deciding for the direction of animal C and to not answer the item from the own, egocentric point of view.

Data Collection

The main study was conducted on the campus of Leuphana University, Lüneburg, Germany for 13 days from mid of May to mid of June 2016. The testing period was at the end of the school year, and all school grades had already been reported. Some classes were two weeks before summer break When they attended the testing day in mid of June.

The test participants were recruited by an invitation to 15 primary schools in the city of Lüneburg to participate in the extracurricular activity „Mathe in Aktion" ('Math in Action') on the campus. The activity was announced in brochures that were sent to the schools. Participating in this activity was free of charge, but the parents and the schools had to give their consent. The children did not receive any financial remuneration for participating in the study. This chapter describes the sample (Section 9.1), the material (Section 9.2) and the experimenters (Section 9.3). Moreover, it specifies the testing procedure (Section 9.4), presents the data documentation (Section 9.5) and descibes the data treatment (Section 9.6) describes the data treatment.

9.1 Participants

The original sample consisted of $n_0 = 261$ fourth graders from 13 different classes. They came from seven different schools, six public schools and one Montessori school. The six public schools are located in different parts of the town and attract

Electronic supplementary material The online version of this chapter (https://doi.org/10.1007/978-3-658-32648-7_9) contains supplementary material, which is available to authorized users.

C. Heil, *The Impact of Scale on Children's Spatial Thought*, Studien zur theoretischen und empirischen Forschung in der Mathematikdidaktik, https://doi.org/10.1007/978-3-658-32648-7_9

a great variety of students with respect to their socio-economic background. About half of the participants attended two large public schools with four parallel classes of which three participated in the study. One school specializes in language-sensitive subject teaching and is attended by students who may have problems in their language development. Another school offers pedagogical afternoon supervision (Table 9.1).

Table 9.1 Description of the Sample for the Main Study

School ID	Trusteeship	Specialization	Nb. of classes	Nb. of students
KB	public	none	2	32
IS	public	none	2	39
MS	private	none	1	16
HG	public	language	1	16
RP	public	none	3	68
RF	public	pedagogical supervision	1	24
HK	public	none	3	66

Six children included in the sample were refugee children with a very basic understanding of German. Three children did not receive the permission to participate in the small scale spatial abilities test (written material) but in the large scale spatial abilities test.

The children in the original sample were aged from 9 years 0 months to 12 years 10 months[1] with a mean age of $M = 9.87$ years and $SD = .53$[2]. The sample consisted of 117 boys and 141 girls (3 children did provide information in this respect).

---

[1] Children aged 12 years or older were either refugee children or children who had attended the primary school longer as usual due to learning difficulties.

[2] On the basis of the exact age, calculated using the children's birthday.

9.2 Materials

For the main study, the revised materials outlined in Section 6.2.5 (p. 218) and Table B.1 in the appendix (p. 430) were used:

- The small-scale spatial abilities test was printed and bound in a small booklet (see appendix A).
- The material for the large-scale spatial abilities test was prepared for every single student (see ESM D.2, pp. 61).
- The CBT and the PTSOT-C were prepared as described in Sections 8.1 (p. 235) and 8.2 (p. 240). They were mounted on two separate tables that allowed the children to sit in front of each test.

9.3 Experimenters

Whereas I conducted the written test in every class, the map-based orientation test was supervised by extra 35 experimenters (Table 9.2). They were recruited from a university course and were among institute colleagues. The group included 14 main experimenters who were present on at least three days during data collection testing eight to twelve students each[3]. 21 additional experimenters were present on one or two days testing two to six children each.

Table 9.2 Experimenters in the Main Study

Role	Nb. of supervised classes	Total nb. of supervised children
1 leading PhD student	13	37
14 main experimenters	3–4	143
21 extra experimenters	1–2	81

All experimenters were trained to perform the map-based orientation test with respect to the manual in two sessions that lasted 90 minutes each. The experimenters studied the manual before starting their training. After a short demonstration, they practiced each item with other students from the experimenter group. After the training, the students obtained a videotaped version showing how to collect data as

---

[3]For each class, a team of myself and seven main or extra experimenters supervised the whole class.

well as a questionnaire that helped them to assess whether or not they understood the manual correctly or not. Prior to data collection, seven of the other experimenters and I reviewed the main issues of the test manual. In addition, the 14 main experimenters received a constant feedback on their work during the study and kept a record of critical issues in the manual from supervision to supervision.

9.4   Procedure

Each of the 13 classes was invited to spend a day of 4.5 hours at the campus of the university. After a short welcome, the day was structured as follows: The children started with the small scale spatial abilities test which lasted 60 minutes in total. After the first test, all the children had a break of at least fifteen minutes. Then, the children attended an interactive unit on probabilistic theory which served as a pilot testing of another PhD project. The unit consisted of small interactive tasks that were solved in groups and contained sufficient breaks for cognitive recovery. In parallel, the children were taken out of the session separately by the experimenters who tested them individually in the map-based orientation test. The order of testing was pre-defined by the number of the seating positions that were spontaneously taken by the children at the beginning of the day and could therefore be considered as being a random order. During the whole day, the child had a self-defined code that served to match observations from the written test to observations in the map-based orientation test. At the end of the day, each child obtained a certificate of attendance before farewell.

During the whole day, the role of the teacher was to accompany the children without taking an active role such as assisting in one of the two tests. During the breaks, the teachers had the duty to supervise. During the maths activities, the teacher observed the children, took pictures or motivated them. All teachers were given the written material with the children the tasks and the day spent at the university once being back at school.

The cognitively demanding tests were embedded into challenges to motivate the children. For the written test, a class-wise challenge in solving the spatial puzzle tasks was introduced. The children were told that they could win the challenge as a class, whenever every child gave the very best in the written test (without copying). On the basis of the main score, the best class was identified and were awarded medals at the last school day. The map-based orientation test was motivated as being an individual treasure hunt on the campus which was based on some spatial puzzle tasks with the map. At the end of the treasure hunt, there was a box with sweets as remuneration. Both scenarios (challenge among spatial puzzlers and treasure hunt) were already announced in the invitational brochure. The children were well informed and motivated when arriving at the campus and they were keen on winning the contest and finding the treasure.

9.4.1 Testing Procedure for the Small-Scale Spatial Abilities Test

I administered the paper-and-pencil test in all classes in the same room at university. The test and three tasks were introduced according to the manual in about 15 minutes. During that phase, the children were allowed to ask questions. After completing the cover sheet of the booklet with their personal code, their birthday and sex, the children started the test. The booklets were collected after 45 minutes. Altogether, the whole procedure lasted for about 60 minutes depending on the number of questions asked by the children.

The children were not allowed to talk, ask any questions on the tasks or copy from their neighbor during completion time of the written material. If they finished their test booklet before the deadline, they were asked to check their answers to achieve the maximum points in the challenge. If there was time left after controlling the results, each child was offered to spend the rest of the time by either drawing or completing extra material.

The six refugee children who did not understand the German instructions in the test material well enough and the three children who did not have the declaration of consent were given the material to uphold equal rights for every attendee. They solved the tasks, but their booklets were not collected for the raw data set.

9.4.2 Testing Procedure for the Large-Scale Spatial Abilities Test

The map-based orientation test was administered individually without imposing any time limit. The children were taken out of the probabilistic theory session one-by-one. The experimenter started by asking some ice-breaker questions. In a first testing step, the child performed the PTSOT-C and the Corsi Block Test. After completing both tests, the children were taken outside the university building.

While the children walked down the stairs in the campus building they were told they were part of a very special treasure hunt. Once arrived at the starting point, the 'treasure map' (map of the campus) was handed out. Next to the treasure map were six stickers, three circular stickers ($\varnothing = 8\,mm$) and four arrows ($7\,mm \times 14\,mm$), each one in yellow, blue, red and orange. Then, the children were turned to the north, holding their map in alignment with the environment. The children were instructed to keep the map in the same position as it had been distributed for the whole treasure hunt. A small discussion was conducted to point out representational correspondences between the environment and the map, emphasizing the starting point and buildings around the starting point.

In a second introductory instruction each child was shown a green scarf that was fixed to a rain spout of a building. First, the child was asked to show with their arm the corresponding direction of the scarf and to use the answering instrument later. The child was then turned 180 degrees (to the south) and the orange arrow was stuck to the map to indicate the viewing direction. Then, the pointing procedure to the scarf was repeated. A small discussion on how the pointing direction changed when the orientation has changed as well as a comment that 'the map does not perfectly align anymore to the environment' was the closing of the introductory part of the procedure. At this point, the experimenter decided whether or not the child understood the answering instrument and the concept of the viewing direction. If the child had trouble, the introductory session was repeated. After that, the child started the map-based orientation test. After the test, the child's map was collected and the experimenter prepared the test material for the next child.

The same general procedure was used for all children, independently of the experimenter. The standardized test documentation material and the test manual ensured an almost[4] standardized procedure. All experimenters talked to the children to settle them into a comfortable position. Experimenters did not help the children to complete the treasure hunt and avoided to give feedback after completing the task. When leading the children to a flag they walked in front of the children. Whenever the child was placing a disk, the experimenter shadowed the child.

During the testing procedure, time problems occurred due to the following reasons: (1) The classes were late in the morning, but had to leave in time, which resulted in a shortened overall time ≤ 4.5 hours; (2) the children had to leave before the official ending due to private reasons; and (3) the children needed more time than expected in the map-based orientation test. All three reasons led to the fact that some children were not able to complete the whole test, thus producing patterns of missing values that did not occurre due to a lack of cognitive abilities but as a matter of fact of the general organization of data collection.

---

[4]Due to the large environment and the involvement of many experimenters, the notion 'almost' is used here, see also a discussion on the limitations of this study at the end of this book (Section 13.2, p. 389).

9.5 Documentation

The data was continuously documented during data collection. Since the small-scale spatial abilities test was presented in a booklet, the documentation process was straight forward: the completed booklets were collected after the test was finished. In every class, it was further documented whether the test was proceeded according to the manual and whether there were any exceptions such as children who did not speak German well enough to understand the tasks.

The PTSOT-C, the Corsi Block Test and the large-scale spatial abilities test were documented separately by the experimenter. In regard to the PTSOT-C, the experimenter documented the number the child pointed to using the answering instrument (see ESM E.1, p. 68). In regard to the Corsi Block Test, the experimenter documented which number series was completed correctly or wrong in the first and second trail (see ESM E.2, p. 70). During the map-based orientation test, the observations of the items and all extra variables were documented as outlined in Section 5.1.4 (see also ESM D.2, p. 62). The documentation of the single items of the large-scale spatial abilities test is summarized in Table 9.3.

Table 9.3 Documentation of the Map-Based Orientation Test

Task	Items	Documentation
MapRot (pointing to unseen landmarks)	2 orientations with 3 directions each	report the chosen number on the answering instrument
Dots (self-position)	3 locations	stickers (colored circles) on the paper map
Dir (self-orientation)	3 directions	stickers (colored arrows) on the paper map
HP (pointing to the start)	3 directions	report the chosen number on the answering instrument
MapUse (pointing to unseen landmarks at the flags)	3 directions	report the chosen number on the answering instrument
Disks (place-finding)	3 disks	mark the place where the disk was located on the answering instrument
MentalMap (pointing to unseen landmarks without the map)	9 directions	report the chosen number on the answering instrument

In addition, the experimenter reported on the basis of self-elicitation when they did not test according to the test manual, whether unpredictable complications occurred, or the child knew the campus. The documentation of the map-based orientation test therefore allowed both to document the observations for each child and to post-evaluate whether or not the experimenter did forget any items (which would result in a missing value).

9.6   Data Treatment

9.6.1   Encoding and Scoring

Since the data of the paper-and-pencil test was not collected from the six refugee children and the three children without a declaration of consent, there was a raw data set of $n = 252$ test booklets. For ensuring the data quality, the observations in the booklets were encoded twice with respect to the Small-Scale Spatial Abilities Test scorebook. For every single item it was encoded whether or not the child accomplished it (Do-variable) and it was further assigned a raw variable according to the coding manual. Since the coding manual proposed clear categories for every solution within the outcome space, the encoding process was highly objective. Different raw codes of the two raters were verified in a subsequent step and integrated into a revised version, the final raw data set. Then, the raw variables were scored using a SPSS (CORP 2017) syntax that implemented the scoring as declared in the test manual.

The data of the map-based orientation instrument of all 252 children was encoded twice with respect to the Large-Scale Spatial Abilities Test scorebook. Differences in the codes were controlled and a corrected version of the respective variable was integrated into the final raw data set, which was then scored according to the manual using SPSS.

A first analysis of the data set from the map-based orientation test revealed that the experimenter of one child did not work according to the test manual since the child did not understand the representational concepts of a map at all. Furthermore, one child explained having thorough knowledge of the campus. Those two data sets were excluded which resulted in a sample of $n = 250$ cases in the data set.

9.6.2   Treatment of Missing Data

Missing data occurred due to two reasons: first, due to the lack of time in the organizational framework, and second, due to ignorance of the instruction manual by the experimenter. Missing data that occurred due to the first reason typically showed patterns of series of missing values in multiple subsequent tasks. Missing values that occurred due to the second reason showed random patterns since they corresponded to items that were forgotten to ask by the experimenters which was documented by blank cells in the map-based orientation test documentation paper. Consequently, the mechanism for missing patterns was not due to a lack of cognitive abilities but due to a number of random reasons. It was hypothesized that the missing data mechanism should therefore be MCAR or MAR.

The percentage of missing values per child was analyzed. For further data treatment (imputation of missing data), all those children who had more than 50% missing values in the map-based orientation test were detected. As a consequence, the data of ten children were excluded, and the data sets of all other 240 children maintained.

Multiple imputations were computed in R (R CORE TEAM 2018) using the package *NPBayesImpute* (SI & REITER 2013; Q. WANG, MANRIQUE-VALLIER, REITER, HU, & WANG 2014). During imputation, the variables sex, a forward span and backward span of the Corsi Block test, the sum score of the PTSOT-C and all sum scores of the units in the written test were included as auxiliary variables. Estimation values for all item scoring variables were subsequently aggregated into sum scores. The post-treated data set contained $n = 240$ individual sum scores for the measures of the small-scale and large-scale spatial abilities test.

Part III
Results and Discussion

Children's Small-Scale Spatial Abilities

10

This chapter presents the empirical results concerning children's small-scale spatial abilities as measured by the paper-and-pencil test proposed in Chapter 6. It presents the results concerning the quality of individual measures (Section 10.1), concerning empirically separable latent abilities (Section 10.2), and the role of visuospatial working memory (Section 10.3). Section 10.4. presents the results of the measure PTSOT-C. The final Section 10.5 discusses the findings with respect to the literature presented in Part I.

10.1 Analyses of Items and Measures

This section presents the results concerning the question of whether or not the paper-and-pencil tests were able to measure children's small-scale spatial abilities in a differentiated way. This section summarizes the results of item statistics and descriptive analyses of all items and measures but does not present detailed descriptive results for each of them.

The analyses relied on a sample of $n = 240$ children, thus including all the cases of the sample. In subsequent analyses, a reduced sample of $n = 215$ children was also used. During item analysis, the descriptive results on items and measures were compared to each other in both samples. Results only differed at the level of the second decimal place which was considered to be sufficiently small. For the sake of shortness, only the analyses of the full sample ($n = 240$) are presented.

Electronic supplementary material The online version of this chapter (https://doi.org/10.1007/978-3-658-32648-7_10) contains supplementary material, which is available to authorized users.

C. Heil, *The Impact of Scale on Children's Spatial Thought*, Studien zur theoretischen und empirischen Forschung in der Mathematikdidaktik, https://doi.org/10.1007/978-3-658-32648-7_10

10.1.1 Summary of Item Analyses

In a first step, item analyses were conducted to examine to which extent single items were able to differentiate between individual performances. Both CTT descriptive statistics (item facility and discrimination) and IRT statistics (weighted MNSQ and item difficulty) were computed for the individual measures that consisted of multiple items and interpreted using the benchmarks provided in Section 5.2.1. All analyses were computed using a free trial version of *ConQuest 4* (M. L. WU ET AL. 2015). The results are summarized in Table 10.1.

All items except for PFT5 and 2DMR2 had item facility values between .20 and .80, thus indicating that none of the items were too easy or too difficult for the children in the sample. A comparison of item facilities within the same measure showed that some measures consisted of an item set of approximately the same facility, whereas some measures had items with a broad range of item facilities. The measures Claudia, Boxes, and Meadow consisted of an item set featuring homogeneously medium item facilities (.45 to .56, .52 to .65, and .50 to .55, respectively), and the measure Ben consisted of items with homogeneously medium to high item facilities (.58 to .76). The measures LR, 2DMR, 3DMR, and PFT consisted of items covering a broad range of item facilities (.40 to .73, .39 to .81, .34 to .77, and .18 to .80, respectively). 2DMR consisted of two pairs of items having the same item facility (medium and high). These observations were confirmed by item difficulties computed during IRT analysis.

Except for the items 3DMR3 and 3DMR4, all items had discrimination values $\geq$ 0.3.[1] Other than for PFT1 and 3DMR2 in which low discrimination values (.32 and .30, respectively) were explainable by high item facilities (.80 and .77, respectively), an explanation of the low discrimination values for PFT1 and 3DMR2 was not straight forward.

Furthermore, a Rasch model was fitted to the data of each measure. Except for item Meadow3, the fit indices of Rasch analyses suggested that the Rasch model explained the data well enough. In other words, it was most likely that the data did not deviate too much from the assumptions of the model, a result that was demonstrated by acceptable weighted MNSQ values (MNSQ $\leq$ 1.3 and MNSQ in the 90% confidence interval) for almost all items (Table 10.1) The item LRstep1 showed a conspicuous value (MNSQ = 1.22) but was still within the acceptable range (see Section 5.2.1, for further elaboration).

---

[1] Item discrimination values were computed for each task/measure separately and not for the whole scale.

Table 10.1 Summary of Items Statistics for the Items Used the Main Study (n = 240)

Measure	Item label	Facility (CTT)	Discrimination (CTT)	weighted MNSQ + CI (IRT)	Difficulty (IRT)
LR	LRstep1	.73	.48	1.22 (0.78, 1.22)	−2.21
	LRstep2	.58	.68	0.91 (0.79, 1.21)	−0.80
	LRstep3	.52	.71	0.89 (0.79, 1.21)	−0.26
	LRstep4	.40	.62	0.96 (0.78, 1.22)	0.83
2DMR	2DMR1	.80	.55	0.88 (0.81, 1.19)	−1.61
	2DMR2	.81	.49	1.07 (0.82, 1.18)	−1.51
	2DMR3	.39	.41	1.01 (0.80, 1.20)	1.90
	2DMR4	.50	.35	1.06 (0.83, 1.17)	1.19
Claudia	Claudia2	.53	.56	1.06 (0.83, 1.17)	−0.27
	Claudia3	.45	.62	0.99 (0.82, 1.18)	0.25
	Claudia4	.52	.63	0.94 (0.83, 1.17)	−0.24
	Claudia5	.56	.55	1.08 (0.83, 1.17)	−0.55
	Claudia6	.45	.54	1.07 (0.82, 1.18)	0.25
3DMR	3DMR1	.42	.36	0.99 (0.88, 1.12)	0.40
	3DMR2	.77	.30	0.98 (0.83, 1.17)	−1.51
	3DMR3	.60	.18*	1.11 (0.88, 1.12)	−0.49
	3DMR4	.50	.28*	1.06 (0.89, 1.11)	0.02
	3DMR5	.34	.40	0.94 (0.87, 1.13)	0.82
Boxes	Boxes1	.52	.53	1.12 (0.83, 1.17)	−0.09
	Boxes2	.65	.65	0.92 (0.80, 1.20)	−0.92
	Boxes3	.55	.67	0.97 (0.82, 1.18)	−0.31
PFT	PFT1	.80	.32	1.06 (0.78, 1.22)	−2.30
	PFT2	.66	.43	1.04 (0.83, 1.17)	−1.14
	PFT3	.32	.49	1.05 (0.82, 1.18)	1.26
	PFT4	.37	.61	0.89 (0.83, 1.17)	0.89
	PFT5	.18	.59	0.84 (0.77, 1.23)	2.44
	PFT6	.29	.48	1.02 (0.81, 1.19)	1.49
Ben	Ben2	.60	.59	1.10 (0.80, 1.20)	−0.79
	Ben3	.75	.72	0.83 (0.75, 1.25)	−2.23
	Ben4	.58	.59	1.12 (0.81, 1.19)	−0.54
	Ben5	.71	.63	1.11 (0.77, 1.23)	−1.81
	Ben6	.76	.72	0.82 (0.74, 1.26)	−2.37

(Continued)

Table 10.1 (Continued)

Measure	Item label	Facility (CTT)	Discrimination (CTT)	weighted MNSQ + CI (IRT)	Difficulty (IRT)
Meadow	Meadow1	.50	.54	1.15 (0.82, 1.18)	0.02
	Meadow2	.55	.72	0.85 (0.82, 1.18)	−0.36
	Meadow3	.50	.47	1.29 (0.82, 1.18)[†]	0.08
	Meadow4	.51	.66	0.93 (0.82, 1.18)	−0.04
	Meadow5	.54	.69	0.87 (0.82, 1.18)	−0.26
	Meadow6	.51	.63	0.96 (0.82, 1.18)	−0.01

*denotes discrimination values < 0.3, [†]denotes weighted MNSQ-values close to 1.3

As a result of the item analyses, two single items were withdrawn from subsequent analyses. 3DMR3 was withdrawn due to its low discrimination value but 3DMR4 was kept to guarantee that the corresponding sum score of 3DMR had at least five categories and could be treated as a continuous variable during subsequent analyses (Section 5.2.2). Furthermore, Meadow3 was withdrawn since it probably was not explained well enough by the Rasch model. Since the corresponding item set of Meadow consisted of items with homogeneous item facility, no impact was expected by withdrawing a single item both from a statistical and content-relation point of view.

Since the remaining items could be assumed to be explainable by the Rasch model, it was concluded that they most likely reflected one underlying latent ability and hence, the sum scores for all measures were subsequently computed. Moreover, the score for the measures Emil (Section 6.2.5) and Cruise (Table 6.12) was computed. In total, ten sum scores reflecting the total scores of the ten tasks included in the paper-and-pencil test were computed for further analyses.

10.1.2 Descriptive Statistics

Since this study intended to use multivariate methods that assume normally distributed data without too strong floor or ceiling effects, the measures were further analyzed from a descriptive point of view using *IBM SPSS 25* (CORP 2017). Table 10.2 shows the descriptive statistics for each measure. All measures had at least five distinct categories and were therefore treated as continuous variables in the subse-

Table 10.2 Descriptive Statistics of Measures in the Paper-and-Pencil Test (n = 240)

Measure	Range	Mean	SD	Skew	Kurtosis	Min[%]	Max[%]
LR	[0,4]	2.23	1.54	−.14	−1.49	18.8	33.8
2DMR	[0,8]	5.00	2.1	−.62	−0.17	5.4	12.5
Claudia	[0,4]	2.22	1.48	−.14	−1.44	16.3	29.2
3DMR	[0,4]	2.03	1.24	−.06	−1.05	10.8	14.6
Boxes	[0,6]	3.44	1.97	−.44	−1.00	12.9	15.8
PFT	[0,6]	2.62	1.77	.42	−0.76	10.8	9.2
Emil	[0,6]	2.80	2.22	.32	−1.41	16.3	23.3
Ben	[0,5]	3.40	1.80	−.92	−0.54	16.3	38.8
Cruise	[0,4]	1.82	1.59	.12	−1.61	32.9	20.8
Meadow	[0,5]	2.62	1.96	−.14	−1.54	20.0	25.8

quent analyses.[2] None of these measures departed considerably from normality (all skews $\leq$ |2| and all kurtosis values $\leq$ |7|, see Section 5.2.2).

The distribution of the single measures was investigated in detail since one goal of this study was to report to which extent measures were suitable for testing individual differences among the children in a sufficiently differentiated manner. Half of the measures were particularly platykurtic (kurtosis $<$ |1.4|). To investigate whether this occurred due to a particularly flat normal distribution or due to bimodality effects, the percentage of children who received no point at all or achieved the maximum number of points per measure was further considered. Those percentages are reported in the last two columns of Table 10.2. For LR, Claudia, Emil, Cruise, and Meadow, they show that a particularly high percentage of children (over 15% and up to 34%, respectively) scored at the extreme values of the scale which indicated that the sum scores might have been distributed in a bimodal way. In other words, the corresponding measures were likely to lack differentiation of performances among both low and high achievers in those tasks.

The measure Ben was particularly highly skewed (−.92) and featured a mean value (3.40) that was well above the theoretical expectation of 2.5 points. In addition, 38.8% of the children achieved five points. In line with observations from analyses at item level showing that all items featured medium to high facility values, the

---

[2]As outlined in Section 5.2.2, variables were considered to be continuous rather than categorical in the statistical analyses when featuring more than 4 distinct categories. From a content-related point of view, these distributions rather represent grouped samples from a normal distribution.

measure Ben seemed to be subject to ceiling effects. The same result was found for LR and Claudia (33.8% and 29.2% of the children achieved the maximum number of points, respectively).

For the sorting tasks Emil and Cruise, the corresponding sum scores were also distributed in a bimodal way. The measure Cruise featured a particular high percentage of children who achieved no point at all (32.9%) and the mean value (1.82) was slightly beyond the theoretical expectation value of 2 points. The measure was therefore assumed to be subject to floor effects or showed clear tendencies that it did not differentiate well enough between different performances.

A series of post-hoc analyses were computed both for comparing descriptive statistics and distributions of sum scores to understand whether or not these results were robust for the whole sample or sensitive to group effects such as sex differences. Significant sex differences in mean values were found for all measures except for LR, 2DMR, and Cruise (Table F.1 in ESM, p. 85). χ^2-tests concerning differences in the distribution of sum scores revealed that except for LR, significant different distributions of sum scores were found for boys and girls (see ESM F.1, pp. 74, for the distributions of sum scores). Consequently, the observed bimodal distributions were probably due to effects of superposition of those distributions. Hence, the measures Claudia, Emil, Ben, and Cruise tended to yield floor effects for girls, and ceiling effects for boys. Since this study did not intend to examine sex differences in detail, these results were not further considered in the subsequent analyses.

10.1.3 Reliabilities

Table 10.3 shows Cronbach's α for all measures that consisted of a set of different items but not for the picture sorting measures Emil and Cruise.

Table 10.3 Cronbach's α for the Small-Scale Spatial Abilities Measures in the Main Study

LR	2DMR	Claudia	3DMR	Boxes	PFT	Emil	Ben	Cruise	Meadow
.81	.66	.80	.54	.77	.75	–	.84	–	.84

Most of the measures had at least an acceptable internal consistency which was reflected by values for Cronbach's α that ranged from .75 to .84. In contrast, the measures 2DMR and 3DMR featured lower (.66 and .54) reliabilities.

Conclusion

Descriptive analyses at the level of items and sum scores showed that all measures in the paper-and-pencil test had at least acceptable characteristics from a psychometrical point of view. With the possible exception of two critical items, all items used in the measures had an acceptable item facility and item discrimination. The fit indices revealed that except for one, all items might be modeled by the Rasch model within their respective measure.

Results of the descriptive analyses indicated that the sum scores of half of the measures were distributed in a bimodal way. They indicated that some sum scores showed higher percentages of children who did either not achieve a point at all nor scored with the maximum number of points, a result that was probably due to effects of sex differences in the distribution of sum scores. The measure Ben was characterized by ceiling effects for the whole sample and the measure Cruise by floor effects. 3DMR was the only measure with consistently low discrimination values and a poor value concerning internal consistency.

10.2 On the Dissociation of Two Subclasses of Small-Scale Spatial Abilities

This section presents the results concerning the analysis of correlational patterns between the individual measures. Those specified an assumed linear relation at the manifest level and suggested which latent factors were likely to be empirically separable. As stipulated by the theoretical model (Section 5.1.2), the measures 3DMR, 2DMR, and PFT were assumed to reflect a first class of spatial abilities, object transformation abilities (OB). The measures LR, Claudia, Boxes, Emil, Ben, Cruise, and Meadow were assumed to reflect a second class of small-scale spatial abilities, egocentric perspective transformation abilities (EGO). Confirmatory factor analyses (CFAs) tested the hypothesized two-factor model and quantified the proportion of variance shared among the OB and EGO.

10.2.1 Class Clustering Effects

Since the data were collected class-wise, the intraclass coefficients (ICs) for every task were computed in a preliminary step. They determined possible biases that may have occurred in the subsequent analyses due to an underlying multilevel structure in the data (B. O. MUTHÉN 2002, p. 90). Table 10.4 shows the results of the *Mplus* (L. K. MUTHÉN & MUTHÉN 2004) analysis.

Table 10.4 Intraclass Coefficients for the Small-Scale Spatial Abilities Measures in the Main Study

LR	2DMR	Claudia	3DMR	Boxes	PFT	Emil	Ben	Cruise	Meadow
.037	.026	.053	.011	.087	.044	.011	.011	.060	.020

Seven out of the ten measures had an IC $\rho_{IC} \leq .050$, a result that showed that variance in these measures was mainly due to individual differences but not due to a considerable effect of belonging to a certain class (GEISER 2012). The tasks Claudia ($\rho_{IC} = .053$), Cruise ($\rho_{IC} = .060$), and Boxes ($\rho_{IC} = .087$) showed a clearer effect of class clustering. Consequently, for these measures, effects of belonging to a certain class may have had an influence on their corresponding variances. In regard to Boxes, for example, 8.7% of the variance was due to the effect of belonging to a certain class, a result that could probably be due to the fact that tasks like Boxes were posed in some but not all school books. This explanation, however, did not explain the class effects found for Claudia and Cruise.

10.2.2 Analysis of Pearson Correlations

To approach relations of the measures on a manifest level, pairwise Pearson correlations were computed to quantify the amount of linear relationship under the assumption of continuous variables (Table 10.5).

All correlations are positive for the sample with 240 children (Table 10.5 above the diagonal). Following J. COHEN's (1988) suggestions, moderate correlations were observed among measures that were intended to reflect OB (.39 to .49) and also moderate correlations among the measures Claudia, Emil, Boxes and Meadow that were intended to reflect EGO (.32 to .47). However, these measures correlated only weakly with the measures Ben and Cruise (.24 to .33). In particular Cruise correlated to the same extent with 2DMR, 3DMR, and PFT (.24 to .37) as with Claudia, Boxes, Emil, and Meadow (.24 to .31). LR, a measure that was intended to reflect EGO, correlated to a high extent with 2DMR, 3DMR, and PFT (.36 to .41) and with other measures of EGO (.25 to .39). The results further indicate a wide range of correlational strength between all OB-measures and EGO-measures (.21 to .52) with the correlation of Boxes with PFT being the strongest one (.52).

Table 10.5 Pearson Correlations between the Paper-and-Pencil Measures (n = 240 above the diagonal and n = 215 below the diagonal)

	1.	2.	3.	4.	5.	6.	7.	8.	9.	10.
1. LR	–	.36	.28	.36	.39	.41	.34	.25	.28	.30
2. 2DMR	.28	–	.26	.39	.39	.43	.23	.23	.24	.21
3. Claudia	.19	.18	–	.30	.45	.45	.40	.27	.24	.32
4. 3DMR	.28	.31	.17	–	.44	.49	.36	.26	.27	.24
5. Boxes	.32	.30	.35	.31	–	.52	.47	.33	.33	.46
6. PFT	.32	.33	.35	.38	.42	–	.39	.42	.37	.43
7. Emil	.26	.14	.30	.25	.39	.27	–	.28	.25	.34
8. Ben	.17	.15	.17	.14	.24	.33	.19	–	.20	.33
9. Cruise	.21	.18	.14	.18	.25	.29	.16	.12	–	.31
10. Meadow	.23	.11	.22	.10	.37	.32	.24	.25	.24	–

Since results from the preliminary descriptive analyses suggested that the majority of the measures tended to differentiate not well enough among low and high achievers, a subsample was computed in which all those children were withdrawn who were in the lower and upper 5%-percentile in relation to the total score of all ten tests. By withdrawing them, all those children who had essential problems in solving the tasks (i. e., scoring extremely low in all of the tasks) or whose performances were not measured differentiated enough (i. e., extremely high scores in all of the measures) were excluded from the sample. The reduced sample consisted of n = 215 children.

Table 10.5 shows the recomputed Pearson correlations for the reduced sample below the diagonal. Results showed the same correlational patterns as outlined for the full sample in Table 10.5, but correlations were on average .08 to .10 lower than those reported for the full sample for most of the measures. 3DMR and PFT showed higher deviations. Differences between Pearson correlations in the original and reduced sample differed up to |.14| (3DMR with Meadow) and |.12| (PFT with Emil).

10.2.3 Confirmatory Factor Analysis

CFAs were conducted using the R package *lavaan* (Rosseel 2012) to investigate the observed patterns of correlations in detail, in particular to examine whether object manipulation spatial abilities were separable from egocentric perspective transformation abilities. To avoid sensitivity of the analysis to extreme outliers, the analyses were computed on the basis of the reduced sample with n = 215 children. Since the preliminary analysis revealed no considerable class-clustering effect, the influences of class-wise testing were neglected by performing a single-level analysis with an estimator using robust standard errors. All analyses were run using the MLR and WLSMV estimator. Both estimators use robust standard errors, and WLSM is particularly suitable when dealing with measures that are distributed in a bimodal way (Section 5.2.2). Since MLR allows for computation of AIC and BIC for model comparison, the results using MLR are reported here and the ones using WLSMV are reported in the ESM.

As in the pilot study, observations at the manifest level (Table 10.5) raised the question of whether LR reflected EGO, as hypothesized, or OB. To investigate this question, a preliminary CFA model was computed, in which the measure was allowed to load on both factors of spatial abilities. Results indicated that the measure loaded only on OB significantly ($p = .043$), a finding that indicated that only OB contributed to explain variance in LR above what has been explained by EGO and not vice versa. Consequently, LR was assumed to be a measure reflecting OB in the subsequent analyses.

In the preliminary model, both Cruise and Ben had poor factor loadings (.38 and .39, see COMREY & LEE 1992). Consequently, the latent factor EGO explained less than 20% of variance in these measures. Moreover, both measures had either a strong floor effect (Cruise) or a strong ceiling effect (Ben), that is, they were not able to distinguish between low and high-performers in these tasks. Since EGO was still measured in a broad way, even without LR, Cruise, and Ben, those two measures were also excluded from subsequent analyses. Consequently, the new two-factor model involved eight measures with LR, 2DMR, 3DMR, and PFT reflecting OB and Boxes, Claudia, Emil, and Meadow reflecting EGO.

The fully standardized two-factor model solution is depicted in Figure 10.1. Numbers above the single-headed arrow lines represent the standardized factor loadings. Since all measures were assumed to load on one single factor in this model, these loadings can be interpreted as standardized regression coefficients (R. B. KLINE 2015, p. 301). The shorter arrows that point to each measure represent error terms, that is, the percentage of variance in the measure that is not explained by the factors. The curved, two-headed arrow represents the latent correlation between both factors OB and EGO. The results of the WLSMV-estimation are shown in Figure F.1 (ESM, p. 95). Estimations deviated by |.02| maximum in this model.

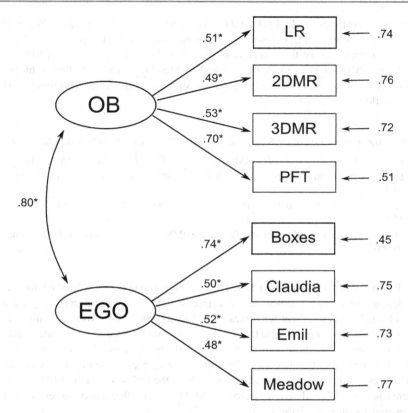

Figure 10.1 Completely standardized solution of the confirmatory factor analysis in the main study revealing two distinct subclasses of small-scale spatial abilities. *indicate significant path coefficients ($p < 0.001$)

The values of the corresponding fit indices are reported in Table 10.6. In line with the benchmarks provided in Table 5.1 (p. 168), all indices suggested that the model explained the empirical data very well. The exact-fit hypothesis was not rejected ($\chi^2(19) = 16.15$, $p = .65$) which indicates that the model did not derive considerably from the data. Moreover, the Root Mean Square Error of Approximation (RMSEA) was .00 with an 95% confidence interval $CI = [.00, .05]$. The corresponding significance test if the value was well beneath .05 produced a nonsignificant $p = .96$. Furthermore, the Bentler Comparitive Index (CFI = 1.00), the

Tucker-Lewis Index (TLI = 1.00), and the Standardized Root Mean Square Residual
(SRMR = .034) were well beneath the commonly used criteria for good fit. Further-
more, an inspection of the residuals showed that they were all low ($\leq |0.1|$) (R.
B. KLINE 2015, p. 278). Only the residual of Meadow correlated slightly higher
with the residual of 3DMR ($-.102$). Those slight deviations between model and
data could be neglected.

Table 10.6 Comparison of the Single-Factor and the Two-Factor Models in the Main Study

Model	χ^2	df	χ^2/df	RMSEA(CI)	CFI	TLI	SRMR	AIC	BIC
1-factor-model	25.62	20	1.28	.03 [.00, .08]*	.97	.96	.045	6443	6496
2-factor-model	16.15	19	0.85	.00 [.00, .05]*	1.00	1.00	.034	6434	6491

*indicates if $p_{RMSEA} \leq .05$ is non-significant. The RMSEA is shown with its 90 % confidence
interval (CI).

In line with COMREY AND LEE (1992), the estimated factor loading for the
indicators were fair for almost all items (factor loadings of .48 to .53), very good
for PFT (.70) and excellent for Boxes (.74). Consequently, for those two indicators
only the latent factors were able to explain a considerable amount of variance which,
in turn, showed that those indicators reflect well the corresponding construct. For
all other measures, the amount of explained variance explained by either OB or
EGO was much lower. The reliabilities for both the OB-scale and the EGO-scale
were ($\omega_{OB} = .64$, and $\omega_{EGO} = .65$, see MCDONALD (1999)) and the set of eight
measures featured a reliability of $\omega_{total} = .76$.

The two-factor model solution indicated that both factors of spatial abilities
were highly correlated factors ($r = .80$), thus sharing 64 % of variance. To examine
whether both factors were identical, an alternative, single-factor model was computed in which all measures loaded on one factor. The fit indices of this model
are also shown in Table 10.6. The overall chi-square statistic for the model was
non-significant which demonstrated that the model did not deviate considerably
from the data, and the values for all other fit indices still met the criteria for good
fit (Table 5.1). A comparison between the two models, however, revealed a better
fit of the two-factor model solution. Consequently, both models could explain the
data, but the two-factor model featured better fit indices. Significance of this observation was confirmed by the $\chi 2$-difference test comparing the fit of both models
($\chi^2(1) = 7.16$, $p = .007$).

Both latent factors were closely related but should better be modeled as distinct ones. This result was also confirmed by the scatterplot mapping the latent estimates performances in all measures of OB against those in EGO for each child (Figure 10.2).

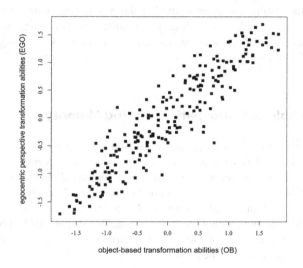

Figure 10.2 Comparison of estimates of performances in the latent standardized constructs OB and EGO in the main study

To examine whether sex had an influence on the extent of the dissociation between the two classes of spatial abilities, a group-wise CFA was computed and analyses of measurement invariance were conducted in a final step (see also ESM F.1, pp. 85). The results indicated partial weak invariance when fixing the loading of the measure Emil to be equal in both groups ($\Delta\chi^2(5) = 9.07$, $p = .110$). Strong partial invariance was not attained ($\Delta\chi^2(6) = 15.35$, $p = .020$). Consequently, except for Emil, both constructs OB and EGO were measured robustly to sex differences. In other words, the factors themselves and their latent correlation did not differ, but the intercepts (corresponding to group-specific mean values) were sex-sensitive.

Conclusion

Moderate manifest correlations were found among all measures included in the small-scale spatial abilities test. A subsequent confirmatory analysis explained correlational patterns in the data by fitting a two-factor model derived from the literature. This model included eight different measures except for Ben and Cruise. As suggested by the hypothesized model and confirmed by the results, object manipulation abilities and egocentric perspective transformation abilities were highly correlated (.80), yet being separable latent factors. Results indicated that a single-factor model could also explain the correlational patterns.

10.3   The Role of Visuospatial Working Memory

This section presents results concerning the relation between visuospatial working memory capacities (VSWM) and small-scale spatial abilities. Since measures of visuospatial working memory were intended to be integrated into the CFA, there was the need to control whether these measures were normally distributed without showing too strong floor or ceiling effects. Moreover, since there was a possible range of different scorings for both forward and backward span that had been proposed in Section 8.1.3, results of a first psychometric analysis indicated which two measures were appropriated to be chosen.

10.3.1 Descriptive Analysis of Different Corsi Block Test Measures

To analyze whether the measures of VSWM proposed in Table 8.1 could be assumed to be normally distributed data without too strong floor or ceiling effects, a descriptive analysis was performed (Table 10.7). Since all measures featured at least seven distinct categories, they were treated as continuous variables in the subsequent analyses. None of these measures departed considerably from normality (all skews $\leq |2|$ and all kurtosis values $\leq |7|$).

An analysis of the descriptive values for the forward span measures Maximum Forward Span (MFS) revealed that all children achieved at least four points out of nine, thus managing to reproduce a sequence of four blocks, and some of the children were able to reproduce sequences of length 8 in the second trail. Results of the

Table 10.7 Descriptive Statistics of All Scores Used for the Corsi Block Test (n = 240)

Measure	MinMx	Range	Mean	SD	Skew	Kurtosis
MFS	[2,8]	[4,8]	6.13	0.83	.11	−.17
MBS	[2,6]	[4,6]	5.50	0.65	−.89	−.29
AFS	[2,8]	[4,8]	5.63	0.79	.13	.15
ABS	[2,6]	[3,6]	5.14	0.68	−.60	.04
MTFS	[2,70]	[18,66]	38.40	10.37	.46	−.36
MTBS	[2,40]	[14,40]	32.20	7.44	−.45	−.95
ATFS	[2,70]	[14,66]	33.10	9.00	.52	.33
ATBS	[2,40]	[11,40]	28.36	7.21	−.17	−.71

descriptive statistics revealed that MFS, Average Forward Span (AFS), Maximum Total Forward Span (MTFS), and Average Total Forward Span (ATFS) were neither subject to floor nor to ceiling effects. Both measures that relied on the evaluation of the achieved span values without taking into consideration the number of trails used, MFS and AFS, were normally distributed with a mean value that was slightly above the theoretical mean value (6.13 for MFS, 5.63 for AFS) and low positive skew values (.11 and .13, respectively), a result that indicated that the corresponding distribution was almost standard normal with a slight effect of being right-tailed. These results were further supported by small values for kurtosis (−.17 and .15, respectively). Both measures that relied on the evaluation of the achieved span values while taking into consideration the number of trails used, MTFS and ATFS were also normally distributed with a mean value that was around the theoretical mean value (38.40 and 33.10, respectively) with an effect of being right-tailed (.46 and .52, respectively). The corresponding absolute values of their kurtosis values (−.36 and .33, respectively) were higher than in the case of MFS and AFS due to the definition of the measures, but still pointed to quasi-normality without the distribution being to flat or too pointed.

In comparison to the results of the forward span, an analysis of the descriptive values for the backward span measures Maximum Backward Span (MBS) revealed that children achieved at least three points out of six, thus managing to invert a sequence of three blocks. However, a considerable number of children scored with the maximum of points possible which was supported by high mean values for MBS, Average Backward Span (ABS), Maximum Total Backward Span (MTBS), and Average Total Backward Span (ATBS), a result that pointed toward possible ceiling effects to consider in the backward measures. More specifically, an analysis of the skew values revealed that all measures of backward span were negatively skewed which indicated that the corresponding distribution was a left-tailed normal distribution. Particularly high skew values were found for measures that did not take into consideration the number of trails used, MBS ($-.89$) and the ABS ($-.60$), and lower values for measures that are defined upon the number of trails used, MTBS ($-.45$) and the ATBS ($-.17$) which indicated that MBS and ABS were subject to ceiling effects, whereas the effect was lower for the more complex measures MTBS and ATBS. The corresponding kurtosis values ($-.95$ and $-.71$, respectively) indicated that the corresponding distribution was still normal, but widely distributed.

Results of a CTT analysis underpinned the results for each measure (Table 10.8). In correspondence with the results outlined above, all backward span measures showed high item facility values (.71 to .92) which indicated that they were not able to differentiate among children with high VSWM capacity well enough. In regard to the forward measures, item difficulty was higher for both measures defined upon the achieved sequence length (.77 and .70 for MFS and AFS, respectively) and showed medium item difficulty for the more complex measures MTFS and ATFS (.55 and .47, respectively).

Concerning the discrimination values for the measures, forward measures showed higher values (.65 to .78) than backward measures (.55 to .69) with AFS (.78) and ABS (.69) having the highest discrimination values, respectively. However, those high discrimination values corresponded to high facility values (.70 and .86, respectively) which did not prove the quality of the indicators. Analyzing the facility-discrimination values revealed that the forward measure ATFS had a high discrimination value (.73) while being medium difficult (.47), and that the backward measure ATBS had an acceptable discrimination value (.57) while having the lowest facility value among all backward measures (.71).

Table 10.8 Summary of Statistics for the CBT Items in the Main Study

Measure	Facility (CTT)	Discrimination (CTT)
MFS	.77	.73
MBS	.92	.68
AFS	.70	.78
ABS	.86	.69
MTFS	.55	.65
MTBS	.81	.55
ATFS	.47	.73
ATBS	.71	.57

Since results of the preliminary descriptive analysis indicated that both measures were normally distributed without showing ceiling effects, subsequent analyses on the relation of spatial working memory to small-scale spatial abilities relied on the measures ATFS and ATBS, thus referring to scorings that took into account the averaged performance in sequence repetition/inversion and the number of trails used to do so.

10.3.2 Analysis of Pearson Correlations

Table 10.9 shows the correlations between the two chosen measures of VSWM, ATFS and ATBS, and the eight measures of spatial abilities. All correlations were positive. Correlations of the forward measure ATFS ranged from .10 to .32, thus representing weak associations according to J. COHEN'S (1988) conventions and only four of them were found to be significant at the level $p = .05$ with Bonferroni-correction for 16 multiple tests. Correlations of the backward measure ATBS ranged from .19 to .33, thus representing weak to medium associations, and all of them were significant.

Table 10.9 Correlations of Single Measures with the Two Measures of Visuospatial Working Memory Capacity (n = 215)

	LR	2DMR	Claudia	3DMR	Boxes	PFT	Emil	Meadow
ATFS	.32*	(.18)	(.10)	(.12)	.26*	.24*	(.18)	.25*
ATBS	.26*	.22*	.21*	.23*	.33*	.31*	.19*	.32*

*$p < 0.05/16 = 0.003$ (two-tailed p with Bonferroni-correction)

A comparison of the correlations with respect to the two different subsets of spatial tasks, OB tasks and EGO tasks did not reveal any important differences in the extent of correlation. For the OB-tasks, 2DMR and 3DMR correlated only in a weak manner with the VSWM-tasks whereas LR and PFT correlated in a medium manner. For the EGO tasks, the labyrinth-tasks Claudia and Emil featured only weak associations, whereas Boxes and Meadow had medium associations with both measures of VSMW.

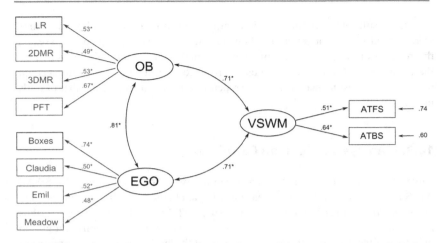

Figure 10.3 Completely standardized solution of the confirmatory factor analysis including two distinct subclasses of small-scale spatial abilities and visuospatial working memory. *indicate significant path coefficients ($p < .001$)

10.3.3 An Extended Model Involving VSWM

To investigate patterns of correlations among all measures of small-scale spatial abilities and VSMW, in particular, to investigate the extent to which each subclass of small-scale spatial abilities overlapped or not with VSWM, a single-level confirmatory factor analysis that neglected the effects of class clustering ($\rho_{IC}(ATFS) = .004$ and $\rho_{IC}(ATBS) = .006$) was computed. To conduct the analysis, a latent factor VSWM[3] was included into the two-factor CFA model of small-scale spatial abilities (Figure 10.1, p. 267). Figure 10.3 shows the fully standardized three-factor model based on MLR estimations using the reduced data set with n = 215. An analysis of the fit indices indicated that the model explained the data very well. The exact-fit hypothesis was not rejected, $\chi^2(32) = 29.98$, $p = .57$, a result that showed that the model did not derive considerably from the data. Moreover, the RMSEA was .00 with an 95% confidence interval $CI = [.00, .05]$. The test that the value was well beneath .05 was non-significant ($p = .97$). Furthermore, the

---

[3]To test for possible sex differences in the VSWM measures, a MANOVA was computed with ATFS and ATBS as dependent variables and sex as independent variable. Multivariate tests showed no significant effect for sex ($F(2, 237) = .097$, $p = 0.91$; $Wilk's \Lambda = .999$, partial $\eta^2 = .00$).

CFI = 1.00, the TLI = 1.00, and the SRMR = 0.039 were well beneath the commonly used criteria for good fit (Table 5.1).

In line with the patterns found in Pearson correlations, the results of the confirmatory factor analysis indicated that VSWM correlated with both subclasses of small-scale spatial abilities in the same amount, sharing $.71^2 = .50$ (50%) of variance. 40.4% and 25.5% of the variance of ATFS and ATBS could be explained by the latent factor VSWM, a result that indicated that both measures reflected the latent factor to a good and very good extent.

> **Conclusion**
> An analysis of the measures that reflect visuospatial working memory capacity (VSWM) revealed that the Average Total Forward Span (ATFS) and the Average Total Backward Span (ATBS), that is, measures that take into account both the number of trials and all items completed during the test, were suited best to measure children's VSWM in a differentiated way. A confirmatory analysis of correlational patterns revealed that VSWM shared a considerable amount of variance (50%) with both subclasses of small-scale spatial abilities.

10.4 Item Analysis for the PTSOT-C

Since the PTSOT-C was performed in an individual setting, it was first tested whether the experimenter had an effect on the results in the test. The experimenters were classified into three groups (Cathleen Heil, experienced experimenters, extra experimenters, see Section 9.3). The corresponding performances (sum scores of all eight items) of children who were instructed by those experimenters, were also devided into three groups. There was no significant difference between groups as determined by one-way ANOVA (F(2,237) = 1.64, p = .20). Consequently, all observations from all three experimenter groups were pooled in one data set for subsequent analysis.

To test whether the preliminary adopted version of the PTSOT can be included as a measure of egocentric perspective transformation abilities, the test and its items were analyzed in detail. A preliminary analysis of the distribution of the relative frequencies of the sum scores for the measure PTSOT-C (Figure 10.4) revealed that

(1) 14% of the children did not score a single point,
(2) over 80% of the children scored three or less points,
(3) none of the children reached seven or eight points.

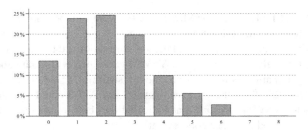

Figure 10.4 Sum scores for PTSOT-C (n = 252)

Based on those observations, the PTSOT-C was observed to contain items that were too difficult for the children in the sample. Single items as well as the measure itself, were not able to differentiate between children with higher or lower egocentric perspective transformation abilities since it was subject to strong floor effects.

Results of a descriptive analyses for the measure including all items supported this observation (Table 10.10). Indeed, the low mean value indicated that the children scored, in the average, only with a quarter of the scores that were possible to achieve.

Table 10.10 Descriptive Statistics for PTSOT-C (n = 252)

Measure	Range	Median	Mean	SD	Skewness	Kurtosis	Cronbach's α
PTSOT-C	[0,8]	2.00	2.17	1.52	.52	−.27	.43

Table 10.11 shows a summary of CTT and IRT item statistics. All item facilities except for PTSOT1 were below medium difficulty with PTSOT4, PTSOT7, and PTSOT8 being particularly difficult (item facility $\leq$ .20). Except for two items, the item discrimination values were below .30. Item discrimination values could, at least in some (PTSOT4, PTSOT7), but not all cases (e. g., PTSOT1) be explained by the corresponding low item facilities. Moreover, Cronbach's alpha was found to be low (.43), a result that indicated that the intercorrelations among test items were not sufficiently high.

Table 10.11 Summary Statistics for the PTSOT-C Items in the Main Study (n = 252)

Item no.	Item label	Facility (CTT)	Discrimination (CTT)	weighted MNSQ + CI (IRT)	Difficulty (IRT)
1	PTSOT1	.60	.11*	1.07 (0.90, 1.10)	−0.52
2	PTSOT2	.38	.18*	1.03 (0.90, 1.10)	0.57
3	PTSOT3	.32	.27*	0.98 (0.88, 1.12)	0.86
4	PTSOT4	.15	.10*	1.04 (0.78, 1.22)	1.93
5	PTSOT5	.25	.19*	1.01 (0.86, 1.14)	1.20
6	PTSOT6	.29	.30	1.00 (0.87, 1.13)	1.06
7	PTSOT7	.07	.05*	1.01 (0.62, 1.38)	2.80
8	PTSOT8	.10	.34	0.95 (0.68, 1.32)	2.48

*denotes discrimination values < 0.3

Conclusion

The PTSOT-C was not considered as a measure in subsequent analyses. The low discrimination values for all items and their item difficulties demonstrated that the items of the PTSOT-C need to undergo further revision. A preliminary version of the PTSOT-C was not suitable to reflect fourth grader's abilities in a differentiated way. The strong floor effects and the low test reliability that were found for the PTSOT-C further supported these results.

10.5 Discussion of the Empirical Findings

To answer the research question of this study (Section 5.1.1), a set of small-scale spatial abilities tasks in paper-and-pencil format were developed. In line with the literature, this study assumed that small-scale spatial abilities can be modeled to involve two subclasses, with object manipulation abilities being partially dissociable from egocentric perspective transformation abilities. These two subclasses were operationalized by a set of reflective measures. This final section discusses the findings concerning the individual measures and the paper-and-pencil test with respect to the theoretical framework outlined in Part I.

Measures Included in the Paper-and-Pencil Test

The results of this study showed that it is possible to measure children's small-scale spatial abilities in a differentiated way using paper-and-pencil tasks. Similar to previous studies (e. g., BERLINGER 2015; GRÜSSING 2012; JOHNSON & MEADE 1987), a range of spatial tasks that had been developed for adults were adopted and administered to children. Results indicated that not only adapted but also self-developed measures were able to reflect a wide range of individual differences in performance.

Consistent with previous studies, the results demonstrated large individual differences in children's performances in most of the small-scale spatial tasks (e. g., DIEZMANN & LOWRIE 2009; GRÜSSING 2012; VANDER HEYDEN ET AL. 2016). In particular, PFT, LR, Boxes, Claudia, and Meadow emerged as reliable measures allowing to differentiate well between children of high and low small-scale spatial abilities. This might be due to the fact that the chosen spatial tasks required the children to perform multistep solutions (e. g., folding, punching and unfolding of a piece of paper, see LINN & PETERSEN 1985).

Another explanation might be that the correct encoding and control of reference systems during mental manipulation might reflect individual differences. Indeed, all tasks needed to perform a set of mental transformations correctly, that, in turn, required the controlled use of different frames of reference, for example, when imagining the appearance of the whole object array in the perspective tasks (e. g., HUTTENLOCHER & PRESSON 1979; NEWCOMBE 1989). The findings are therefore in line with previous findings on children which demonstrated that by the age of 8, children are able to switch their frame of reference according to the task but not with the same success (ALLEN 1999a). In conclusion, the results of this study suggest that children are still challenged by these kinds of paper-and-pencil tasks proposed although it has been demonstrated that small-scale spatial abilities undergo substantial development during the age of primary school (e. g., LIBEN 2006).

The paper-and-pencil test was objective since it was conducted due to a standardized test manual, involved only items of closed item format, and the results were scored according to a fully standardized scorebook. Similar to other studies, however, some of the measures suffered from poor reliability (e. g., JOHNSON & MEADE 1987) or involved items that suffered from low item discrimination values (e. g., GRÜSSING 2012). In particular both mental rotation tests 2DMR and 3DMR had poor reliability values, a finding that is in line with GRÜSSING (2012) results for these measures. A possible explanation might be that there was no time limit for all spatial tasks. In particular for 2D mental rotation tasks, however, it has been emphasized that it is the speed of mental rotations or the rate of rotations in a certain time limit that accounts for individual differences (JUST & CARPENTER 1985; MUMAW ET AL. 1984; PELLEGRINO & KAIL 1982). Although this might explain the low reliability for 2DMR, this is not necessarily the case for 3DMR since tasks involving multistep solution processes (LOHMAN 1982) or involving complex 3D stimuli (MUMAW & PELLEGRINO 1984) have not been found to be speed-dependent. An inspection of the descriptive item statistics of 3DMR, however, revealed that for item 3DMR3 and 3DMR4, it was probably too obvious that the distractors were reflections of the original snake. Consequently, children did not need to perform three-dimensional mental rotations. To be used in other studies, the measure 3DMR need careful re-analysis. Coloring of some of the cubes could be one approach to make the 3DMR items easier for children. Although the cognitive mechanisms behind this finding are not well elaborated, it has been shown to facilitate performances in mental rotation in adults (KHOOSHABEH & HEGARTY 2008) which could be one possibility to improve the measure.

Another unexpected finding was that although the test proved to be construct valid, only two of the measures (PFT and Boxes) had high factor loadings. Strictly speaking, only in those two measures, the underlying latent construct was able to explain sufficient variance. Estimates indicated that up to three quarters of the variance was not explained by the underlying model for all other measures. Consequently, it was not clear whether those measures reflected the underlying construct well enough (R. B. KLINE 2015). Since this study did not use other approaches to control for the validity of individual tests, it remained unclear, at least from a psychometric point of view, whether those tasks measured what they intended to. From a content-related point of view, most of the tasks were developed close to others that had been proposed in the literature and could, at least from a face validity point of view, be considered to be valid (e. g., BERLINGER 2015; GRÜSSING 2012).

Dissociation of Two Subclasses of Small-Scale Spatial Abilities
The empirical findings of this study indicate that a two-component model distinguishing between object-based transformation abilities and egocentric perspective transformation abilities is a suitable way of modeling children's small-scale spatial abilities among other possible models reported in the literature (e. g., LINN & PETERSEN 1985). This finding is in line with previous studies in adults (HEGARTY ET AL. 2006; HEGARTY & WALLER 2004; KOZHEVNIKOV & HEGARTY 2001), primary school children (VANDER HEYDEN ET AL. 2016), and kindergarden children (FRICK 2019) that suggested that object manipulation abilities are dissociable from egocentric perspective transformation abilities. The results of this study showed that measures that required children to mentally manipulate objects to solve them did

not reflect the same latent construct as measures that required children to mentally transform their imagined self. Results of the CFA demonstrated that a model assuming a dissociation (Figure 10.1, p. 267) explained the data significant better than a model that was based on the assumption that small-scale spatial abilities are an undifferentiated construct. Similar to most of the studies outlined above, however, a strong latent correlation between the subclasses was found (.80).

One possible explanation of this dissociation might be that the specific cognitive processes to solve each subclass of tasks operate on distinct neural subsystems (ZACKS & MICHELON 2005; ZACKS ET AL. 2000). Object-based transformation abilities, that is, abilities that allow an individual to imagine the varying locations of objects in the environment while keeping the own viewpoint on the mental situation invariant in the environment, seem to rely on cognitive processes that allow to manipulate changing relations between the object-to-environment and self-to-environment frame of reference. They are isomorphic to the actual physical movement of an object (e. g., SHEPARD & METZLER 1971; WIMMER ET AL. 2016). In contrast, egocentric perspective transformation abilities, that is, abilities that allow an individual to imagine transformations of the self to reason about spatially invariant objects, seem to rely on cognitive processes that allow to manipulate changing relations between the self-to-object and self-to-environment frame of reference (ZACKS & MICHELON 2005). Performing egocentric transformations has been described as a cognitive effortful mental spatial transformation that allows children to update a representation of oneself after or by imagining movements of the self by rotations and translations (EPLEY, MOREWEDGE, & KEYSAR 2004). They are isomorphic to cognitive processes that govern the planning and simulation of prospective movements (PARSONS 1987b) and have been suggested to emerge from mental emulations of movements that have previously experienced and internalized in real space (KESSLER & THOMSON 2010). Although the underlying cognitive processes were not explicitly measured in this study, this explanation finds support both in the literature on adults (e. g., HEGARTY & WALLER 2004; ZACKS & MICHELON 2005) and children (CRESCENTINI ET AL. 2014).

Taking a developmental perspective, a possible dissociation might be explained by the fact that egocentric perspective transformations are rather embodied. As children quite early acquire the abilities to distinguish between left and right in their development, this might lead to egocentric automatisms that need to be controlled when solving perspective taking tasks (EPLEY, MOREWEDGE, & KEYSAR 2004). To overcome these, additional life experience from social perspective taking and cognitive control are necessary to overcome the initial taken viewpoint. This might not be the case when manipulating lifeless objects (CRESCENTINI ET AL. 2014). A possible dissociation between those different but similar cognitive abilities (mental operations with visuospatial stimuli) does, however, not necessarily imply that there is no or very low latent correlation at all between them. Innate or socio-environmental factors may have influenced the individual's abilities to perform both types of spatial transformations and can be very similar for both subclasses (CRESCENTINI ET AL. 2014; HEGARTY & WALLER 2004).

The high latent correlation between both subclasses of small-scale spatial abilities indicates that they have a considerable amount of shared variance. These findings suggest that although both subclasses rely on specific types of mental transformations (object-based and egocentric ones), both subclasses of spatial abilities seem also to rely on a set of common cognitive processes. Those might be the encoding of the presented visuospatial information and the maintenance of these information in (visuospatial) working memory. The shared variance that was found in this study might therefore reflect individual differences in these common processes. This also explains the high latent correlations that were found for each subclass with VSWM (.71). Indeed, encoding and maintenance of visuospatial information have been suggested in the literature to be controlled in the VSWM (e.g., MIYAKE ET AL. 2001).

A second possible interpretation of the high latent correlation found might lie in the use of strategies while solving the spatial tasks. Although strategy-homogeneity was assumed in this study, previous literature suggested that children and adults solve tasks with individual strategies (e.g., BARRATT 1953). It is further known that strategy use might be dependent on individual preferences in dealing with spatial information (e.g., KOZHEVNIKOV 2007) and the level of being challenged by a task, which might have led to the fact that children solved all tasks with the same strategy (for example using analytic strategies by mentally rotating the inputs, by changing the own position, or by consistently guessing). In particular for perspective taking tasks, it has been shown that children (GRÜSSING 2002; VANDER HEYDEN ET AL. 2017) and adults (KESSLER & THOMSON 2010; KOZHEVNIKOV & HEGARTY 2001) might use mental rotation strategies to solve tasks that should measure spatial perspective taking. This explanation can be found in the fact that the latent factors were not able to explain variance in the individual performances to a sufficient extent for some measures. The factor representing the abilities to perform a transformation of the imagined self, for example, was merely able to explain a quarter of the variance in both labyrinth tasks. This finding might be due to the fact that not all children solved the tasks with the intended strategy but might have rather used mental rotations of the labyrinth plan. One example is the task LR. An unexpected finding of this study was that LR did not reflect the abilities to take imagined viewpoints on a map to decide on whether a turn is to the left or to the right but rather reflected the subclass of object-based transformation abilities.

Consequently, individuals might have solved this task by mentally turning the map to perform left/right-decisions and did not transform the imagined self.

The findings of this study enrich previous findings in the literature on children's small-scale spatial abilities. They extend the findings of VANDER HEYDEN ET AL. (2016) both for the age at which a dissociation between object and perspective transformation is empirically demonstrable and concerning the extent of dissociation. Their dissociation model between object and viewer transformation abilities of 11 to 12 year olds only marginally outperformed the same model for 9 to 10 year olds. The findings of this study provide further empirical support that the dissociation might be found for children aged from 9 to 12, and that both subclasses of small-scale spatial abilities remain highly correlated (.80 in this study vs. .75 in their study). The fact that VANDER HEYDEN ET AL. (2016) investigated the dissociation using spatial tasks with manipulable material to measure viewer transformation abilities indicates a certain independence of the empirical findings of both studies towards the concrete tests used. Consequently, VANDER HEYDEN ET AL.'S (2016) findings and the ones from this study demonstrate that the dissociation between object and viewer transformation abilities is most likely to emerge by the age of 9 to 10 and manifests until the age of 12 years.

Moreover, the findings of this study provide a strong support for the extrinsic-intrinsic differentiation in dynamic spatial abilities in children as it was proposed in the theoretical 2 x 2-typology by NEWCOMBE AND UTTAL in several studies (e. g., DAVIS & SPATIAL REASONING STUDY GROUP 2015; NEWCOMBE 2018; UTTAL ET AL. 2013). The results of this study therefore complement the study of MIX ET AL. (2018) who were not able to show the extrinsic-intrinsic dissociation consistently in kindergarden children, third graders and sixth graders. The findings of this study, however, may not be considered to be a test of the whole 2 x 2-typology because only dynamic spatial abilities were reflected by the measures used.

The Role of Visuospatial Working Memory
The results of this study (Figure 10.3, p. 274) show that visuospatial working memory capacity as measured by the Corsi Block Test was strongly related to both subclasses of small-scale spatial abilities to the same extent. Consequently, both subclasses of small-scale spatial abilities seemed to be involved to an equal extent to the controlled storage and maintenance of visuospatial information in this subsystem of working memory for further processing. The result that there seems to be a close relation between individual differences in visuospatial working memory and those in spatial abilities is in line with previous findings in the literature (e. g., MIYAKE ET AL. 2001). Some models of small-scale spatial abilities, for example,

acknowledge the importance of visuospatial working memory by including a factor representing working memory capacity (e. g., CARROLL 1993; EKSTROM 1976).

At the level of measures, ATFS and ATBS differentiated well between subjects of low and high capacity of visuospatial working memory. Measures of maximum forward span were in line with normative studies that proposed that the average forward span is about six blocks by the end of primary school (e. g., FARRELL PAGULAYAN ET AL. 2006). Simple measures of backward span showed ceiling effects, a result that could be explained by the fact that the maximum number of blocks included in the design was limited to six. In addition, a lot of children reported playing with a digital App that requires them to perform backward-span tasks. Consequently, simply measuring the maximum sequence did not help to differentiate between low and high-performers well enough. More complex measures, however, that involved the length of the sequence and the number of trials required were able to do so.

The extent to which both factors of small-scale spatial abilities are related to visuospatial working memory are similar to those found in studies with adults (MIYAKE ET AL. 2001). They are similar both at the level of zero order correlations (.19 to .33 in this study vs. .21 to .39 in their study), for individual tasks such as the PFT (.31 compared to .25), and at the latent level (.71 and .71 vs. .54 to .69). Deviations might be due to the fact that this study measured the construct of VSWM by using one test (CBT with two different span measures) instead of including four different ones as MIYAKE ET AL. (2001) did. Both of these studies relied on another set of small-scale spatial abilities tests.

More specifically, this study found significant correlations between working memory capacity and tasks requiring object-based transformation, which is similar to previous studies with adults (KAUFMAN 2007). Although the extent of the relations that were identified (.22 to .31) was smaller than in adults (.47 to .51), which might probably be due to the quality of the measures used, one might suggest that capacity of maintenance of spatial information in working memory and performances in solving spatial tasks requiring object-based transformations are also related for children. Indeed, all tasks that reflected this subclass of small-scale spatial abilities required the children to maintain a mental representation of all the parts of an object, and the interrelation to other parts or other objects in working memory while simultaneously transforming these by rotation or folding (KAUFMAN 2007).

The finding that visuospatial working memory capacity is correlated with the abilities to transform the imagined self as reflected by a set of perspective taking tasks requiring the children to control conflicting frames of reference is in line with previous correlational studies with children (VANDER HEYDEN ET AL. 2017).

Although the extend of interaction cannot be compared since VANDER HEYDEN ET AL. (2017) did not provide correlations but computed the relative predictive importance of visuospatial working memory measures while controlling for age and intelligence, one might agree their results are consistent with the findings of this study. These empirical findings might be explained by the cognitive mechanisms that have been suggested to underlie the solution of perspective taking tasks. To complete these, children have to inhibit what they perceived from their own viewpoint, but the same time they have to make this view mentally available to perform the egocentric transformation using higher-order control and computation processes in working memory (EPLEY, MOREWEDGE, & KEYSAR 2004; YU & ZACKS 2017).

Children's Large-Scale Spatial Abilities

11

This chapter presents the empirical results concerning children's large-scale spatial abilities as measured by the map-based orientation test proposed in Chapter 7. It presents the results concerning the psychometrical quality of individual measures measures (Section 11.1), about empirically separable latent abilities (Section 11.2), and the role of visuospatial working memory (Section 11.4). The final Section 11.4 discusses the findings with respect to the literature presented in Part I.

11.1 Analyses of the Items and Measures

This section presents the results of analyses of all measures included in the map-based orientation test. All analyses were computed using a free trial version of *ConQuest 4* (M. L. WU ET AL. 2015) and *IBM SPSS 25* (CORP 2017). First, detailed analyses are presented for all measures, involving an analysis or the CTT and IRT item statistics. Summaries of the analyses can be found in Section 11.1.2 (p. 303) and Section 11.1.3 (p. 304). The descriptive analyses were based on the data without missings, thus applying pairwise deletion mechanism, and they were interpreted using the benchmarks provided in Section 5.2.1.

Electronic supplementary material The online version of this chapter
(https://doi.org/10.1007/978-3-658-32648-7_11) contains supplementary material, which is available to authorized users.

11.1.1 Detailed Analysis of Individual Measures

MapRot

An analysis of the relative frequencies of the sum scores of MapRot (Figure 11.1) revealed that

(1) almost 60% of the students achieved at maximum one point,
(2) only one student achieved the maximum score.

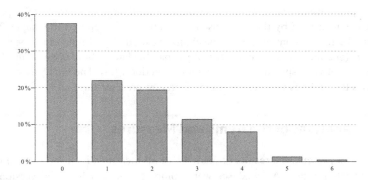

Figure 11.1 Sum scores for MapRot (n = 237)

Both results showed that the measure MapRot was subject to floor effects. The measure could therefore not differentiate well enough between low and high achievers in a map rotation task. The items seemed to be too difficult for the children in the sample which might not only have been due to the lack of cognitive abilities but also due to problems in understanding the task. A supplementary analysis of the descriptive values of the measure (Table 11.1), in particular the high skewness, supported these observations.

Table 11.1 Descriptive Statistics for MapRot (n = 237)

Measure	Range	Median	Mean	SD	Skewness	Kurtosis
MapRot	[0,6]	1	1.36	1.39	.80	−0.22

Summary statistics for item difficulties and other item statistics were computed (Table 11.2). All item facilities were below .30, a result that showed that the items were difficult because less than 30% of the children were able to solve them. The item MapW3 had a facility value of .10. In addition, all items had low discrimination values. The one of item MapW3 was particularly low (.20), a result that could be explained by the low facility value of the item. The values of the weighted MNSQ indicated the individual items in the measure to satisfy the criterion of Rasch scalability. Results of the IRT analysis provided further evidence that the items of the measure were very difficult. Indeed, an analysis of the corresponding person-item-map (Figure 11.2) revealed that items 1 to 5 were equally difficult and item six was even more difficult. With respect to the person scorings, the item-person map showed that the first five items were appropriate to differentiate among very high achievers only but not among low achievers.

For subsequent analyses, the item MapW3 was withdrawn from the measure due to its bad psychometric properties. Although the items MapS1 and MapW2 had also problematic discrimination values, they were kept within the measure to provide a wider spectrum of cognitive tasks. For the remaining five items, item statistics were recomputed (see Table 11.18 in the summary, p. 303).

Table 11.2 Summary of Items Statistics for MapRot (n = 237)

Item no.	Item label	Facility (CTT)	Discrimination (CTT)	Weighted MNSQ + CI (IRT)	Difficulty (IRT)
1	MapS1	.28	.24*	1.06 (0.85, 1.15)	1.20
2	MapS2	.28	.38*	0.94 (0.85, 1.15)	1.23
3	MapS3	.26	.40*	0.93 (0.84, 1.16)	1.36
4	MapW1	.23	.31*	1.01 (0.83, 1.17)	1.56
5	MapW2	.21	.27*	1.00 (0.82, 1.18)	1.74
6	MapW3	.10	.20*	1.02 (0.69, 1.31)	2.70

*denotes discrimination values < 0.3

Figure 11.2 Person-item
map for MapRot. Each 'X'
represents 2.2 cases

Homepointing

Summary statistics for the measure Homepointing (HP) were computed. An analysis of the relative frequencies of the sum scores (Figure 11.3) revealed that

(1) about 20% of the children did not achieve a single point,
(2) about 50% of the children scored with a single point,
(3) only 6% of the children achieved the full score.

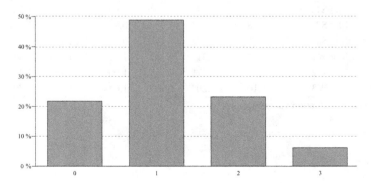

Figure 11.3 Sum scores for HP (n=211)

These results indicated that not all of the children have understand the task or it was too challenging, thus leading to a considerable amount of zero-scores. One item seemed to be less difficult, thus allowing half of the children to score with at least one single point. However, the small percentage of children scoring with three points showed that the items of the measure seemed to be very difficult. These observations were confirmed in the corresponding descriptive statistics for the measure (Table 11.3). In particular the positive skew value (.41) pointed toward a right-skewed distribution which was centered around the score of 1.

Table 11.3 Descriptive Statistics for HP (n = 211)

Measure	Range	Median	Mean	SD	Skewness	Kurtosis
HP	[0,3]	1	1.14	0.83	.41	−.29

To investigate the measure at the level of single items, a summary of item statistics was computed (Table 11.4). As stipulated before, one item (Start1) emerged to be easier (item facility of .67) than the two others (.25 and .22, respectively) which could be explained by the fact that for the first item, the children had been walking only a short distance, whereas children had to walk a longer route and performed different tasks meanwhile for completing the two others. All three items had considerable low discrimination values, a result that suggested that all three items were not useful to distinguish between low and high achievers. The values of the weighted MNSQ indicated that the individual items in the measure to satisfy the criterion of Rasch scalability. These observations were confirmed in the corresponding person-item-map (Figure 11.4).

Table 11.4 Summary of Items Statistics for HP (n = 211)

Item no.	Item label	Facility (CTT)	Discrimination (CTT)	Weighted MNSQ + CI (IRT)	Difficulty (IRT)
1	Start1	.67	.08*	1.02 (0.88, 1.12)	−0.81
2	Start2	.25	.11*	1.01 (0.83, 1.17)	1.28
3	Start3	.22	.18*	1.00 (0.82, 1.18)	1.40

*denotes discrimination values < 0.3, †denotes conspicuous MNSQ-values with a T-Value $T \geq 2$

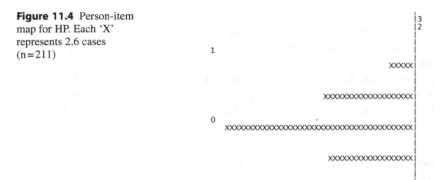

Figure 11.4 Person-item map for HP. Each 'X' represents 2.6 cases (n = 211)

For the subsequent analyses, the measure HP was withdrawn due to its problematic psychometric properties. In this initial conception, the task did not measure performances of the children in a differentiated way.

Moreover, most of the experimenters reported that they forgot to cover the map while asking for the starting point. In this case, children could use the map to indicate the direction. This was problematic; the measure could no longer be treated as a reflecting SurveyMap since it turned out to become a comprehension task. It was therefore similar to ActiveMapUse, but asked for the starting point as a particular landmark.

ActiveMapUse

An analysis of the relative frequencies of the sum scores of ActiveMapUSe (Figure 11.5) revealed that

(1)   over 25% of the children did not achieve a single point,
(2)   about 40% of the children achieved one point,
(3)   under 10% of the children achieved the full score.

These results indicated that the task was either difficult to understand for a large number of children or over-challenged them which yield scores of zero points. Whenever children achieved points, they mostly scored with one point, a result that indicated that one item was rather easy for them. The third observation suggested that one item was difficult to master. Descriptive statistics were computed to underpin these observations (Table 11.5).

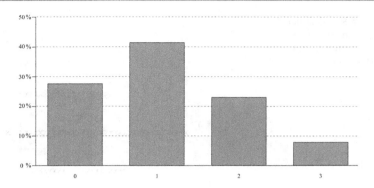

Figure 11.5 Sum scores for ActiveMapUse (n = 217)

Table 11.5 Descriptive Statistics for ActiveMapUse (n = 211)

Measure	Range	Median	Mean	SD	Skewness	Kurtosis
ActiveMapUse	[0,3]	1	1.11	0.90	.43	−.52

Item statistics (Table 11.6) showed that all items had low to medium item facilities (.22 to .57), a result that demonstrated that all items were of medium to high difficulty. The item MapUseRed was particularly difficult (.22). All three items had low discrimination values (.15 to .21), a result that indicated that the items were not able to distinguish well between low and high achievers. The values of the weighted MNSQ indicated the individual items in the measure to satisfy the criterion of Rasch scalability. The corresponding item difficulties showed that the item MapUseRed was a very difficult one that was barely able to differentiate among students of different performances (see also Figure 11.6).

Table 11.6 Summary of Items Statistics for ActiveMapUse (n = 217)

Item no.	Item label	Facility (CTT)	Discrimination (CTT)	Weighted MNSQ + CI (IRT)	Difficulty (IRT)
1	MapUseYellow	.57	.15*	1.01 (0.89, 1.11)	−0.25
2	MapUseBlue	.36	.21*	1.00 (0.88, 1.12)	0.79
3	MapUseRed	.22	.16*	0.99 (0.82, 1.18)	1.46

*denotes discrimination values < 0.3

Figure 11.6 Person-item map for ActiveMapUse. Each 'X' represents 2.3 cases (n = 217)

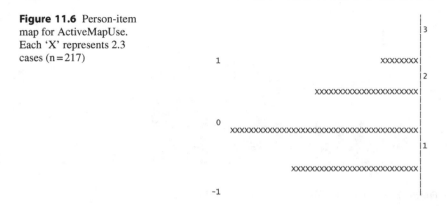

For subsequent analyses, the measure ActiveMapUse (all three items) was withdrawn due to its problematic psychometric properties. In its initial conception the task did not measure performances of the children in a differentiated manner.

Towards an Integrated Measure of Landmark-Oriented Map Use
Withdrawing both HP and ActiveMapUse would have resulted in a loss of information concerning the measurement of map-based pointing tasks as one type of comprehension tasks. Consequently, a measure integrating all three items from HP and ActiveMapUse was designed, denoted as ActiveHP. This could be done from a content-related perspective since d the task HP was often forgotten to be conducted without a map uring data collection and therefore represented a pointing task to the unseen landmark 'start'.

An analysis of the relative frequencies of the sum scores (Figure 11.7) revealed that

(1)   under 10% of the children did not achieve a single point,
(2)   the distribution was approximately a right-skewed normal distribution,
(3)   under 10% of the children achieved five points or the full score.

These results showed that parceling the two previous measures into one single measure represented children's performances in a much more differentiated manner. In the new measure, a smaller number of children scored with zero points, a large number of children scored with one to four points, and few of them scored high or with six points. These observations were confirmed by the descriptive statistics for the new measure (Table 11.7).

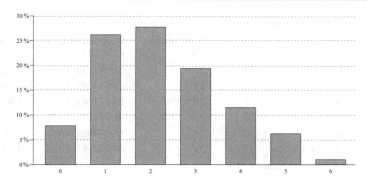

Figure 11.7 Sum scores for ActiveHP (n = 191)

Table 11.7 Descriptive Statistics for ActiveHP (n = 211)

Measure	Range	Median	Mean	SD	Skewness	Kurtosis
ActiveHP	[0,6]	2	2.24	1.38	.47	−.36

Item statistics were recomputed (Table 11.8) to analyze the item properties for each of the three items. The item facilities did not change because they were independent of their composition into a new measure. Item discrimination values remained ≤ 0.3 but they were higher as compared to the cases of the individual measures HP and ActiveMapUse. The values of the weighted MNSQ indicated the individual items in the measure to satisfy the criterion of Rasch scalability. The corresponding item difficulties showed that the measure ActiveHP had three very difficult items that were suitable to distinguish among high-achievers but involved only one item with medium difficulty to differentiate among students scoring moderately (see also Figure 11.8).

An investigation of the composite measure revealed that all items still suffered from low discrimination values. With its six items, however, the measure was able to differentiate in a sufficient way between students of different performances. For this reason, all items were kept and integrated into the measure ActiveHP for further analyses. Besides the effect of improving some of the psychometrical properties of the items, parceling two measures into one achieved a decrease in the number of indicators for further analyses, which was particularly desirable for the planned multivariate analyses.

Table 11.8 Summary of Items Statistics for ActiveHP (n = 191)

Item no.	Item label	Facility (CTT)	Discrimination (CTT)	Weighted MNSQ + CI (IRT)	Difficulty (IRT)
1	Start1	.57	.21*	0.99 (0.90, 1.10)	−0.32
2	Start2	.36	.15*	0.98 (0.89, 1.11)	0.68
3	Start3	.22	.20*	1.00 (0.83, 1.17)	1.44
4	MapMensa	.57	.25*	1.01 (0.87, 1.13)	−0.82
5	MapBib	.36	.24*	1.01 (0.84, 1.16)	1.29
6	MapGarden	.22	.21*	0.99 (0.82, 1.18)	1.42

*denotes discrimination values < 0.3

Figure 11.8 Person-item map for ActiveHP. Each 'X' represents 1.6 cases (n = 191)

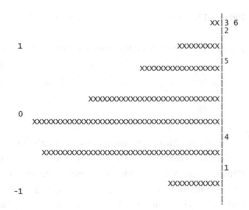

Dots

An analysis of the relative frequencies of the sum scores of Dots (Figure 11.9) revealed that

(1) almost 40% of the children did not achieve a single point,
(2) the majority of the children achieved one to three points,
(3) about 10% of the children achieved four to five points.

These results indicated that all items concerning 'finding the locations after walking a pre-defined way in real space' were difficult for the children in the sample. Although

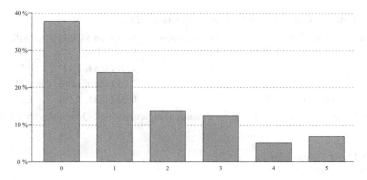

Figure 11.9 Sum scores for Dots (n = 233)

the polytomous scoring also honored less precise solutions with one point, almost two fifth of the children were not able to manage the task Dots. Consequently, the measure was subject to floor-effects. An analysis of the corresponding descriptive statistics, in particular the high positive skew (.91), confirmed these observations (Table 11.9).

Table 11.9 Descriptive Statistics for Dots (n = 233)

Measure	Range	Median	Mean	SD	Skewness	Kurtosis
Dots	[0,5]	1	1.44	1.55	.91	−.22

To analyze the psychometric properties of each of the three items, CTT and IRT item statistics were computed (Table 11.10). The item facility values were found to be low (.25 to .39), a result that showed that all three items were above medium difficulty. The item discrimination values were also low, yet acceptable (≥ 0.3). The values of the weighted MNSQ indicated the individual items in the measure to satisfy the criterion of Rasch scalability. The corresponding analysis of item difficulties revealed that all three items were indeed suitable to differentiate among medium achievers but not among low or high achievers (Figure 11.10). For subsequent analyses, all items were kept due to their acceptable psychometric properties.

Table 11.10 Summary of Items Statistics for Dots (n = 233)

Item no.	Item label	Facility (CTT)	Discrimination (CTT)	Weighted MNSQ + CI (IRT)	Difficulty (IRT)
1	DotYellow	.39	.38	1.01 (0.87, 1.13)	0.59
2	DotBlue	.25	.41	1.05 (0.80, 1.20)	1.21
3	DotRed	.27	.49	0.99 (0.81, 1.19)	1.12

*denotes discrimination values < 0.3, †denotes conspicuous MNSQ-values with a T-Value $T \geq 2$

Figure 11.10 Person-item map for Dots. Each 'X' represents 8.8 cases (n = 233)

Dir

An analysis of the relative frequencies of the sum scores of Dir (Figure 11.11) revealed that

(1) about 12% of the children did not achieve a single point,
(2) the majority of the children scored with one point,
(3) almost 30% of the children scored full.

These results indicated that this measure was understandable for most of the children and featured at least one simple item that could be completed by many of them. The third observation suggested that all three items were not too difficult for the children in the sample since more than a quarter of them were able to score with three points. An analysis of the corresponding descriptive statistics of the measure (Table 11.11) confirmed these observations. In the average, students scored above the expectation value (1.67) and the distribution was widely spread which was indicated by a high kurtosis value (−1.14).

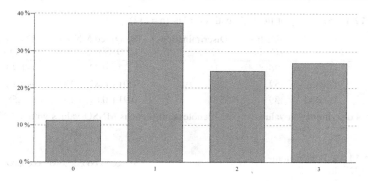

Figure 11.11 Sum scores for Dir (n = 232)

Table 11.11 Descriptive Statistics for Dir (n = 232)

Measure	Range	Median	Mean	SD	Skewness	Kurtosis
Dir	[0,3]	2	1.67	0.95	.01	−1.14

Item statistics (Table 11.12) revealed that the item facilities ranged from low to high (.39 to .76), thus indicating that there were easy (DirYellow) and more difficult items (DirRed). Concerning the item discrimination values, the ones of DirYellow and DirBlue were critical ($\leq$ .03). This could, in the case of DirYellow, be explained by the high corresponding facility value. Results of the Rasch analysis indicated that according to the fit indices, all three items did not deviate significantly from the assumptions of the Rasch model and could therefore be assumed to be Rasch scalable. The corresponding item difficulties showed that the measure Dir consisted of three items of different difficulty that could differentiate among low, medium and high performers. This observation was confirmed in the corresponding person-item map (Figure 11.12).

For the subsequent analyses, the first item DirYellow should have ideally be withdrawn from the measure due to its problematic psychometric properties. This would, however, have resulted in a measure that consisted only of two items, thus being highly categorical which was unwanted for reasons of parameter estimation during subsequent latent analyses. Consequently, the item was kept in the measure.

Table 11.12 Summary of Items Statistics for Dir (n = 232)

Item no.	Item label	Facility (CTT)	Discrimination (CTT)	Weighted MNSQ + CI (IRT)	Difficulty (IRT)
1	DirYellow	.76	.21*	1.07 (0.82, 1.18)	−1.49
2	DirBlue	.52	.28*	1.01 (0.87, 1.13)	−0.12
3	DirRed	.39	.39*	0.91 (0.87, 1.13)	0.58

*denotes discrimination values < 0.3, †denotes conspicuous MNSQ-values with a T-Value $T \geq 2$

Figure 11.12 Person-item map for Dir. Each 'X' represents 5.6 cases (n = 232)

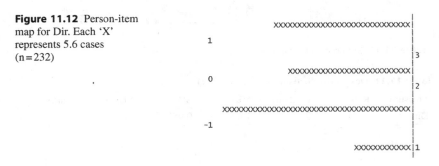

Disks

An analysis of the relative frequencies of the sum scores of Disks (Figure 11.13) revealed that

(1) more than 10% of the children did not score a single point,
(2) the distribution of sum scores was approximately equally distributed for scores up to four points,
(3) almost 30% of the children scored with the maximum number of points.

These results showed that the measure was understandable for a large number of children. Indeed, the polytomous scoring, which honored also less precise solutions with one point, allowed for large differentiation among individual performances. Altogether, the measures consisted of three items that were sufficiently simple to be mastered by a large number of children, thus leading to a slight floor effect. These observations were confirmed in the descriptive values of the measure (Table 11.13).

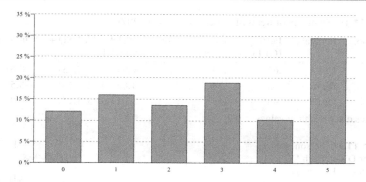

Figure 11.13 Sum scores for Disks (n = 207)

Table 11.13 Descriptive Statistics for Disks (n = 207)

Measure	Range	Median	Mean	SD	Skewness	Kurtosis
Disks	[0,5]	3	2.87	1.77	−.19	−1.32

CTT and IRT psychometric item characteristics were computed (Table 11.14) to analyze the measure at the level of items. Two items showed a higher item facility value (.62 and .63), a result that indicated that they were slightly easier to complete than the third one (.48) with medium difficulty. All three discrimination values were above .30. The item discrimination values, however, were still low (.33 to .41), thus showing that they were not good enough to differentiate between high and low achievers. The values of the weighted MNSQ indicated the individual items in the measure to satisfy the criterion of Rasch scalability.

The corresponding item difficulties showed that one item featured medium difficulty and two slightly below medium difficulty. This result indicated that the three items were thus able to differentiate among medium achievers. However, there was no item to differentiate among low and no item to differentiate among high achievers. These observations were confirmed in the person-item map (see Figure 11.14). For subsequent analyses, the measure was kept unmodified due to the acceptable psychometric properties.

Table 11.14 Summary of Items Statistics for Disks (n = 207)

Item no.	Item label	Facility (CTT)	Discrimination (CTT)	Weighted MNSQ + CI (IRT)	Difficulty (IRT)
1	Disk1	.62	.33	0.99 (0.87, 1.13)	−0.61
2	Disk2	.63	.38	1.02 (0.82, 1.18)	−0.50
3	Disk3	.48	.41	1.01 (0.83, 1.17)	0.03

*denotes discrimination values < 0.3

Figure 11.14 Person-item map for Disks. Each 'X' represents 6.1 cases (n = 207)

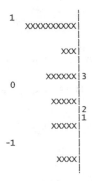

MentalMap

There were nine different items in MentalMap, three of them measuring memorizing landmarks that were not directly experienced but subject to other tasks (Restaurant, Library, Garden), three of them measuring memorizing landmarks that were passively experienced (yellow, blue, and red flag) and three of them measuring memorizing landmarks that were actively experienced (Disk1, Disk2, and Disk3).

Summary statistics for the measure including all nine were computed first. An analysis of the relative frequencies of the sum scores (Figure 11.15) revealed that

(1) no child achieved 7, 8, or 9 points,
(2) about half of the children scored with maximal two points,
(3) the distribution of sum scores was a right-skewed normal distribution.

These results suggested that most of the items must have been very difficult for the children, yielding low scores and the fact that some possible scores were not achieved at all. To further analyze these observations, item statistics were computed for each item in each of the three groups (items reflecting the abilities to cognitively map locations on the map, items reflecting the abilities to cognitively map flag

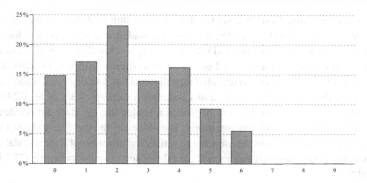

Figure 11.15 Sum scores for MentalMap (n = 214)

locations, and items reflecting the abilities to cognitively map disk locations). Table 11.15 shows the corresponding item statistics.

Eight of the nine items featured low item facility values and were rather difficult to solve for the children in the sample. Two of these items had a facility value ≤ 0.2 (MMap1 and MMap3). Only one item (MYellow) featured medium difficulty.

An item discrimination analysis for the first group of items showed that all three items had problematic discrimination values ≤ 0.3. Those items were therefore not able to differentiate well enough among high- and low achievers in this task.

Table 11.15 Summary of Items Statistics for MMap, MFlag, MDisk (n = 223/n = 224/n = 214)

Item no.	Item label	Facility (CTT)	Discrimination (CTT)	Weighted MNSQ + CI (IRT)	Difficulty (IRT)
1	MMap1	.19	.11*	1.00 (0.80, 1.20)	1.71
2	MMap2	.31	.14*	1.02 (0.87, 1.13)	0.95
3	MMap3	.14	.25*	0.99 (0.75, 1.25)	2.08
4	MYellow	.67	.29*	1.07 (0.83, 1.17)	−1.01
5	MBlue	.35	.49*	0.91 (0.85, 1.15)	0.55
6	MRed	.35	.34*	1.06 (0.84, 1.16)	0.89
7	MDisk1	.38	.28*	1.03 (0.87, 1.13)	0.57
8	MDisk2	.35	.32*	0.98 (0.86, 1.14)	0.77
9	MDisk3	.32	.24*	1.02 (0.86, 1.14)	0.93

*denotes discrimination values < 0.3

An item discrimination analysis for the second group of items (items reflecting the abilities to cognitively map flag locations) revealed that all items had low discrimination values with one item (MYellow) having a value slightly ≤ 0.3. An item discrimination analysis for the third group of items (items reflecting the abilities to cognitively map disk locations) revealed that all items had low discrimination values with two of the three items having a discrimination value ≤ 0.3 (.28 and .24). An IRT analysis for each of the three item groups revealed that the data did not deviate significantly from the model assumptions. The values of the weighted MNSQ indicated the individual items in the measure to satisfy the criterion of Rasch scalability. The corresponding item difficulties reflected the observations concerning CTT item difficulties.

For further analyses, the first group of items was withdrawn due to their extremely bad psychometric properties. Two different measures from the second and the third group of items (MFlag and MDisk) were computed. This split up was necessary for reasons of the planned subsequent CFA. The CFA analyses should include measures reflecting the factor SurveyMap, for reasons of model identification, at least two measures had to be included. Since the measure HP did not reflect these abilities due to the fact that the map was not covered by the experimenters during pointing to the start, two measures were needed to ensure model identification.

To avoid having not enough categories in each of the two measures, all items were kept, although some of these featured problematic psychometric properties. Table 11.16 and Table 11.17 show the corresponding descriptive statistics for the two measures. As suggested by the corresponding item analysis, children achieved better results in the measure MFlag than in the measure MDisk. Both measures are positively skewed.

Table 11.16 Descriptive Statistics for the Measure MFlag (n = 224)

Measure	Range	Median	Mean	SD	Skewness	Kurtosis
MFlag	[0,3]	1	1.42	1.05	.16	−1.17

Table 11.17 Descriptive Statistics for the Measure MDisk (n = 214)

Measure	Range	Median	Mean	SD	Skewness	Kurtosis
MDisk	[0,3]	1	1.07	1.00	.58	−0.73

11.1.2 Summary: Psychometric Properties of All Items

After analyzing all measures in detail, item statistics were recomputed for all those items that were included in the seven measures of large-scale spatial abilities (Table 11.18). The results from the detailed analyses in Section 11.1.1 may be briefly summarized as follows:

Table 11.18 Summary of Items Statistics for the Map-Based Orientation Test

Measure	Item label	Facility (CTT)	Discrimination (CTT)	Weighted MNSQ + CI (IRT)	Difficulty (IRT)
MapRot	MapS1	.28	.24*	1.04 (0.85, 1.15)	1.21
	MapS2	.28	.37*	0.95 (0.85, 1.15)	1.23
	MapS3	.26	.40*	0.96 (0.84, 1.16)	1.37
	MapW1	.23	.30*	0.99 (0.83, 1.17)	1.57
	MapW2	.21	.25*	1.06 (0.81, 1.19)	1.75
ActiveHP	Start1	.57	.21*	0.99 (0.90, 1.10)	−0.32
	Start2	.36	.15*	0.98 (0.89, 1.11)	0.68
	Start3	.22	.20*	1.00 (0.83, 1.17)	1.44
	MapUseYellow	.57	.25*	1.01 (0.87, 1.13)	−0.82
	MapUseBlue	.36	.24*	1.01 (0.84, 1.16)	1.29
	MapUseRed	.22	.21*	0.99 (0.82, 1.18)	1.42
Dots	DotYellow	.39	.38*	1.01 (0.87, 1.13)	0.59
	DotBlue	.25	.41*	1.05 (0.80, 1.20)	1.21
	DotRed	.27	.49*	0.99 (0.81, 1.19)	1.12
Dir	DirYellow	.76	.21*	1.07 (0.82, 1.18)	−1.49
	DirBlue	.52	.28*	1.00 (0.87, 1.13)	−0.11
	DirRed	.39	.39*	0.91 (0.87, 1.13)	0.58
Discs	Disk1	.62	.33*	0.99 (0.87, 1.13)	−0.61
	Disk2	.63	.38*	1.02 (0.82, 1.18)	−0.50
	Disk3	.48	.41*	1.01 (0.83, 1.17)	0.03
MFlag	MYellow	.67	.29*	1.07 (0.83, 1.17)	−1.01
	MBlue	.35	.49*	0.91 (0.85, 1.15)	0.55
	MRed	.35	.34*	1.06 (0.84, 1.16)	0.89
MDisk	MDisk1	.38	.28*	1.03 (0.87, 1.13)	0.57
	MDisk2	.35	.32*	0.98 (0.86, 1.14)	0.77
	MDisk3	.32	.24*	1.02 (0.86, 1.14)	0.93

*denotes discrimination values < 0.3

(1) Items featured facility values from .21 to .76 with half of the items being rather difficult, featuring a facility ≤ 0.4.

(2) The corresponding item discrimination values were low, in particular in the measure ActiveHP and MDisk.

(3) Within each measure, the values of the weighted MNSQ indicated the individual items in the measure to satisfy the criterion of Rasch scalability.

11.1.3 Summary: Descriptive Statistics of the Measures

Since it was intended to use multivariate methods in subsequent analysis that assume normally distributed data without too strong floor or ceiling effects, descriptive statistics of the final measures were recomputed (Table 11.19). In total, the test consisted of four measures with six or more categories, which were considered to be continuous variables in the subsequent analysis. None of the measures in the test departed, from a statistical analysis point of view, considerably from normality (all skews $\leq |2|$ and all kurtosis values $\leq |2|$, but see Section 5.2.2). Three of the measures involved, however, only four categories which made the conclusion of denoting measures as involving 'normal distributed data' rather abstract. Instead, those results indicated that estimators that do not require a certain distribution should also be considered during subsequent analyses.

Table 11.19 Descriptive Statistics of All Measures in the Map-Based Orientation Test

Measure	n	Range	Mean	SD	Skew	Kurtosis
MapRot	237	[0,5]	1.26	1.30	.76	−0.43
ActiveHP	191	[0,6]	2.24	1.38	.47	−0.36
Dots	233	[0,5]	1.44	1.55	.91	−0.22
Dir	232	[0,3]	1.67	0.99	.01	−1.14
Discs	207	[0,5]	2.87	1.77	−.19	−1.32
MFlag	224	[0,3]	1.42	1.05	.16	−1.17
MDisk	215	[0,3]	1.07	1.00	.58	−0.73

These results can be summarized as follows:

(1) All measures had a mean value that was below the theoretical expectation value, a result that showed that the measures were difficult to solve for the children in the sample.

(2)   Except for Disks, all measures were positively skewed, a result that further demonstrated that the measures tended to be difficult for the children.

(3)   Mean value of Disks was higher than the mean value of Dots.

(4)   Mean value of MFlag was higher than the mean value of MDisk.

To understand whether these results were sensitive to sex differences, a series of post-hoc analyses were computed both for comparing descriptive statistics and distributions of sum scores. Significant sex differences in mean values were found for all measures except for MapRot and Dir (Table F.3 in the ESM, p. 94). χ^2-tests concerning differences in the distribution of sum scores revealed that significant different distributions of sum scores were found for boys and girls (ESM F.1, pp. 87) for the measures ActiveHP, Dir, MFlag, and MDisk. Since this study did not intend to examine sex differences in greater detail, these results were not further considered in the subsequent analyses.

Because the test instrument was based on an experimental setting in real space involving multiple experimenters for data collection, the individual measures were controlled for the influence of experimenters, weather conditions, and the crowdedness of the campus by computing three MANOVAs for each of the experimental conditions, testing the hypothesis whether there would be one or more mean differences between each of the groups:[1]

–   A non-significant first MANOVA for the three different groups of experimenters (PhD student, experienced experimenters, casual experimenters), Pillais' Trace = .10, $F(12, 294) = 1.33$, $p = .198$, indicated that there was no significant influence of the experimenter group on performances in the map-based orientation tasks.

–   There was no significant influence of the weather (sunny/cloudy/rainy) on the dependent variables (Pillais' Trace = .06, $F(12, 290) = 0.78$, $p = .665$).

---

[1] Computing three subsequent analyses on the same sample would normally demand for a alpha-error-correction. Since it was intended to report non-significant results, no correction of the alpha-level was done, that is, all results that were non-significant at a 0.05-level were also non-significant at a 0.05/6-level.

Prior to all MANOVAs, Box's M was computed and tested. The results are not reported here. Non-significance was interpreted to indicate that the covariance matrices between the groups were assumed to be equal for the subsequent analyses.

- There was no significant influence of the campus status (crowded/moderate/empty) on the dependent variables (Pillais' Trace = .07, $F(12, 286) = 0.91$, $p = .534$).

Since children were assumed to have different activity levels and prior experiences with map use at home and at school, robustness of the measures against those variables was tested by applying another three MANOVAs:

- Prior knowledge on maps learned at school had no significant influence on performances in the map-based orientation tasks (Pillais' Trace = .20, $F(35, 700) = 0.84$, $p = .728$).
- Prior knowledge on maps learned at home had no significant influence on the dependent variables (Pillais' Trace = .27, $F(35, 705) = 0.16$, $p = .241$).
- An analysis of how 162 children got to school and how these means of transport related to their performances revealed that 12 of the 162 children were brought to school by car, 11 by bus, 41 on foot, 82 by bike and 6 on a scooter. There was no significant influence of the means of transport to school on the performances in the map-based tasks (Pillais' Trace = .21, $F(35, 730) = 0.92$, $p = .605$).

To sum up, results from the analyses on effects on experimental conditions and child-dependent variables indicated that the measures were robust against these factors. The variances of the dependent variables that were accounted for by the control variables were low (3.2% to 5.2%). Consequently, those control variables were not included in subsequent analyses at the latent level.

11.1.4 Reliabilities

Measures included in the map-based orientation test showed questionable internal consistency (Table 11.20), which was reflected by values for Cronbach's α that ranged from .43 to .60.

Table 11.20 Cronbach's α for the Measures in the Map-Based Orientation Test (Based on Pairwise Deletion of Missing Values)

MapRot	ActiveHP	Dots	Dir	Discs	MFlag	MDisk
.55	.43	.60	.47	.54	.56	.46

Conclusion
Descriptive analyses at the level of items and sum scores for all measures used in the large-scale spatial abilities test showed acceptable characteristics from a psychometrical point of view. The items were found to be rather difficult, and a some of them showed poor discrimination values. As a result, some of the measures did not distinguish between children of high and low abilities at the level of items. The items in the measures satisfied the criterion of Rasch scalability. Results of the analyses also indicated that the outdoor measures were difficult to solve for the children. Finally, reliability was low for all of the measures. In particular, the measures ActiveHP, Dir, and MDisk had very low values (.43, .47, and .46).

11.2   On the Dissociation of Three Subclasses of Large-Scale Spatial Abilities

This section reports results concerning the analysis of correlational patterns in the measures of large-scale spatial abilities. Pairwise correlations specified the relations of the individual measures at the manifest level and latent analyses allowed for interpretation of correlational patterns in terms of empirically separable latent factors that represented subclasses of large-scale spatial abilities. In addition, the results of confirmatory factor analyses quantified the proportion of variance shared among the identified factors and specified the amount of variance explained in the individual measures by the assumed underlying latent factors.

11.2.1 Analysis of Results From Multiple Imputation Procedure

All analyses were based on Multiple Imputation (MI) data sets. An assumption for the multiple imputation procedure was that missing data was missing completely at random (MCAR, see Section 5.2.3). LITTLE's (1988) test resulted in a non-significant $\chi^2(1652) = 1652.09$, $p = 0.495$, a result that showed that the data were most likely to be missing at random. In particular, no systematic patterns in the missing data could be identified from a experimental procedure point of view. The range of missing data was from 1.25% to 10.8%, with a mean percentage of 5.26% (SD = 3.57%). Figure 11.16 shows the percentage of missing values per item.

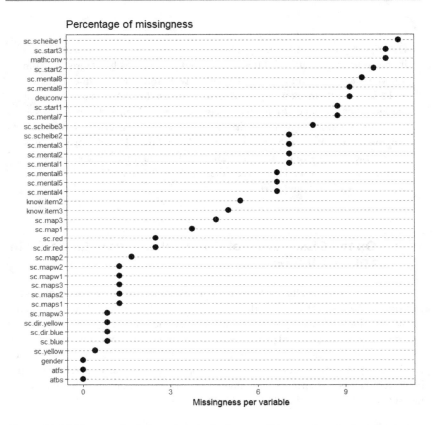

Figure 11.16 Patterns of missing values in the data set of the map-based orientation test

The data set used for this analysis was computed with 30,000 MCMC iterations, 30 imputed data sets, 5,000 MCMC burn-ins and 30 latent classes. To test the robustness of the chosen imputations, a range of imputations for different parameter sets (20 to 40 imputations, 2,500 to 10,000 burn-ins and 5 to 40 latent classes) were computed and compared concerning means and standard deviations of the parameters imputed. The results of this analysis demonstrated that the imputed values varied at maximum 0.01, indicating that the imputations were robust against different parameter sets.

Based on the 30 MI data sets, descriptive statistics were recomputed. Table 11.21 shows a comparison of descriptive statistics from the imputed data (n = 240) with, in parenthesis, descriptive statistics of the data set with complete cases only (n = 191 to 237). A comparison between both values suggested that occurring differences were only marginal (e. g., .00 to .05, for the mean values) with the biggest differences occurring in those measures that featured most of the missing data before imputation (.04 for ActiveHP and .05 for Discs, compare also to Figure 11.16). Consequently, it was concluded that the multiple imputation treatment of missing data had not changed the descriptive statistics of the measures.

Table 11.21 Descriptive Statistics of All Imputed Measures (in Parentheses Are the Descriptive Statistics for Complete Data Sets)

Measure	Range	Mean	SD	Skew	Kurtosis
MapRot	[0,5]	1.26 (1.26)	1.29 (1.30)	.75(.76)	−0.43 (−0.43)
ActiveHP	[0,6]	2.28 (2.24)	1.37 (1.38)	.45(.47)	−0.36 (−0.36)
Dots	[0,5]	1.44 (1.44)	1.55 (1.55)	.90 (.91)	−0.25 (−0.22)
Dir	[0,3]	1.67 (1.67)	0.99 (0.99)	.01 (.01)	−1.14 (−1.14)
Discs	[0,5]	2.82 (2.87)	1.76 (1.77)	−.15 (−.19)	−1.31 (−1.32)
MFlag	[0,3]	1.42 (1.42)	1.04 (1.05)	.16 (.16)	−1.13 (−1.17)
MDisk	[0,3]	1.06 (1.07)	0.99 (1.00)	.58 (.58)	−0.72 (−0.73)

11.2.2 Class Clustering Effects

Since the data was collected class-wise, the intraclass coefficient (ρ_{IC}) for every measure was computed to determine possible biases in subsequent analyses due to an underlying multilevel structure (B. O. MUTHÉN 2002, p. 90). Table 11.22 shows the results of the *Mplus* (L. K. MUTHÉN & MUTHÉN 2004) analysis.

Table 11.22 Intraclass Coefficients for the Measures in the Map-Based Orientation Test

MapRot	ActiveHP	Dots	Dir	Discs	MFlag	MDisk
.021	.020	.067	.050	.030	.007	.027

Six out of the seven measures had an $\rho_{IC} \leq .050$, indicating that variance in the measure was mainly due to individual differences but not due to a considerable effect of belonging to a certain class (GEISER 2012). Only the task Dots ($\rho_{IC} = .067$) showed a higher effect of class clustering which, however, was not straightforward to explain.

11.2.3 Analysis of Pearson Correlations

All results that are reported in this section are based on the analysis of the MI data set with n = 240 children or, in the reduced version, with n = 215 children (see Section 10.2.2). To approach the possible relations among large-scale spatial abilities measures on a manifest level, pairwise Pearson correlations were computed to quantify the amount of linear relationship under the assumption of continuous variables. Those are reported in Table 11.23. Pearson correlations both for the complete data set using a pairwise deletion process in SPSS (right above the diagonal) and the data set based on multiple imputation using a pooling function for the 30 single correlation matrices (left below the diagonal) are provided in the table.

All correlations were positive. Correlation coefficients from both procedures of computations (MI imputations/pairwise deletion) differed by no more than .30. According to J. COHEN (1988), the correlations had small (.15/.14) to large (.55/.55) effect sizes. The measures MFlag and MDisk correlated moderately with each other (.42/.44). They had small to moderate effect sizes when it came to correlations with the measures MapRot, ActiveHP, and Disks (.15/.14 to .34/.33 respectively) and small to moderate effect sizes when it came to correlations with the measures Dots and Disks (.21/.23 to .34/.36). The measures Dots and Dir correlated highly with each other (.55/.55). They correlated with small to moderate effects with the measures MapRot, ActiveHP, and Disks (.21/.21 to .48/.47). MapRot correlated to ActiveHP with small effect size (.22/.23) and to Disks with moderate effect size (.32/.33), respectively. ActiveHP and Disks were moderately correlated (.29/.33).

Table 11.23 Pearson Correlations between the Measures in the Map-Based Orientation Test (n = 240)

	1.	2.	3.	4.	5.	6.	7.
1. MapRot	–	.23	.21	.22	.33	.14	.27
2. ActiveHP	.22	–	.24	.33	.33	.19	.27
3. Dots	.21	.30	–	.55	.47	.31	.36
4. Dir	.21	.33	.55	–	.31	.29	.23
5. Discs	.32	.29	.48	.32	–	.33	.33
6. MFlag	.15	.23	.30	.28	.32	–	.44
7. MDisk	.25	.28	.34	.21	.34	.42	–

Pearson correlations were recomputed for the reduced sample with n = 215 children for the MI data (Table 11.24, left below the diagonal). Correlations deviated by .00 to .05 from the ones computed on the complete sample with n = 240 cases.

Table 11.24 Pearson (Left Below Diagonal) and Mixed (Right Above Diagonal) Correlations between the Measures in the Map-Based Orientation Test ($n = 215$)

	1.	2.	3.	4.	5.	6.	7.
1. MapRot	–	.23	.20	.21	.31	.16	.25
2. ActiveHP	.23	–	.25	.32	.26	.22	.29
3. Dots	.20	.25	–	.57	.44	.30	.36
4. Dir	.20	.30	.53	–	.28	.29	.23
5. Discs	.31	.26	.44	.27	–	.33	.36
6. MFlag	.15	.21	.29	.27	.33	–	.48
7. MDisk	.25	.28	.35	.20	.33	.42	–

Since some of the measures had only a few categories, assuming them to be continuous while computing Pearson correlations was not exact. To show that this assumption had, however, almost no implications on the correlations computed, the measures MFlag, MDisk, and Dir were treated as categorical, and correlations were re-computed. This resulted in a mixed correlation matrix with Pearson correlations between measures with continuous variables, and Polychoric correlations between all other measures (Table 11.24, right above the diagonal). Treating MFlag, MDisk, and Dir as categorical resulted in correlations that deviated by .00 to .06 to the ones computed assuming those variables to be continuous, with the highest deviation occurring between MFlag and MDisk (.06).

11.2.4 Confirmatory Factor Analyses

To investigate the patterns of correlations found, a series of CFAs were conducted using *R lavaan* (ROSSEEL 2012) with the package *semtools* (JORGENSEN ET AL. 2018). The package fitted a specified model on each of the imputed data sets and computed a pooled version of the model parameter estimates and fit indices afterwards. Since the preliminary analysis revealed no considerable class clustering effect, the influences of testing in classes were neglected by performing a single-level analysis. Moreover, the models were computed using estimators with robust standard errors (MLR, WLSMV), a method to control for possible small influences of class clustering effects.

The deviations in correlations occurring due to the assumption that MFlag, MDisk, and Dir were treated as continuous were neglected because they were sufficiently small. Treating those measures as ordered categorical variables in the subsequent confirmatory analyses would have been the ideal choice but would have made estimations of the corresponding thresholds of the variables necessary, which, in turn, would have increased the number of parameters to estimate. Since the sample size was limited, this was unwanted.

Since three of the seven measures had less than five categories, WLSMV estimation was applied in addition to MLR estimation (see Section 5.2.2). To test whether the computed results were robust with regard to the treatment of missing values, the MI data set based estimations were compared to results based on FIML methods (for MLR estimation only) and to those based on the complete data set with 142 cases. Throughout the analyses, the set with 215 cases was used for testing if a certain model of large-scale spatial abilities could explain the data. This was done to ensure that a possible model that explained the data could subsequently be analyzed also with regard to models describing small-scale spatial abilities. The latter ones were based on the reduced set with 215 cases. To justify the underlying factor structure of large-scale spatial abilities, however, the results were compared to estimations based on multiple imputation data sets with 240 cases.

To test whether the computed results were robust with regard to the estimators used, the models were computed using both MLR and WLSMV estimation. Although WSLMV seemed more appropriate, MLR estimations were not expected to deviate too much from the WLSMV estimations (see Section 5.2.2). In total, each of the hypothesized models were computed for the following seven cases:[2]

(1) complete data only (n = 142) + MLR estimation
(2) complete data only (n = 142) + WLSMV estimation
(3) FIML treatment of missing values (n = 215) + MLR estimation
(4) multiple imputation data sets (n = 215) + MLR estimation
(5) multiple imputation data sets (n = 215) + WLSMV estimation
(6) multiple imputation data sets (n = 240) + MLR estimation
(7) multiple imputation data sets (n = 240) + WLSMV estimation

Two-Factor Model A first theoretical model stipulated that the measures MFlag and MDisk were reflected a first subclass of large-scale spatial abilities, survey mapping (SurveyMap), and the measures MapRot, ActiveHP, Dots, Dir and Disks reflected a second subclass of large-scale spatial abilities, map-environment-self

---

[2]Due to the specific nature of the distribution of the measures and the fact that some of these have only a few categories, the presentation of the results will be based on case 5 if it explains the data well enough.

interactions (MapEnvSelf). Based on these assumptions, a preliminary two-factor model was computed for each of the seven cases outlined above. The corresponding fit indices are reported in Table 11.25.[3]

Table 11.25 Fit Indices for the Two-Factor Models of Large-Scale Spatial Abilities

Model / case	χ^2	df	χ^2/df	RMSEA [CI]	CFI	TLI	SRMR
(1) complete MLR	34.86	13	2.68†	.11 [.07, .15]	.88	.81	.060
(2) complete WLSMV	24.79	13	1.90†	.08 [.03, .13]	.94	.90	.057
(3) FIML	37.24	13	2.86†	.09 [.06, .13]	.90	.84	.054
(4) MI (MLR, 215)	29.21	13	2.24†	.08 [.04, .11]	.93	.88	–
(5) MI (WLSMV, 215)	24.47	13	1.88†	.06 [.02, .10]	.95	.92	–
(6) MI (MLR, 240)	26.63	13	2.04†	.07 [.03, .10]	.95	.92	–
(7) MI (WLSMV, 240)	22.94	13	1.76†	.06 [.01, .09]	.96	.94	–

†indicates that the corresponding χ^2-test for model fit was significant ($p < .05$.)
The RMSEA is shown with its 95% confidence interval [CI].
Pooled fit indices for MI data sets are Satorra-Bentler correction scaled ones.

The results indicated that a two-factor model did not explain the data in all of the cases. The corresponding χ^2-statistic was significant, a result that showed that the model deviated from the data. In addition, the corresponding fit indices did not meet the criteria for good fit (compare Table 5.1).

Three-Factor Model In line with the literature outlined in Part I, an alternative model was a three-factor model assuming that the factor MapEnvSelf could be further differentiated into one describing the abilities to solve production and another describing the abilities to solve comprehension tasks. The model stipulated that MFlag and MDisk reflected a first factor, Survey Map, that Dots and Dir reflected a second factor, Prod, and that MapRot, ActiveHP, and Disks reflected a third factor, Comp.

Table 11.26 shows the fit indices for this three-component model. Except for model (1) and (3), the fit indices of all models indicated that these three-factor models explained the data very well. For each of these models, the corresponding χ^2-statistics was non-significant, a result that indicated that the model did not deviate

---

[3]Pooled values for SRMR were not supported in *semtools*.

considerably from the data. Moreover, values for the RMSEA, the CFI,and TLI met
the criteria for good fit.

Table 11.26 Fit Indices for the Three-Factor Models of Large-Scale Spatial Abilities

Model	χ^2	df	χ^2/df	RMSEA [CI]	CFI	TLI	SRMR
(1) complete (MLR, n = 142)	21.05	11	1.91†	.08 [.02, .13]	.95	.90	.044
(2) complete (WLSMV, n = 142)	14.24	11	1.29	.05 [.00, .11]	.98	.97	–
(3) FIML (MLR, n = 215)	20.41	11	1.86†	.06 [.01, .11]	.96	.93	.035
(4) MI (MLR, n = 215)	14.71	11	1.34	.04 [.00, .09]	.99	1.00	–
(5) MI (WLSMV, n = 215)	11.30	11	1.03	.01 [.00, .07]	1.00	.99	–
(6) MI (MLR, n = 240)	13.86	11	1.26	.03 [.00, .08]	.99	.98	–
(7) MI (WLSMV, n = 240)	11.20	11	1.02	.01 [.00, .07]	1.00	1.00	–

†indicates that the corresponding χ^2-test for model fit was significant ($p < .05$.)
The RMSEA is shown with its 95% confidence interval (CI).
Pooled fit indices for MI data sets are Satorra-Bentler correction scaled ones.

Since pooling the results of CFAs on each of the MI data sets did not allow for
computing robust corrections to the fit indices but only scaled ones, it remained to
test whether those were so biased that model fit was suggested although this was not
the case. To avoid a misinterpretation of the data, robust fit indices were computed
for each of the 30 data sets and model fit was controlled for (see Table F.4 in the
ESM, p. 94 for robust fit indices of the 10 first data sets for case 5). By doing so, it
turned out that the scaled fit indices, in particular the CFI and TLI, were better than
the robust ones. The model fit for all models except for case (1) and (3), however,
could be confirmed. Moreover, controlling the residuals for the different data sets
showed that residuals were all low ($\leq |0.1|$) (R. B. KLINE 2015, p. 278).

Whereas the fit indices belonging to case (1) indicated that it did not explain
the data very well, this was not clear for case (3). For the latter mode, the exact-fit
hypothesis was rejected ($\chi^2(11) = 20.41, p = .040$). However, $\chi^2/df < 2$ was
found, and a rule of thumb indicated model fit (e. g., MOOSBRUGGER & KELAVA
2012, p. 337). In addition, the RMSEA was at the limit of good fit (.06), the CFI
was higher than .95, and the SRMR smaller than .04. However, the TLI was smaller
than the the suggested limit of .95. Since the criteria of good fit in this study were

quite strict in comparison to other studies (e. g., GRÜSSING 2012), model (3) was also considered to explain the data.

To sum up, those results suggested that the three-factor model explained the data very well. This result was robust against possible effects of the chosen estimator, the underlying sample size and even with regard to the fact that imputations were used. This was particularly demonstrated by the fact that the model explained the complete data and the FIML-based data set as well as all of the MI data sets.

Table 11.27 Estimated Latent Correlations for the Three-Factor Model of Large-Scale Spatial Abilities

Model / case	SurveyMap–Prod	SurveyMap–Comp	Prod–Comp
(2)	.63	.64	.64
(3)	.58	.69	.65
(4)	.59	.77	.72
(5)	.59	.77	.72
(6)	.58	.76	.78
(7)	.59	.76	.78

In a subsequent step, the standardized model solutions concerning the estimated latent correlations were compared for all cases that indicated good fit (Table 11.27). Estimations for the latent correlation between SurveyMap and Prod did not considerably vary among all cases (.58 to .63). When it came to estimations of latent correlations with the factor Comp, however, the estimations deviated for those cases that were based on the MI data sets. In other words, deviations were found both for the latent correlation between SurveyMap and Comp (.64/.69 versus .76 to .77) and latent correlation between Prod and Comp (.64/.65 versus .72/.78). This result was not surprising, given the fact, that most of the missing values occurred in comprehension tasks. Since subsequent analyses relied on models that used MI data sets, a higher latent correlation with Comp would result in a higher percentage of shared variance with this factor. This could be controlled for in some of the analyses.

Figure 11.17 shows the fully standardized three-factor model solution of case (5). Since all measures loaded on one single factor, these loadings are interpretable as regression coefficients (R. B. KLINE 2015, p. 301). Latent correlations among the different factors were high, thus indicating that each pair of factors was sharing a considerable amount of variance (35% to 59%).[4]

---

[4] To address possible biases due to the assumption that three of the measures were continuous rather than being categorical (Dir, MFlag, MDisk), a version of this model was re-computed treating them as ordered variables. Results revealed a latent correlation of .77 between Sur-

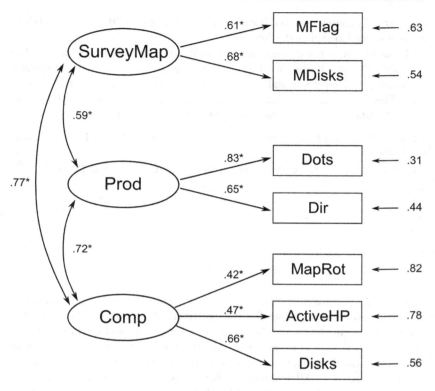

Figure 11.17 Completely standardized solution of the confirmatory factor analysis revealing three distinct subclasses of large-scale spatial abilities. *indicate significant path coefficients ($p < .001$)

In line with COMREY AND LEE (1992), the estimated factor loadings for the indicators were poor (MapRot), fair (ActiveHP), good (MFlag), very good (MDisk, Dir, Disks) and excellent (Dots). Consequently, the underlying latent factors sufficiently explained variance in most of the measures. Prod, for example, explained 69% of variance in Dots, and Comp, for example, explained 43% of variance in Disks. SurveyMap accounted for 46% of variance in MDisk. MapRot was the only measure

veyMap and Comp, of .56 between SurveyMap and Prod, and of .72 between Prod and Comp. Those results indicated that the estimated solutions of model (5) are robust against the assumption of treating all variables as continuous ones.

that was not well explained by the model; only 18% of variance were accounted for by the underlying factor Comp.

The reliabilities[5] for the scales in case (5) were acceptable (McDonald's $\omega_{MentalMap}$ = .60, ω_{Prod} = .74, and ω_{Comp} = .56) and the set of the seven measures featured an acceptable test reliability (ω_{total} = .78).

The fact that the latent factor SurveyMap was more closely related to Comp than to Prod can also be observed in the scatterplots that mapped the latent estimated performances in one subclass of large-scale spatial abilities against those in another for each child (Figure 11.18). The three scatterplots show the relations between all three factors, thereby indicating increasing latent correlations.

Since a conceptual distinction between tasks that could be completed without movement and map-based tasks requiring the movement of the subject was done in the conception of the single map-based orientation tasks, a second third-factor model was fitted to the data. In line with the theoretical conception outlined in Section 7.1, this three-factor model assumed that the factor MapEnvSelf can be differentiated into two factors, with one involving map use without and another involving map use with movement in real space. The model stipulated that MFlag and MDisk reflected a first factor, SurveyMap, that MapRot, and ActiveHP reflected a second factor, StaticMapUse, and that Dots, Dir, and Disks reflected a third factor, DynamicMapUse.

Results of a confirmatory factor analysis revealed that the fit indices did again not meet the criteria for good fit for for all cases (1) to (7). In regard to case (5), for example, the exact-fit hypothesis was rejected $\chi^2(11) = 22.50, p = .021$, thus indicating that the model deviated considerably from the data under WLSMV-estimation. Furthermore, the CFI and TLI were low (.95 and .91, respectively), the RMSEA was 0.07 with a 95% confidence interval of [.03, .11] (fit indices for all other cases are not reported here, but showed similar evidence for misfit). Consequently, a three-factor model that differentiated the factor MapEnvSelf by means of the necessary movement of the child in real space to solve map-based orientation tasks did not explain the data.

---

[5]The computed values are based on a single imputed data set, because there is no package that supports to compute a pooled version of reliability values.

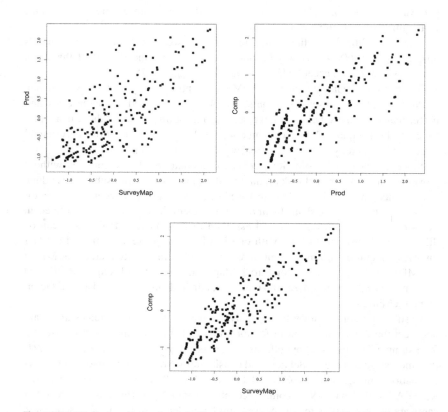

Figure 11.18 Comparison of standardized performances in tasks reflecting SurveyMap, Comp, and Prod in the main study

In a final step, the results were intended to be tested if they were robust against sex-differences. To examine whether sex had an influence on the extent of the dissociation between three subclasses of spatial abilities, a group-wise CFA was to be computed. The group-wise model, however, did not fit the data in the case of the three-dimensional model ($\chi^2(22) = 38.59$ (p=.016), CFI=.93, TLI=.87, RMSEA=.08). Consequently, no conclusion could be drawn concerning sex differences using this kind of analysis.

> **Conclusion**
> Small to moderate correlations among all measures in the large-scale spatial abilities test were found. A confirmatory factor analysis of the correlational patterns indicated that a three-component model explained the data best. The model included all seven measures and indicated that survey mapping (SurveyMap) is partially dissociable from the abilities to solve production tasks (Prod) which were, in turn, dissociable from the abilities to solve comprehension tasks (Comp). This model did not only explain the data sufficiently well, but also was found to be robust against biases from missing data treatment and use of different estimators. A two-component model and an alternative three-component model did not explain the data.

11.3   The Influence of Visuospatial Working Memory

This section presents the results concerning the influence of visuospatial working memory capacity (VSMV) on large-scale spatial abilities. First, Pearson correlations between two measures of VSWM and all of the measures in the map-based orientation test were analyzed on a manifest level. Then, their patterns of correlations with the different map-based orientation tasks were analyzed in an extended three-component model of large-scale spatial abilities involving a factor VSWM.

11.3.1 Analysis of Pearson Correlations

Table 11.28 shows the correlations between the two chosen measures of VSWM, ATFS and ATBS, and the seven measures of large-scale spatial abilities. All correlations were positive. Correlations of the VSWM forward measure ATFS ranged from .06 to .34, thus representing weak to medium relations according to J. COHEN (1988). Only the correlations with Dots, Dir, and Disks were found to be significant at the level $p = .05$ with Bonferroni-correction for 14 multiple tests. Correlations of the VSWM backward measure ATBS ranged from .06 to .33, thus representing weak to medium relations, and again, only the correlations with the measures Dots, Dir, and Disks were found to be significant at the level $p = .05$ with Bonferroni-correction.

Table 11.28 Correlations of the Single Measures of Large-Scale Spatial Abilities with the Two Measures of Visuospatial Working Memory Capacity (n = 215)

	MapRot	ActiveHP	Dots	Dir	Disks	MFlag	MDisk
ATFS	(.13)	(.10)	.34*	.29*	.25*	(.06)	(.13)
ATBS	(.10)	(.06)	.33*	.26*	.24*	(.13)	(.06)

$^*p < .05/14 = .0036$ (two-tailed p with Bonferroni-correction)

11.3.2 An Extended Model Involving VSWM

To investigate patterns of correlations among all measures of large-scale spatial abilities and VSMW, in particular, to investigate the extent to which each subclass of large-scale spatial abilities shared variance with VSWM, a single-level confirmatory factor analysis was computed that neglected the effects of class clustering ($\rho_{IC}(ATFS) = 0.004$ and $\rho_{IC}(ATBS) = .006$).

To conduct the analysis, a latent factor VSWM[6] was included into the three-factor CFA model based on WLSMV-estimation and the multiple imputation data sets (case 5) as depicted in Figure 11.17 (p. 316).

An analysis of the fit indices showed that the model explained the data sufficiently well. The exact-fit hypothesis was not rejected, $\chi^2(21) = 14.26$, $p = .858$, which indicated that the model did not derive considerably from the data. Moreover, the RMSEA was .00 with an 95% confidence interval [.00,.03]. Furthermore, the CFI = 1.00, and the TLI = 1.00 beneath the used criterion for good fit. Fit was also confirmed when considering the robust fit indices for every single of the 30 data sets and the parameter estimates were robust against the choice of the estimator (see Figure F.2 in the ESM, p. 96, for results of MLR estimation).

Figure 11.19 shows the fully-standardized solution of the extended three-factor-model. In line with the results concerning the zero-order Pearson correlations, the results of the latent analysis indicated that visuospatial working memory capacities did not correlate with all subclasses of large-scale spatial abilities to the same amount. The latent correlation with SurveyMap was .26 ($p = .035$). The latent correlation with Comp was .53 ($p < .001$), thus both factors shared 28% of variance. In line with the observations concerning the Pearson correlations, the latent

---

[6]To test for possible sex differences in the VSWM measures, a MANOVA was computed with ATFS and ATBS as dependent variables and sex as independent variable. Multivariate tests showed no significant effect for sex ($F(2,237) = 0.097$, $p = .91$; Wilk's $\Lambda = 0.999$, partial $\eta^2 = .00$).

correlation between VSMV and Prod was the highest one with .71 ($p < .001$), thus both factors shared 50% of variance.

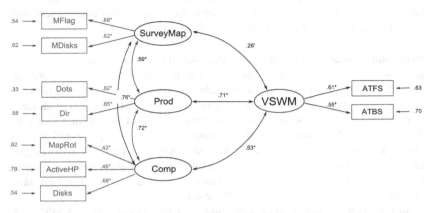

Figure 11.19 Completely standardized solution of the confirmatory factor analysis including three distinct subclasses of large-scale spatial abilities and visuospatial working memory. *indicate significant path coefficients ($p < .001$). 'indicate significant path coefficients ($p < .05$)

Conclusion

Visuospatial working memory capacity related highly, but not to the same extent, to those subclasses of large-scale spatial abilities that involved map use: considerable shared variance was found with the factor representing the abilities to solve production tasks (.71). The shared variance with the abilities to solve comprehension tasks was weaker (.53). The estimated relation to the abilities to encode a survey map was small (.26).

11.4 Discussion of the Empirical Findings

To answer the research question of this study (Section 5.1.1), a large-scale spatial abilities test using map-based orientation format was developed. In line with the literature, this study assumed that large-scale spatial abilities can be modeled to involve at least two subclasses, namely the abilities to encode and use a survey map and the abilities to keep track of the varying map-environment-self relations. The latter subclass was assumed to be dissociable into two further subclasses, namely the abilities to solve production tasks and comprehension tasks. These three subclasses were operationalized by a set of reflective measures. This final section discusses the findings concerning the individual measures in the map-based orientation test with respect to the theoretical framework outlined in Part I.

Items and Measures Included in the Map-Based Orientation Test
In line with previous studies with children of a similar age, the map-based orientation tasks chosen to reflect the latent constructs in this study mimicked the practical use of maps in everyday life (CHRISTENSEN 2011; LIBEN ET AL. 2013). The children were asked to point to unseen landmarks with the help of their map, to indicate the direction of the starting point after moving in real space, to place stickers on a map to indicate their location based on flags shown by an experimenter, to navigate towards pre-defined places that they identified on the map, and, finally, to point to landmarks without using their map. All tasks were integrated into a treasure hunt that consisted of a subsequent mix of all these tasks.

In line with previous studies with children (CHRISTENSEN 2011; KASTENS & LIBEN 2010; LIBEN ET AL. 2013), this study found large individual differences in performances of two kinds of map-based tasks in real space and in pointing tasks based on survey knowledge. These individual differences are not surprising, given that even adults struggle to locate themselves on a map in an unknown environment (LIBEN ET AL. 2008, 2010) or struggle to successfully navigate unknown environments with schematic maps (M. LEVINE 1982). These findings are in line with the developmental literature (HART & MOORE 1973) and they are also in line with empirical findings on children at the beginning of secondary school that tend to hardly master similar tasks in a more complex environment (WRENGER 2015). Although the focus of the discussion will lie on the explanation of problems that the children in the sample faced, the findings of this study also demonstrate that even children were able to complete cognitively challenging tasks with a certain degree of accuracy without having been instructed in these tasks before or having practiced on a similar task.

The most obvious interpretation of the difficulties in map-based tasks found might be that an additional level of difficulty was generated by the demand that the children were not allowed to turn their map and by the fact that the names of the buildings on the campus (which can be seen when navigating it) were not available on the map. Consequently, the children were forced to reason about geometric correspondences while applying a sophisticated level of map interpretation that involved the use of projective and Euclidean concepts rather than reasoning about topological concepts alone, which would have been possible with aligned maps. Indeed, previous studies found that map turning and reasoning with topological cues alone is the dominant strategy used by children during map-based navigation of real spaces (e. g., LIBEN ET AL. 2013; WRENGER 2015). Thus, to solve the tasks, children had to compensate for the misaligned map by either mentally rotating it or by choosing another strategy such as using multispatial cues (KASTENS & LIBEN 2010). This extra rotation or the integration of additional pieces of information might have increased not only the complexity of the tasks but also the amount of information that had to be maintained and transformed in working memory. Even adults struggle when they are not allowed to turn the map (DAVIES & UTTAL 2007), so it is not surprising that many children faced difficulties while completing the tasks.

Some further reasons might have accounted for the children's difficulties in the use of the maps (see KASTENS & LIBEN 2010 for a classification of failure modes in map-reading). First, some children might have failed to understand the concept of a map and the related symbolic correspondences as such. Although the developmental literature suggests that children are able to establish those (e. g., NEWCOMBE ET AL. 2013), some children might have consistently failed to perform tasks such as Dots, Dir, and Disks and might have received low scores.

Second, children might have attended to irrelevant information in the environment and on the map, especially when being asked to locate themselves on a map. That is, children might have identified a small area of grass as the 'garden' on the map or misinterpreted the shape of small paths on the campus as clues on the map. Since the campus was an unknown environment for the children and since it represented a visual complex environment, the amount of visual-spatial information available might have overwhelmed them (KASTENS & LIBEN 2010).

Third, children who considered multiple spatial clues and compensated for the misaligned map in one way or another, might have had problems in transferring those onto the map or the environment, for example by mixing up left/right turnings or having problems with Euclidean concepts such as scale or distance. Children tend to face problems when deriving distances from the map and struggle to understand the concept of scale and how to apply it in real space. This has been articulated in similar studies with more complex maps in a town (WRENGER 2015).

A surprising finding was that children performed very well in tasks requiring them to navigate toward a pre-defined goal with the map (i. e., solving Disks), but struggled to locate themselves on the map (i. e., solving Dots). That is, when actively engaging in a navigation task, they seemed to be able to keep track of the varying relations between the self, the environment, and the map. This was not the case when they were asked to determine their location after passively experiencing the environment or to apply Euclidean concepts such as reasoning about angles (UTTAL 1996). Another explanation might be that they were required to use multispatial clues and projective reasoning when locating themselves on the map. Consequently, the children who only partially or incompletely incorporated location information into their reasoning process might have been unable to locate themself on the map (KASTENS & LIBEN 2010). A tentative interpretation for better performing in Disks than in Dots might be that, knowing their initial position, they could focus on the map alignment only by constantly testing their self-location in space using single topological cues. This might have been sufficient to establish geometric correspondences during navigation. This finding is in line with the result that children performed well in those type of comprehension tasks but not those requiring them to apply projective and Euclidean concepts to make inferences on directions to landmarks.

One explanation for the children's difficulties when solving the pointing tasks at the end of the treasure hunt was that they were not explicitly told to memorize the landmarks encountered during the treasure hunt. However, individuals acquire survey knowledge without directly attending to landmarks, so the task instruction reflected everyday experiences (see CHRISTENSEN 2011, p. 17). Due to the challenging and probably exciting treasure hunt, the children might not have paid as much attention to landmarks as they would have if they had experienced the unknown environment without performing additional tasks. Consequently, they might have struggled to encode landmark and route information in a coordinated frame of reference which would have enabled them to successfully complete the pointing task (HART & MOORE 1973).

The map-based orientation test could be considered to be objective since it was conducted in accordance with a standardized test manual; there were no influences concerning the experimenters and environmental conditions, and the items were designed to be in closed response format and they were scored according to a standardized scorebook (RAMMSTEDT 2010). The test was found to be construct valid and most of the measures were found to be good, very good, or excellent indicators of the latent construct (R. B. KLINE 2015). Except for MapRot, the latent factors were able to explain more than 20% of variance in all of the measures. Consequently, these could be considered to reflect the underlying construct well enough. It might be due to the fact that MapRot was conducted at the very beginning of the map-based orientation test that the children were not yet aware about the fact that the map was unaligned with the environment and this might have caused answers by mostly guessing directions. Since this study did not involve further approaches to investigate validity of the individual tasks, it remains an open question why this test did not well reflect the abilities to solve comprehension tasks.

Moreover, although the results of the field testing of the initial conception of the map-based orientation test demonstrated that individual differences in children's large-scale spatial thought existed, some of the psychometric test characteristics were problematic for most of the measures. Indeed, almost all measures had a low reliability, indicating that the test quality of the individual tests was very limited. This might, of course, be explained by the fact that most measures involved only few items (three to six). In addition, some items within the measure were not sufficiently well able to distinguish between children who performed low or high in certain tasks.

Dissociation of Three Subclasses of Large-Scale Spatial Abilities
The empirical findings of this study suggested that children's large-scale spatial abilities should not be treated as an undifferentiated construct. Results indicated that the three-component model explained the data of this study better than other models. Hereby, a first factor represented the subclass of large-scale spatial abilities that allowed children to make inferences from survey knowledge that they acquired during map-mediated movement in real space. This factor was reflected by two pointing tasks to landmarks encountered before. A second subclass represented those large-scale spatial abilities that allowed children to solve production tasks. Those were reflected by tasks requiring them to indicate their position and viewing direction on the map after movement in real space. A third subclass represented the abilities to solve comprehension tasks. Those were reflected by tasks requiring the children to point to unknown landmarks with the help of the map, to indicate the starting point, and to navigate to pre-defined goals (Figure 11.17, p. 316).

The three-component model proposed is one possible way to model children's large-scale spatial abilities when these are conceptualized involving the use of maps as cognitive tools to grasp real space. The empirical findings are in line with what has been suggested in the theoretical model in Figure 3.11 (p. 100), namely that there seems to be a double dissociation when it comes to describing large-scale spatial abilities. The first dissociation is the one between map-based acquirement of survey knowledge, that is partially, but not completely dissociated from the abilities to keep track of the map-environment-self relation in real space. A second dissociation is the partial one between the two subclasses of abilities required to solve two different kinds of map-based tasks, namely production and comprehension tasks.

There are several possible explanations for the two dissociations found in this study. One possible explanation might be that the individual differences observed at the level of tasks reflect individual differences in the underlying cognitive processes that, in turn, might vary for each of the subclasses of large-scale spatial abilities. Concerning the first dissociation, distinct cognitive processes seem to be involved when processing large-scale spatial information from direct experience and when using it while reasoning with a map. Survey mapping abilities, that is, abilities that allow children to be aware of the changing relations between the self and landmarks during movement in space, involve the constant update of the relations with the egocentric frame of reference, whereas map use requires the children to update both the relations with the egocentric and the intrinsic frame of reference (the map) during movement in space (LOBBEN 2004; OTTOSSON 1987).

Since large-scale environments do not provide any explicit boundaries and are, in the case of unknown environments, full of unfamiliar landmarks and thus not easy to identify, children only perceive a certain part of large-scale spatial information (LIBEN ET AL. 2013). These pieces of information have to be integrated into a cognitive map that is distorted, incomplete, non-metric, and that rather involves ground-level spatial information (e. g., SHELTON & MCNAMARA 2004; SHELTON & PIPPITT 2007). In contrast, children who use a map have to understand which pieces of the multiple spatial information they perceive are actually represented in a map, or they have to extrapolate knowledge represented on the map into their ground-level view. This, in turn, involves complete, metric, and simultaneously perceivable information, which is cognitively processed in a different way than information gained from direct experience (e. g., COLUCCIA ET AL., 2007).

Reasoning from the survey map requires children to make inferences from distorted, piecewise information and is rather different from reasoning about directions from a paper map (see also CHRISTENSEN 2011). This explanation is in line with HEGARTY ET AL.'s (2006) findings who showed that survey knowledge obtained from direct experience is dissociable from survey obtained from visual inputs. Consequently, although both survey mapping and map use deal with the processing of large-scale spatial information that has to be assimilated and maintained in working memory, they provide children with qualitatively different ways of gaining insight into real space.

Concerning the second dissociation, distinct cognitive processes seem to be involved when solving production or comprehension tasks. Although both kinds of tasks required the children to mentally align the map with the environment, solving production tasks required them to perform environment-to-map transformations, whereas comprehension tasks involve map-to-environment transformations. The latter subclass involves the formulation of a spatial hypothesis that is related to landmark information and that requires a child to visualize how two-dimensional information might appear in three-dimensions in the egocentric frame of reference (LOBBEN 2004; OTTOSSON 1987; SEILER 1996). In contrast, the former subclass involves the formulation of a spatial hypothesis that is related to symbolic information on the map and that requires a child to collapse the perceived three-dimensional information perceived from an egocentric perspective into a two-dimensional abstract representation of these (LOBBEN 2004; OTTOSSON 1987; SEILER 1996). In other words, mental transformations of abstract, two-dimensional information into three-dimensional ones by means of visualization seem to be unique cognitive processes involved when solving comprehension tasks, whereas transformation of three-dimensional specific information into two-dimensional abstract ones seem to be unique cognitive processes involved when solving production tasks. Consequently, although both subclasses of large-scale spatial abilities require the individual to use visuospatial working memory to keep the map aligned while transforming both landmark and symbolic information, they allow children to perform qualitatively different types of dimensional transformations of spatial information.

Although this study found evidence for a dissociation of those subclasses of large-scale spatial abilities, the high latent correlations indicated that all subclasses were considerable related and shared a considerable amount of variance. An interesting finding was that SurveyMap was correlated to a higher extent to Comp (.77) than to Prod (.59), a result that was in line with the theoretical model Figure 3.11 (p. 100). In line with what has been argued in the introduction, maps may serve as cognitive tools that allow children not only to reason about the spatial information perceived but also to structure the way they perceive and encode the environment. This is in line with classical findings on cognitive mapping using maps (e. g., THORNDYKE & HAYES-ROTH 1982) and what has been proposed in the literature on map use (OTTOSSON 1987; SEILER 1996). Another explanation would be that both subclasses were measured by pointing tasks that involved an abstract answering instrument (the arrow-and-circle device). In MFlag and MDisk, children had to make inferences from a survey map, whereas when solving MapRot and ActiveHP, they made inferences from a map. A third explanation for the different amount of shared variance between SurveyMap and Comp, and SurveyMap and Prod might be that children had to follow the experimenter to solve production tasks and thus were not fully engaged while moving in the environment. In contrast, when solving comprehension tasks, they were active, also when moving in real space and placing the disks. Being active rather than passive may have influenced the acquisition of the corresponding survey knowledge.

The finding that children's large-scale spatial abilities involving map-use do not only vary with respect to performances in individual measures, but also with respect to the latent structure underlying the correlational patterns of those measures, extends previous findings of (CHRISTENSEN 2011). He found differences in children's performances in production and comprehension tasks also using flag-sticker-tasks and suggested that performances in these tasks are dissociated from the abilities to reason using a mental map (CHRISTENSEN 2011). In contrast to CHRISTENSEN's (2011) study, however, this study found a partial relation of the abilities to encode and use a survey map with the ones to use a paper map. This might be explained by the different study designs used in both studies.

The findings of this study suggested that children perform better in Disk, a comprehension task, than in Dots, a production task. These findings are in line with the ones of CHRISTENSEN (2011). Since the children participating in his study navigated the campus without guidance, the results of both studies indicate that the findings are robust against the specific environment used and concerning the concrete tasks parameters. They are in line with previous findings in language and temporal sequence reasoning, which have shown that completing comprehension tasks can be easier than solving production tasks (e. g., KOERBER AND SODIAN, 2008, as cited in CHRISTENSEN 2011, p. 50).

The Role of Visuospatial Working Memory
This study was probably the first one with children to relate measures of visuospatial working memory to a differentiated model of large-scale spatial abilities. The results of this study indicated (Figure 11.19, p. 321) that visuospatial working memory was considerably related to both of the subclasses of large-scale spatial abilities that involved map use (.71 for Prod, .53 for Comp), but to a smaller extent to the subclass of abilities that required the acquisition of survey knowledge (.26). All subclasses therefore seemed to be involved to a different degree to the controlled storage and maintenance of information in the visuospatial working memory for further processing. Visuospatial working memory seems to be involved in all cognitive large-scale spatial tasks that require children to generate a mental representation from a visual stimulus and to maintain the information in memory while transforming it simultaneously (e. g., ZIMMER ET AL., 2010, p. 13).

The abilities to make inferences on directions to landmarks from survey knowledge shared little variance with visuospatial working memory capacity (6.7%). The finding that there seems to be a relation is in line with previous studies on adults (e. g., ALLEN ET AL. 1996, LABATE ET AL. 2014, SHOLL & FRAONE 2004) and children (e. g., PURSER ET AL. 2012) but the findings of these studies suggested a rather limited extent of involvement. This can be explained by the fact that the acquisition of environmental knowledge has been shown to be related mainly to long-term memory and that the retrieval of spatial information from a cognitive or survey map rather relies on components of working memory such as executive control and visuospatial working memory (PURSER ET AL. 2012; WOLBERS & HEGARTY 2010). Consequently, the extent of involvement of visuospatial working memory might also be influenced by the measure that was chosen to externalize an individual's knowledge. This study used pointing tasks that required egocentric direction pointing, a task which probably involved less information to be maintained in working memory than sorting of pictures (e. g., ALLEN ET AL. 1996) or pointing from an imagined viewpoint in the environment (e. g., ROSSANO & MOAK 1998).

The findings that visuospatial working memory capacity relates to map-based activities in real space is in line with previous findings on adults (e. g., COLUCCIA ET AL. 2007; PLORAN ET AL. 2015). The findings of this study, however, differ from these studies in two major respects. First, concerning the extent of relation, this study found stronger relations than some of the studies on adults. This deviation might be explained by the very particular test settings in their study, involving a highly unfamiliar, unstructured, and complex environment (e. g., PLORAN ET AL. 2015). Second, concerning the kind of visuospatial working memory involved, this study found that working memory as measured by the Corsi Block Test, previously denoted as passive and sequential working memory (COLUCCIA ET AL. 2007), is involved in map use in real space. This extends the findings of COLUCCIA ET AL. (2007) who found that simultaneously working memory is involved when it comes to learning about spatial relations from a map. Considering both studies, one tentative interpretation might be that perceiving information only from the map requires simultaneous working memory, whereas relating this information to the fragmented pieces of them perceived in real space during movement also requires sequential working memory.

Finally, the finding that the visuospatial working memory was correlated higher to Prod than to Comp ($r = .71$ compared to $= .53$) provides evidence for the assumption that both subclasses of large-scale spatial abilities should be regarded as two distinct ones. Indeed, to solve production tasks such as Dots and Dir, children had to perceive and encode different pieces of information in working memory where they had to be maintained for the development of a spatial hypothesis concerning the location on the map. When solving comprehension tasks, children can also rely on spatial information perceived on the map, thereby decreasing the cognitive load of the task (see also CHRISTENSEN 2011, p. 65).

Relations Between Children's Small-Scale and Large-Scale Spatial Abilities

12

This chapter presents the empirical results concerning the relation between children's small-scale spatial abilities as measured by the paper-and-pencil test and large-scale spatial abilities as measured by the map-based orientation test. In addition, it presents results concerning the role of visuospatial working memory capacity. Section 12.1 presents results of correlation and multiple regression analyses that examined possible relations between both classes of spatial abilities at the manifest level. Section 12.2 presents results of structural equation modeling analyses that examined those relations at the latent level. In both sections, results concerning the role of visuospatial working memory are additionally presented. The final Section 12.3 discusses the findings with respect to the literature presented in Part I.

12.1   Results at the Manifest Level

This section presents results concerning the relations of individual performances in tasks that reflect small-scale and large-scale spatial abilities as well as of sets of these tasks. Results of a zero-order correlation analysis are shown in a first step. They indicate whether and to which extent performances in both sets of spatial tasks are related. In a second step, results of subsequent semipartial correlation analyses are presented. They explain patterns of correlations found among different sets of tasks. Finally, results of hierarchical multiple regression analyses are shown. Those relate performances in individual small-scale spatial tasks to performances in

---

Electronic supplementary material   The online version of this chapter (https://doi.org/10.1007/978-3-658-32648-7_12) contains supplementary material, which is available to authorized users.

individual large-scale spatial tasks while controlling for performances in tasks requiring visuospatial working memory capacities and while controlling for sex differences. Results of these suggest which spatial tasks emerged as important predictors in explaining variance in other spatial tasks.

12.1.1 Zero-Order Correlation Analysis

To assess the relationships between performances in individual small-scale spatial tasks and performances in large-scale spatial tasks, pairwise Pearson correlations were computed based on the 30 MI data sets with 215 cases using the package *mitools* (LUMLEY 2006). Moreover, based on the results presented in Chapter 10, the sum score of LR, MR2d, MR3d, and PFT was computed since these measures were found to reflect the abilities to perform object-based transformations (OB). The sum score of Boxes, Claudia, Emil, and Meadow was computed since these measures were found to reflect the abilities to perform transformations of the egocentric viewpoint (EGO). Similarly, based on results presented in Chapter 11, the sum score of MFlag and MDisk (SurveyMap), Dots and Dir (Prod), and MapRot, ActiveHP, and Disks (Comp) were computed. Based on these sum scores of performances, another set of pairwise Pearson correlations between these sum scores and individual tasks were computed, and Pearson correlations were also computed at the level of sum scores (Table 12.1). To avoid multiple testing for 100 times, which would have resulted in a very restrictive correction of the increasing alpha error, no significance levels were computed for the reported correlations at this moment of analyses.

Except for one $(-.02)$, all correlations among individual measures were positive. In line with J. COHEN's (1988) suggestions concerning the interpretation of correlation coefficients, most of them were low (.07 to .29), and six of them were moderate (.30 to .35). Overall, the highest correlations were found among the small-scale spatial tasks LR, PFT, Boxes, and Meadow and those large-scale spatial tasks that involved the use of a map. The strongest correlations were found between performances in the task LR and Dots (.34) and between performances in the tasks Boxes and Disks (.35). Except for PFT (.13) and Boxes (.12), no correlations were found among performances in small-scale spatial tasks and performances in the tasks MFlag. Except for Claudia (.08), correlations among performances in MDisk and performances in EGO tasks were low (.21 to .24).

Concerning the correlations among performances in sets of large-scale spatial tasks and performances in individual small-scale spatial tasks (last three columns

Table 12.1 Pooled Pearson Correlations between Individual and Clustered Measures of Small-Scale and Large-Scale Spatial Abilities (n = 215)

	MFlag	MDisk	Dots	Dir	MapRot	Active HP	Disks	Survey Map	Prod Map	Comp
LR	.06	.12	.34	.16	.21	.16	.30	.11	.30	.32
MR2d	.08	.15	.20	.08	.14	.20	.16	.14	.17	.23
MR3d	-.02	.06	.18	.16	.09	.23	.16	.02	.20	.23
PFT	.13	.18	.29	.22	.22	.30	.32	.19	.30	.40
Boxes	.12	.22	.30	.19	.22	.22	.35	.20	.29	.37
Claudia	.10	.08	.17	.14	.03	.18	.21	.11	.18	.21
Emil	.07	.24	.22	.10	.10	.14	.18	.18	.19	.20
Meadow	.09	.21	.28	.15	.07	.14	.25	.18	.26	.22
OB	.10	.19	.36	.21	.24	.31	.33	.17	.34	.41
EGO	.14	.28	.35	.21	.16	.24	.35	.25	.33	.36

in Table 12.1), most of the correlations were low (.11 to .29) and five of them were moderate (.30 to .40). Overall, correlations among performances in tasks that required memorizing of landmarks without the map and performances in individual small-scale spatial tasks were, if existent, low (.11 to 20). Except for two (LR, PFT), all correlations between performances in tasks that required children to solve production tasks and performances in singe small-scale spatial tasks were low (.17 to .29). There were moderate correlations between performances in comprehension tasks and performances in the two OB tasks LR (.32), PFT (.40), and Boxes (.37) and smaller ones with all other small-scale spatial tasks.

Concerning the correlations among performances in sets of small-scale spatial tasks and performances in individual large-scale spatial tasks (last two rows in Table 12.1), most of them were low (.10 to .28). Five were found to have moderate effects (.31 to .36), four of these occurring in correlations with the task Dots and Disks. Correlations among performances in OB tasks with performances in two tasks that required children to solve comprehension tasks was higher than in the case for EGO tasks (MapRot .24 compared to .16, and ActiveHP .31 compared to .24). Correlations for both performances in OB and EGO tasks with performances in the tasks Dots and Disks were about equally high (.36 compared to .35, and .33 compared to .35). Correlations with the task MFlag were low (.10 to .14), and performances in the task MDisk was correlated higher to performances in tasks that required egocentric perspective transformations (.28), then to tasks that required the children to perform object-based transformations (.19).

These observations concerning performances in individual tasks or pairwise correlations with performances of sets of tasks were also observed in the results concerning correlations at the level of parceled sum scores in these sets of tasks (bottom right in Table 12.1). There were moderate correlations between performances in EGO tasks and performances in SurveyMap tasks (.25) that were higher than correlations with performances in OB tasks (.17). Performances in both sets of small-scale spatial tasks correlated moderately to production and comprehension tasks, but correlations were higher with production tasks (.34/.33) than with comprehension tasks (41/.36). In the latter case, OB tasks were correlated higher to comprehension tasks (.41) than EGO tasks (.36). This was only marginally the case when it came to correlations with performances in production tasks.

In summary, these results suggested that children's performances in the two subsets of small-scale spatial tasks were not equally related to children's performances in the three subsets of large-scale spatial tasks. To investigate this result in greater detail, semipartial correlation analyses were conducted in a subsequent step.

12.1.2 Semipartial Correlation Analyses

Since the sets of spatial tasks in the paper-and-pencil test were related to each other and since the sets of spatial tasks in the map-based orientation test were related to each other, some of the correlations reported in Table 12.1 were assumed to be spurious. In other words, they might have occurred due to the effect of shared variance among the sets of tasks (SIMON 1954; VÖLKL & KORB 2018). Consequently, interpreting the zero-order-correlations without further detangling the effects of confounding variables would have resulted in misleading conclusions.

Indeed, the pairwise correlations between OB and EGO ($r = .49$), between SurveyMap and Prod ($r = .39$), between SurveyMap and Comp ($r = .44$), and between Prod and Comp ($r = .45$) showed that the sum scores of all these sets of spatial tasks shared a considerable amount of variance that should be considered in subsequent analyses.

To control for shared variance and to measure unique contributions of individual variables (i.e., sets of spatial tasks) in explaining variance of other variables (i.e., another set of spatial tasks), semipartial correlation analyses were conducted (e.g., KRAHA, TURNER, NIMON, ZIENTEK, & HENSON 2012; PITUCH & STEVENS 2015). More specifically, those allowed for measuring the extent of unique contribution of one variable while controlling for shared variance with other ones when it came to accounting for variance of a dependent variable. Semipartial correlation analyses is therefore close to regression analyses and implies the use of causal directions. As there are no clear suggestions from the literature, specifying both causal directions, performances in small-scale spatial tasks predicting performances in large-scale spatial tasks and vice versa, was reasonable. Consequently, semipartial correlation analyses were computed for both causal directions. The 'small-to-large' direction is reported first.

Performances in OB and EGO Tasks as Unique Predictors of Performances in Large-Scale Spatial Tasks

Since OB and EGO shared a considerable amount of variance (24%), the extent to which performances in each of the sets of small-scale spatial tasks explained unique variance in performances in each of the three sets of large-scale spatial tasks was measured. First-order[1] semipartial correlations were computed using the package *ppcor* (KIM 2005). To do so, the correlation between EGO and OB with each sum score of large-scale spatial tasks was computed while controlling for shared variance

---

[1] The order of the semipartial correlation refers to the number of shared variances that are partialled out.

between EGO and OB (Table 12.2). To control for α-error inflation due to multiple testing, the significance-level was adopted using HOLM's (1979) correction for 28 (Table 12.2) + 12 (Table 12.3) = 40 multiple tests.[2]

Partialling out the shared variance between performances in EGO and OB tasks yielded the following three results:

(1) Performances in OB tasks made a significant unique contribution to the prediction of variance in performances in the tasks Dots and ActiveHP.
(2) Performances in EGO tasks made a significant unique contribution to the prediction of variance in performances in Disks.
(3) Neither the performances in OB nor in EGO tasks made a significant unique contribution to the prediction of variance in performances in the tasks MFlag, MDisk, Dir, or MapRot.

Table 12.2 First-Order Semipartial Correlations for OB and EGO with Individual Large-Scale Spatial Tasks (n = 215)

	MFlag	MDisk	Dots	Dir	MapRot	ActiveHP	Disks
OB	.10	.19	.36***	.21*	.24**	.31***	.33***
EGO	.14	.28***	.35***	.21*	.16	.24**	.35***
Residual OB-EGO	.04	.05	.20*	.11	.16	.20*	.17
Residual EGO-OB	.08	.19	.19	.10	.04	.09	.20*

$*p < 0.1$ $**p < 0.05$ $***p < 0.01$ (Holm-correction for 40 multiple comparisons)

To generalize these results, first-order semipartial correlations between sets of small-scale spatial tasks and sets of large-scale spatial tasks were computed (Table 12.3). Performances in the two sets of small-scale spatial tasks, EGO and OB tasks both predicted performances in almost all large-scale spatial tasks not only at the level of individual tasks but also at the level of sets of them. The total variance of performances in SurveyMap tasks explained was 6.4%. After partialling out shared variance between performances in EGO and OB tasks, neither performances in EGO nor OB emerged as a significant predictor of performances in this subset of large-scale spatial tasks. A similar result was found when it came to explaining variance in performances in Prod tasks. The total variance explained was 15.2%,

---

[2]Holm corrections were used because they are less restrictive than Bonferroni ones.

but neither performances in EGO nor OB tasks were found to make a significant unique contribution to explaining variance. The total variance in performances in Comp tasks explained was 20.5%. When controlling for shared variance, only performances in OB tasks made a significant unique contribution to explaining variance in performances in Comp tasks (6.3% of unique contribution).

Table 12.3 First-Order Semipartial Correlations for OB and EGO with SurveyMap, Prod, and Comp (n = 215)

	SurveyMap	Prod	Comp
OB	.17	.34***	.41***
EGO	.25**	.33***	.36***
Residual OB-EGO	.05	.19	.25**
Residual EGO-OB	.16	.17	.17

$*p < 0.1$ $**p < 0.05$ $***p < 0.01$ (Holm-corrected values for 40 multiple comparisons)

Performances in SurveyMap, Prod, and Comp Tasks as Unique Predictors of Performances in Small-Scale Spatial Tasks

Since SurveyMap, Prod and Comp were sharing a considerable amount of variance (15,2% to 20,2%), the extent was computet to which performances in each of the sets of large-scale spatial tasks explained unique variance in performances in each individual small-scale spatial task. To do so, second-order semipartial correlations were computet, that is, the correlation between SurveyMap, Prod and Comp with each sum score of small-scale spatial tasks while controlling for shared variance with the two other variables (Table 12.4).

Partialling out the shared variance yielded the following three results:

(1) Performances in SurveyMap tasks made no significant unique contribution to the prediction of variance in performances in small-scale spatial tasks.
(2) Performances in Prod tasks made a significant unique contribution to the prediction of variance in performances in Emil and Meadow.
(3) Both performances in Prod and Comp tasks made significant unique contributions to the prediction of variance in performances in LR, PFT, and Boxes. In the tasks PFT and Boxes, Comp predicted a higher degree of unique variance than Prod.

Table 12.4 Second-Order Semipartial Correlations for SurveyMap, Prod and Comp with Individual Small-Scale Spatial Tasks (n = 215)

	LR	2DMR	3DMR	PFT	Boxes	Claudia	Emil	Meadow
SurveyMap	.11	.14	.02	.19	.20	.11	.18	.18
Prod	.30***	.17	.20	.30***	.29***	.18	.19	.26***
Comp	.32***	.23**	.23**	.40***	.37***	.21*	.20	.22**
Residual SurveyMap	.02	.08	−.04	.08	.11	.05	.13	11
Residual Prod	.25***	.12	.16	.23**	.21*	.14	.21*	.21*
Residual Comp	.23**	.16	.17	.30***	.27***	.15	.12	.13

$*p < .1 \ **p < .05 \ ***p < .01$ (Holm-corrected values for 60 multiple comparisons)

To generalize these results at the level of sum scores of sets of tasks, second-order semipartial correlations between sets of small-scale spatial tasks and sets of large-scale spatial tasks were computed (Table 12.5). Performances in the three sets of large-scale spatial tasks, SurveyMap, Comp, and Prod tasks, predicted performances in small-scale spatial tasks not only at the level of individual tasks but also at the level of sets of them.

The total variance of performances in OB tasks explained was 20.4%. After partialling out shared variance between all sets of large-scale spatial tasks, performances in Prod and Comp made a significant unique contribution to explain the variance in performances in OB tasks. Comp (9.6% of unique contribution) had slightly higher unique contribution than Prod (7.3% of unique contribution). The total variance of performances in EGO tasks explained was 17.0%. Again, when

Table 12.5 Second-Order Semipartial Correlations for SurveyMap, Prod, and Comp with OB and EGO (n = 215)

	OB	EGO
SurveyMap	.17	.25***
Prod	.34***	.33***
Comp	.41***	.36***
Residual SurveyMap	.06	.15
Residual Prod	.27***	.26***
Residual Comp	.31***	.25***

$*p < .1 \ **p < .05 \ ***p < .01$ (Holm-corrected values for 60 multiple comparisons)

controlling for shared variance, both performances in Prod and Comp tasks made a unique contribution to explain variance in performances in EGO tasks. Unique contributions had about the same extent (6.8% compared to 6.3%).

12.1.3 Hierarchical Multiple Regression Analyses

In a subsequent step, this study aimed to investigate how strong performances in individual small-scale spatial tasks are related to performances in each of the large-scale spatial tasks and vice versa. It also examined the role of visuospatial working memory and controlled for robustness against possible influences of sex differences. Interpreting the zero-order correlations (Table 12.1) exclusively would have provided a limited view to understand the relations of individual tasks. Because both the small-scale and large-scale spatial tasks were correlated with each other, that is, there were effects of multicollinearity, zero-order correlations were likely to be biased and, probably, spurious (e. g., KRAHA ET AL. 2012). Computing semipartial correlation analyses for all the individual measures would have been one means to control for multicollinearity (e. g., KRAHA ET AL. 2012). This would have, in particular when searching to investigate in addition the role of the predictors visuospatial working memory and sex, resulted in multiple testings of the same example. Corrections for alpha errors such as provided by the Holm corrections would have led to very restrictive levels for determining significance.

This study used series of hierarchical multiple regression analyses to address the question of how performances in individual spatial tasks, visuospatial working memory capacity, and sex predicted performances in other spatial tasks. Multiple regression analyses allow for testing a model that describes how variance in the dependent variable (i. e., performances in one spatial task) might be explained by more than one predictor (e. g., ROSS & WILLSON 2017). Results of these analyses do not only quantify the amount of variance explained in the dependent variable but provide standardized regression weights for each of the predictors (denoted as β weights) and their significance level. Since they are estimates of the predictors' extent of contribution when holding the influence of all other predictors constant KRAHA ET AL. (2012), they can provide one means to specify the importance of certain predictors, ideally if being combined with other post-hoc analyses.[3] The focus of the analyses here was to investigate whether or not predictors were significant

---

[3] Ideally, one would consider semipartial correlation analyses and dominance analyses as possible post-hoc analyses (see KRAHA ET AL. 2012; NATHANS, OSWALD, & NIMON 2012, for further elaboration). Since determining the importance of individual spatial tasks was a

to gain first insights into the role of individual spatial tasks, the role of visuospatial working memory as well as the role of sex as predictors. In addition, beta weights were tentatively interpreted to state which one of the predictors might have been the most important one.

Conducting hierarchical multiple regression analyses allowed for entering the predictors in three blocks, an approach that resulted in three subsequent models that were tested (e. g., Ross & Willson 2017). In line with the theoretical findings, a first block of variables involved performances of spatial tasks. When testing models that explained variance in large-scale spatial tasks, small-scale spatial tasks were entered as predictors and vice versa, large-scale spatial tasks were entered as predictors when it came to explaining variance in small-scale spatial tasks. A second block of variables involved performances in the Corsi Block Test. A third block of variables involved sex. Because predictors were entered in blocks, it was possible to compare the subsequent models and to identify whether or not the explained variance increased significantly after entering a block of variables.

Hierarchical multiple regression analyses were performed for all small-scale and large-scale spatial tasks using the same analysis scheme applied to the 30 MI data sets:

1. In a preliminary analysis (**Step 0**), a set of small/large-scale spatial tasks that significantly predicted variance in a large/small-scale spatial task was identified using the package *relaimpo* (Grömping 2006).
2. In **Step 1** of the hierarchical multiple regression analysis, those spatial tasks were entered into the analysis of a first model. Unstandardized and standardized regression coefficients and their significance were computed.
3. In **Step 2**, the measures ATFS and ATBS were entered into the first model. The extended, second model was analyzed and the regression statistics were computed.
4. In **Step 3**, the variable sex was entered into the second model. The extended, third model was analyzed and the regression statistics were computed.

Hierarchical multiple regression analyses were computed for the sample involving 240 cases, since there was no need to relate the results to subsequent analyses. The fact that all findings at the latent level were independent of the estimator and the basic sample used, no impact was expected on the intended qualitative interpretations. The regressions were computed on the data set with 30 imputations, using the R

---

minor research goal of this study, their role was investigated within the limited methodical framework outlined.

package *mitools*. This package allowed for computing pooled regression coefficients and pooled values for R^2. However, by the time of computing the statistics, there was no possibility to compute pooled F-scores for the overall model fit statistics.

Predictions of performances in large-scale spatial tasks by performances in small-scale spatial tasks and vice versa were to be expected from the literature. Consequently, models were computed for each of the individual tasks in both sets of spatial tasks. The 'small-to-large' direction is presented first, that is, results concerning the prediction of performances in large-scale spatial tasks by means of performances in small-scale spatial tasks, visuospatial working memory capacity, and sex.

Hierarchical Multiple Regression Models Predicting Performances in Large-Scale Spatial Tasks
Table 12.6 summarizes the results concerning the hierarchical multiple regression analyses for predicting performances in large-scale spatial tasks. Small-scale spatial tasks that were found to be significant predictors are shown in the second column. In addition to calculating the explained variance (R^2) at each step of the hierarchical regression analyses, Table 12.6 also specifies the corresponding effect sizes (f^2). In line with J. COHEN's (1988) conventions for interpreting effect sizes, all $f^2 \geq .02$ were interpreted as small, $f^2 \geq .15$ as medium, and $f \geq .35$ as large effects.

Table 12.6 Explained Variance and the Corresponding Cohen's f^2 in Large-Scale Spatial Tasks by Adding Predictors Stepwise (n = 240)

Task	Step 1				Step 2			Step 3	
	Small-scale predictors	R^2	f^2	VSWM	R^2	f^2	Sex	R^2	f^2
MapRot	LR, Boxes	7.0%	.08	ATFS, ATBS	7.4%	.08		8.2%	.09
ActiveHP	3DMR, PFT	15.1%	.18		15.2%	.18		19.8%	.25
Dots	LR, PFT, Meadow	23.9%	.31		27.5%	.38		28.2%	.39
Dir	3DMR, PFT	9.0%	.10		15.4%	.18		15.8%	.19
Disks	LR, PFT, Boxes	22.4%	.29		23.7%	.31		24.4%	.32
MFlag	Boxes	1.5%	.02		3.1%	.03		6.1%	.07
MDisk	Emil, Meadow	7.6%	.08		8.2%	.09		11.0%	.13

The hierarchical multiple regression analyses conducted yield some important results:

Model 1   Except for 2DMR and Claudia, all small-scale spatial tasks were found to be significant predictors in one or other preliminary analyses. When entering them into a first model, small (MapRot, Dir, MFlag, and MDisk) to medium (ActiveHP, Dots, Disks) effect sizes were found when it came to explain variance in these large-scale spatial tasks.

Model 2   Entering ATFS and ATBS as predictors increased R^2 in all models, but it was lower than 1% for MapRot, ActiveHP, and MDisk. In regard to the tasks Dots and Dir, second models involving ATFS and ATBS in addition to the small-scale spatial task predictors led to an important increase in R^2. The second model for Dots had large, and the model for Dir had medium effect size.

Model 3   Entering sex as another predictor increased R^2 in almost all models, but it was lower than 1% for MapRot, Dots, Dir and Disks. The third models, involving all three blocks of predictors were able to predict variance in large-scale spatial tasks. Small (MapRot, MFlag, MDisk), medium (ActiveHP, Dir, Disks) and large (Dots) effect sizes were found.

This section exemplarily presents results of two hierarchical multiple regression analyses. First, the results of analyses of the task Dots are presented to provide an example in which some of the small-scale spatial tasks as well as measures of visuospatial working memory capacities emerged as significant predictors. These results were independent of possible influences of sex differences. Second, results of these analyses of the task MFlag are presented to provide an example in which the variable sex emerged not only as a significant predictor when it came to explain performances in MFlag but also influenced the possible importance of other predictors. The summaries of hierarchical multiple regression analyses predicting performances in all other small-scale spatial tasks are provided in ESM, Section F.3, pp. 100.

Table 12.7 (p. 343) summarizes the results of the hierarchical multiple regression analyses for Dots. Significant predictors are marked in bold face. An adjusted value (R^2_{adj}) is reported in addition to the amount of variance explained by the predictors (R^2, see also in Table 12.6). R^2_{adj} is more sensitive to useless predictors entered into the model. Results of the computed models and model comparisons are described here based on the adjusted R^2-values and in the following summaries of analyses.[4]

---

[4]In Table 12.6 the range of effect sizes is independent of the chosen kind of R^2.

Table 12.7 Summary of Multiple Hierarchical Regression Analyses for Predicting Performances in Dots (n = 240)

	Step 1			Step 2			Step 3		
R^2	23.9%			27.5%			28.2%		
R^2_{adj}	22.9%			25.9%			26.3%		
$\Delta R^2_{adj}(p)$				3.0% (.010)			0.4% (.133)		
	B	β	p	B	β	p	B	β	p
LR	.26	**.25**	.000	.21	**.21**	.001	.20	**.20**	.001
PFT	.17	**.19**	.003	.13	**.15**	.020	.12	**.14**	.031
Meadow	.15	**.19**	.002	.11	**.14**	.027	.10	**.12**	.049
ATFS				.01	.12	.072	.01	.12	.052
ATBS				.02	**.15**	.009	.02	**.16**	.005
Sex							.27	.09	.118

In the first model, the three significant predictors LR, PFT, and Meadow accounted for 22.9% of the variance in Dots with LR emerging to be likely an very important predictor ($\beta = .25$).[5] Prediction was significantly improved by 3.0% ($p = .010$) by adding the predictors ATFS and ATBS in a second step. In the extended, second model, ATBS emerged as yet another significant predictor. Moreover, adding ATFS and ATBS as predictors in Step 2 led to a decrease of the relative importance of the three spatial task predictors. They remained, however, significant at the 0.05-level. The addition of the variable sex in Step 3 led to a marginal but not significant model improvement of R^2_{adj} by 0.4% ($p = .133$). Sex was just not found to be a significant predictor.

The third, final model for predicting performances in Dots accounted for more than a quarter of the variance in performances in this task which can be considered a large effect (J. COHEN 1988). The model proposed that variance could be explained by performances in a set of small-scale spatial tasks, VSWM capacity, and Sex, with LR, PFT, Meadow, and ATBS being significant predictors.

Table 12.8 summarizes the results of the hierarchical multiple regression analyses for MFlag. In the first model, the significant predictor Boxes accounted for 1.1% of the variance in MFlag. Prediction was marginally but not significantly improved by 0.8% ($p = .149$) by adding the predictors ATFS and ATBS in a second step. In the extended, second model, ATBS emerged as a significant predictor. Moreover, adding ATFS and ATBS as predictors in Step 2 led to a decrease of the relative

---

[5]A β of .25 means that one might expect a change of .25 (in standardized metric) in performance in Dots when increasing LR by one standardized unit.(KRAHA ET AL. 2012).

importance of Boxes and the predictor became non-significant at the .05-level. The addition of the variable sex in Step 3 led to a significant model improvement of R^2_{adj} by 2.7% ($p = .007$). Sex was found to be a significant predictor.

Table 12.8 Summary of Multiple Hierarchical Regression Analyses for Predicting Performances in MFlag (n = 240)

	Step 1			Step 2			Step 3		
R^2	1.5%			3.1%			6.1%		
R^2_{adj}	1.1%			1.9%			4.6%		
$\Delta R^2_{adj}(p)$				0.8% (.149)			2.7% (.007)		
	B	β	p	B	β	p	B	β	p
Boxes	.07	**.12**	.049	.04	.08	.250	.02	.04	.599
ATFS				.00	−.02	.834	.00	−.01	.947
ATBS				.01	**.14**	.040	.01	**.16**	.020
Sex							.37	**.18**	.008

The third, final model for predicting performances in MFlag accounted for less than 5% of the variance in performances of this task which can be considered a small effect (J. COHEN 1988). The model proposed that variance could be explained by performances in a set of small-scale spatial tasks, VSWM capacity, and sex, with ATBS and sex being significant predictors.

Results of the hierarchical multiple regression analyses for MFlag showed that some of the significant predictors of the first model (significant large-scale spatial tasks in Table 12.6) did not remain significant in the extended second and third models. In other words, these predictors were not robust against possible interaction effects with measures of visuospatial working memory capacity and sex that were entered in step 2 and 3. The predictors for Dots, however, were robust against influences of visuospatial working memory and sex since they all remained significant predictors in the third model. Table 12.9 shows a summary of the results of the third (Step 3) multiple hierarchical regression models for all large-scale spatial tasks. All predictors that were significant ($p \leq .05$) and nearby significant ($p \leq .10$) after entering all three blocks of predictors.

A comparison of Table 12.9 with Table 12.6 (p. 341) showed that in the cases for MapRot, Dots, and Disks, small-scale spatial tasks that were found to be significant predictors in a first model, remained significant predictors when controlling for influences of visuospatial working memory and sex. When it came to model performances in Dir, the predictors 3DMR and PFT did not remain significant predictors when entering measures of visuospatial working memory in the analyses (see also Table F.15 in the ESM, p. 101). When it came to model performances in ActiveHP, MFlag, and MDisk, the small-scale spatial tasks predictors did not remain significant predictors when entering the variable sex in the analyses (see e. g., Table 12.8). Performances in LR were found to significantly predict performances in Dots. PFT was a predictor for ActiveHP, Dots, and Disks. Boxes was a predictor for MapRot, and Disks. Meadow was a predictor for Dots.

Table 12.9 Summary of Hierarchical Multiple Regression Analysis for all Large-Scale Spatial Tasks (n = 240)

Task	Summary	Significant ($p \leq .05$)	Nearby significant ($p \leq .1$)
MapRot	Table F.13	Boxes	LR
ActiveHP	Table F.14	PFT, Sex	3DMR
Dots	Table 12.7	LR, PFT, Meadow, ATBS	ATFS
Dir	Table F.15	ATFS, ATBS	
Disks	Table F.16	PFT, Boxes	LR
MFlag	Table 12.8	ATBS, Sex	
MDisk	Table F.17	Sex	Meadow

Hierarchical Multiple Regression Models Predicting Performances in Small-Scale Spatial Tasks

Table 12.10 summarizes the results concerning the hierarchical multiple regression analyses for predicting performances in small-scale spatial tasks. Large-scale spatial tasks that were found to be significant predictors are shown in the second column. In addition to showing the explained variance (R^2) at each step of the hierarchical regression analyses, the table also specifies the corresponding effect sizes (f^2). In line with J. COHEN's (1988) conventions, all $f^2 \geq .02$ were interpreted as small, $f^2 \geq .15$ as medium, and $f \geq 0.35$ as large effects.

The hierarchical multiple regression analyses conducted yield some important results:

Model 1 Solely the large-scale spatial tasks Dots, Disks, and ActiveHP were found to be significant predictors in the preliminary analyses. When entering them into a first model, small (2DMR, 3DMR, Claudia, Emil) to medium (LR, PFT, Boxes,

Meadow) effect sizes were found when it came to explain variance in performances in these small-scale spatial tasks.

Model 2   Entering ATFS and ATBS increased R^2 in all models. In regard to the task 2DMR, a second model involving the predictor Dots as well as both measures of VSWM had a small effect size. In regard to the tasks PFT and Boxes, a second model involving ATFS and ATBS in addition to Dots, Disks, and Active HP as predictors even had large effect size. The corresponding second models had medium effect sizes for all other small-scale spatial tasks.

Table 12.10 Explained Variance and the Corresponding Cohen's f^2 in Small-Scale Spatial Tasks by Adding Predictors Step by Step (n = 240)

Task	Model 1			Model 2				Model 3	
	Large-scale predictors	R^2	f^2	VSWM	R^2	f^2	Sex	R^2	f^2
LR	Dots, Disks	18.3%	0.22	ATFS, ATBS	22.7%	0.29		23.2%	0.30
2DMR	Dots	6.3%	0.07		10.0%	0.11		10.1%	0.11
3DMR	ActiveHP	9.4%	0.10		15.4%	0.18		18.0%	0.22
PFT	Dots, Disks, ActiveHP	24.4%	0.32		28.8%	0.40		28.7%	0.40
Boxes	Dots, Disks, ActiveHP	22.0%	0.28		26.8%	0.37		27.9%	0.39
Claudia	Disks, ActiveHP	11.5%	0.13		14.4%	0.17		17.8%	0.22
Emil	Dots, ActiveHP	11.8%	0.13		14.0%	0.16		22.0%	0.28
Meadow	Dots, Disks	14.6%	0.17		20.7%	0.26		22.9%	0.30

Model 3   Entering sex as another predictor increased R^2 in almost all models, but not to the same extent. Entering this predictor did not better predict 2DMR than before (no increase in R^2), but increased R^2 for Emil, for example. The third models, involving all three blocks of predictors, were able to predict variance in the small-scale spatial tasks; small (2DMR), medium (LR, 3DMR, Claudia, Emil, Meadow) and large (PFT, Boxes) effect sizes were found.

This section exemplarily presents results of two hierarchical multiple regression analyses. First, results of analyses of the task Boxes are presented to provide an example in which some of the large-scale spatial tasks and measures of visuospatial working memory capacities emerged as significant predictors, independently of possible influences of sex differences. Second, results of these analyses of the task Emil are presented to provide an example in which the variable sex emerged not only as a significant predictor when it came to explain performances in Emil but also influenced the possible importance of other predictors. The summaries of hierarchical multiple regression analyses predicting performances in all other small-scale spatial tasks are provided in the ESM, Section F.3, pp. 96.

Table 12.11 Summary of Multiple Hierarchical Regression Analyses for Predicting Performances in Boxes (n = 240)

	Model 1			Model 2			Model 3		
R^2	22.0%			26.8%			27.9%		
R^2_{adj}	21.0%			25.2%			26.1%		
$\Delta R^2_{adj}(p)$				4.2% (.001)			0.9% (.059)		
	B	β	p	B	β	p	B	β	p
Dots	.26	**.20**	.006	.16	.12	.080	.15	.11	.100
Disks	.29	**.26**	.000	.25	**.22**	.002	.23	**.21**	.003
ActiveHP	.22	**.15**	.012	.21	**.15**	.009	.17	**.12**	.046
ATFS				.01	.09	.189	.01	.10	.151
ATBS				.03	**.20**	.002	.03	**.21**	.001
Sex							.44	.11	.058

Table 12.11 summarizes the results of the hierarchical multiple regression analyses for Boxes. In the first model, the three significant predictors Dots, Disks, and ActiveHP accounted for 21.0% of the variance in Boxes with Disks emerging to be likely to be an very important predictor ($\beta = .26$). Prediction was significantly improved by 4.2% ($p = .001$) by adding the predictors ATFS and ATBS in a second step. In the extended, second model, ATBS emerged as yet another significant predictor. Moreover, adding ATFS and ATBS as predictors in Step 2 led to a decrease of the relative importance of the three spatial task predictors and Dots become non-significant at the 0.05-level. The addition of the variable sex in Step 3 led to a marginal but not significant model improvement of R^2_{adj} by 0.9% ($p = .059$). Sex was not found to be a significant predictor.

The third, final model for predicting performances in Boxes accounted for more than a quarter of the variance in performances in this task which can be considered to be a large effect (J. COHEN 1988). The model proposed that variance could be explained by performances in a set of large-scale spatial tasks, VSWM capacity, and sex. Disks, ActiveHP, and ATBS were found to be significant predictors.

Table 12.12 presents the detailed results of the hierarchical multiple regression analyses for Emil. In the first model, the two significant predictors Dots, and ActiveHP accounted for 11.1% of the variance (R^2_{adj}) in Emil. Dots emerged as the most important predictor ($\beta = .26$). Prediction was slightly but not significantly improved by 1.4% ($p = .053$) by adding the predictors ATFS and ATBS in a second step. In the extended, second model, ATBS emerged as yet another significant predictor. Moreover, adding ATFS and ATBS as predictors in Step 2 led to a decrease of the relative importance of the two spatial task predictors. Those predictors remained, however, significant. The addition of the variable sex in Step 3 led to a significant model improvement of R^2_{adj} by 7.8% ($p = .000$). Sex was found to be the most important and significant predictor ($\beta = .30$). In contrast, adding the variable sex to the model led to a considerable decrease in the relative importance of Dots and ActiveHP, and ActiveHP became non-significant.

The third, final model for predicting performances in Emil accounted for a fifth of the variance in performances in this task which can be considered a medium effect (J. COHEN 1988). The model proposed that variance could be explained by performances in a set of large-scale spatial tasks, VSWM capacity, and sex. Dots, ATBS, and sex were found to be significant predictors.

Table 12.12 Summary of Multiple Hierarchical Regression Analyses for Predicting Performances in Emil (n=240)

	Step 1			Step 2			Step 3		
R^2	11.8%			14.0%			22.0%		
R^2_{adj}	11.1%			12.5%			20.3%		
$\Delta R^2_{adj}(p)$				1.4% (.053)			7.8% (.000)		
	B	β	p	B	β	p	B	β	p
Dots	.37	**.26**	.000	.28	**.20**	.004	.23	**.16**	.014
ActiveHP	.26	**.16**	.011	.26	**.16**	.012	.12	.07	.246
ATFS				.01	.06	.326	.01	.08	.188
ATBS				.02	**.13**	.035	.03	**.16**	.018
Sex							1.33	**.30**	.000

Results of the hierarchical multiple regression analyses for Boxes and Emil showed that some of the significant predictors of the first model (significant large-scale spatial tasks in Table 12.10) did not remain significant in the extended second and third models. In other words, these predictors were not robust against possible interaction effects with measures of visuospatial working memory capacity and sex that were entered in step 2 and 3. Table 12.13 shows a summary of the results of the third (Step 3) multiple hierarchical regression models for all small-scale spatial tasks. All predictors that were significant ($p \leq .05$) and nearby significant ($p \leq .10$) after entering all three blocks of predictors.

Table 12.13 Summary of Results of the Hierarchical Multiple Regression Analyses for All Small-Scale Spatial Tasks (n = 240)

Task	Summary	Significant ($p \leq .05$)	Nearby significant ($p \leq 0.1$)
LR	Table F.7	Dots, Disks, ATFS	ATBS
2DMR	Table F.8	ATBS	Dots
3DMR	Table F.9	Active HP, ATBS, Sex	
PFT	Table F.10	Disks, ActiveHP, ATBS	Dots
Boxes	Table 12.11	Disks, ActiveHP, ATBS	Sex
Claudia	Table F.11	Disks, ActiveHP, ATBS, Sex	
Emil	Table 12.12	Dots, ATBS, Sex	
Meadow	Table F.12	Dots, ATFS, ATBS, Sex	Disks

A comparison of Table 12.13 with Table 12.10, p. 346 showed that those large-scale spatial tasks that were found to be significant predictors in a first model, remained significant predictors even when controlling for influences of VSWM and sex in most of the cases. Only when it came to model variances in performances in Boxes, the task Dots did not remain significant (see also Table 12.11). When it came to model variances in performances in Emil, ActiveHP did not remain significant in the third model (see also Table 12.12). Performances in Dots were found to significantly predict performances in LR, Emil, and Meadow. Performances in Disks were found to significantly predict performances in LR, PFT, Boxes, and Claudia. Performances in ActiveHP were found to significantly predict 3DMR, OFT, Boxes, and Claudia. Performances in 2DMR were predicted by performances in large-scale spatial tasks to a very limited extent only. Performances in tasks of VSWM emerged as significant predictors of performances in all small-scale spatial tasks. Sex emerged as significant predictor in the models of 3DMR, Claudia, Emil, and Meadow.

Conclusion
Correlation and regression analyses showed that there was an empirical relation between individual performances in small-scale spatial tasks and performances in large-scale spatial tasks. Those relations were particularly strong for some of the relations between performances in LR, PFT, Boxes and Meadow with performances in Dots, Disks, and ActiveHP. Relations to performances in spatial tasks at another scale of space were particularly low for performances in 2DMR, Dir, MFlag, and MDisk. Visuospatial working memory seemed to be a predictor for all small-scale spatial tasks and the production tasks Dots and Dir. Correlation and semipartial correlation analyses at the level of sets of tasks revealed that performances in OB tasks explained unique variance in Prod and Comp tasks. Vice versa, performances in Prod and Comp tasks explained unique variance in both OB and EGO tasks. No unique contributions in explaining variance in performances of SurveyMap or by OB and EGO tasks were found.

12.2   Results at the Latent Level

This section presents results concerning the relation of small-scale and large-scale spatial abilities at the latent level, which is free of measurement errors. It also specifies results concerning the role of visuospatial working memory capacity. To investigate these issues, different structural equation models (SEMs) were tested that included latent variables representing spatial abilities at different scales of space. This approach did not only allow to determine whether or not small-scale and large-scale spatial abilities are related and, if so, to which extent, but also to investigate how different subclasses of these are related and whether visuospatial working memory could be considered to be a mediator.

12.2.1 Results From Structural Equation Modeling

SEM analyses were conducted using *lavaan* (ROSSEEL 2012) and *semtools* (SEMTOOLS CONTRIBUTORS 2015). Those packages allowed for an estimation of these models based on each of the 30 multiple imputation (MI) data sets and for computation of pooled parameter estimates and fit indices. Several models were tested and interpreted in the SEM analyses using the benchmarks provided in Section 5.2.2. The result of the confirmatory factor analyses (CFA) was that small-scale spatial abilities are best described in a two component model (Section 10.2) involving the subclasses object-based transformation abilities (OB) and egocentric perspective transformation abilities (EGO). A single-factor model, however, could also be used to describe small-scale spatial abilities (involving the factor 'Small'). Moreover, CFAs revealed that large-scale spatial abilities are best described in a three-factor model (Section 11.2), involving the subclasses survey mapping abilities (SurveyMap), abilities to solve production (Prod), and comprehension (Comp) tasks using environment-to-map and map-to-environment transformations, respectively. To investigate relations between small-scale and large-scale spatial abilities, two possible models (denoted as 'Full SEM' and 'Reduced SEM', see Table 12.14) were fitted to the data set with 215 cases preferably using the WLSMV estimator to deal with categorical data and bimodal distributions of sum scores (Section 5.2.2).

Similar to the approach to identifying a model to describe large-scale spatial abilities (Section 11.2), both the full and the reduced model also were estimated using MLR, to show robustness against the choice of the estimator. Moreover, WLSMV estimations were performed on two data sets of two different sample sizes (215 versus 240) to show robustness against the limitation of the sample size (Table 12.14). The full data sets with 142 cases only was not considered since the SEMs to be estimated involved a large number of parameters that needed to be estimated.

Table 12.14 Models Tested During Structural Equation Model Analyses

Model	Number of factors		Estimator	n
	Small-scale	Large-scale		
Full SEM I	2	3	WLSMV	215
Full SEM II	2	3	MLR	215
Full SEM III	2	3	WLSMV	240
Reduced model I	1	3	WLSMV	215
Reduced model II	1	3	MLR	215
Reduced Model III	1	3	WLSMV	240

Figure 12.1 shows the results of a SEM analysis relating measures of small-scale spatial abilities to measures of large-scale spatial abilities (Full SEM I). The exact-fit hypothesis for the model was not rejected, $\chi^2(80) = 69.38$, $p = .796$. $\chi^2/df = .87$, which was below the value of 2.0, that was considered as an indicator of good fit in the literature. The RMSEA was .00 with a 95% confidence interval [.00, .026]. The CFI and TLI were both 1.00, measure of how well does the model fit better than the baseline model. All indicated indices suggested good model fit (compare to Table 5.1). Since *semtools* does not provide robust and scales estimated for the fit indices, good fit was checked for all individual 30 data sets (see Table F.6, p. 96 in the ESM, for fit indices of the first 5 data sets).

All estimates of latent correlations were found to be significant. For each of the latent constructs, the estimated model reproduced the results of the individual CFAs. In other words, the latent correlation between OB and EGO was .79 as found in the CFA (Figure F.1, p. 95 in ESM), and the latent correlations between subclasses of large-scale spatial abilities were found to be .57, .71, and .76 (in comparison to .59, .72, and .77 found in the CFA, Figure 11.17, p. 316).

The significant latent correlations between both classes of spatial abilities indicated that both subclasses of small-scale spatial abilities, OB and EGO, correlated with all three subclasses of large-scale spatial abilities, SurveyMap, Prod and Comp. They did, however, not correlate to the same extent: First, low to moderate correlations were found between OB and EGO with SurveyMap (.27 and .39, respectively), but strong latent correlations were found with Prod and Comp (.47 to .71). Thus, the variance shared with measures that reflected the abilities to encode landmarks was lower (7% to 15%) than the shared variance with measures that reflected the abilities to update map-environment-self relation during movement in real space (22% to 50%).

Conversely, latent correlations with each of the subclasses of large-scale spatial abilities differed for both subclasses of small-scale spatial abilities. Latent correlation between OB and SurveyMap was lower (.27) than the one with EGO (.40). Latent correlations between OB and EGO with Prod were about the same (.50 and .47, respectively.) This was not the case when it came to correlations with Comp. Latent correlations between OB and Comp (.71) were higher than the ones with EGO (.63).

The results of the full SEM suggested that small-scale spatial abilities are partially but not completely dissociated from large-scale spatial abilities. To test whether the data supported the partial dissociation model (Figure 5.1, p. 155), the corresponding 95% confidence intervals were computed for each of the standardized parameter estimates (Table 12.15). Results were inconsistent with a model describing spatial abilities at both scales of space as a unitary construct since the 95% confidence

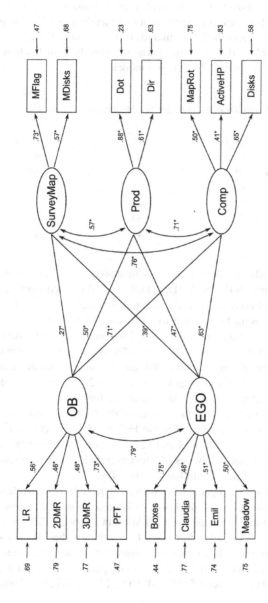

Figure 12.1 Completely standardized solution of the full SEM analysis relating two subclasses of small-scale spatial abilities to three subclasses of large-scale spatial abilities. *indicate highly significant path coefficients ($p < .001$). `indicate significant path coefficients ($p < .05$)

intervals did not include 1. They were also inconsistent with a model describing spatial abilities at both scales of space as a completely dissociated construct since the 95% confidence intervals did not include 0. To conclude, these empirical results indicated small-scale and large-scale spatial abilities do most likely relate to each other according to the partial dissociation model.

Table 12.15 95% Confidence Intervals for the Standardized Latent Correlation Estimates (n=215)

Correlation	Std. Estimate	CI_{low}	CI_{upper}
SurveyMap – OB	.27	.07	.48
SurveyMap – EGO	.39	.20	.58
Prod – OB	.50	.35	.65
Prod – EGO	.47	.27	.66
Comp – OB	.71	.54	.88
Comp – EGO	.63	.43	.83

To investigate whether the parameter estimates depended on the estimator or the sample used, the models Full SEM II and Full SEM III were computed. The exact-fit hypothesis for the Full SEM II model was not rejected, $\chi^2(80) = 72.85$, $p = .70$. The CFI and TLI were both found to 1.00. The RMSEA was .00 [.00, .03]. In addition, all individual data sets were found to have similar fit indices (see Table F.5 in ESM, p. 95, for the first five data sets). The parameter estimates for the latent correlation between OB and EGO was .80, and thus in line with the estimated latent correlation in the CFA as shown in Figure 10.1 (p. 267). The parameter estimates for the latent correlations among large-scale factors were found to be .59, .72, and .77 and were in line with the results of the CFA as shown in Table 11.27 (p. 315). The exact-fit hypothesis for the Full SEM III model was not rejected, $\chi^2(80) = 68.56$, $p = .81$. The CFI and TLI were both found to 1.00. The RMSEA was .00 [.00, .02]. Again, estimates of latent correlations between small-scale and large-scale factors reproduced findings of the CFAs.

Table 12.16 shows a comparison between estimates of latent correlations between subclasses of small-scale and subclasses of large-scale spatial abilities for the full SEMs I to III. Estimates for the same 30 MI data sets with 215 cases were found to be independent of the estimator used. Estimates for the MI data set with 215 cases and the one with 240 cases were slightly different. Estimates with Survey Map were lower (.27 compared to .25, .39 compared to .36), but estimates with Prod and Comp were higher (.50/.57, .47/.55, .71/.76, .63/.70). Those deviations were, however, sufficiently small, to consider the estimates to be independent of the sample used.

Table 12.16 Standardized Latent Correlation Estimates for Full SEM I to III

Correlation	Full SEM I (WLSMV, n = 215)	Full SEM II (MLR, n = 215)	Full SEM III (WLSMV, n = 240)
OB–SurveyMap	.27	.29	.25
EGO–SurveyMap	.39	.40	.36
OB–Prod	.50	.51	.57
EGO–Prod	.47	.48	.55
OB–Comp	.71	.72	.76
EGO–Comp	.63	.65	.70

Since a single-factor model was found to also describe small-scale spatial abilities, a first reduced model was computed in which a single factor representing small-scale spatial abilities (Small) was related to the three large-scale spatial factors. The exact-fit hypothesis model was not rejected, $\chi^2(84) = 80.95$, $p = .574$. Both CFI and TLI were 1.00 and the RMSEA was .00 [.00, .04]. The fit indices indicated therefore good model fit. Figure 12.2 shows the completely standardized solution of the reduced model. The factor Small is reflected by all OB and EGO measures and the factors SurveyMap, Prod, and Comp are reflected by the measures as shown in Figure 12.1. For the sake of simplicity and clearness of presentation, their loadings are omitted in Figure 12.2.

Latent correlation among subclasses of large-scale spatial abilities reproduced the results of the CFA (Figure 11.17). There were moderate latent correlation between Small and SurveyMap (.36 with a 95% confidence interval of [.17, .54]), a strong relation to Prod (.52 with a 95% confidence interval of [.37,.67]), and a strong relation to Comp (.72 with a 95% confidence interval of [.56,.87]). Similar to the results in the full model (Figure 12.1), the model reproduced the result that latent correlations with SurveyMap were smaller than latent correlations with Prod, which were, in turn, smaller than the ones with Comp, a result that was robust against influences of the estimator used and the sample used (Table 12.17). Furthermore, the model supports the partial dissociation between small-scale spatial abilities and subclasses of large-scale spatial abilities, since the corresponding confidence intervals to the estimates neither contain .00 nor 1.00.

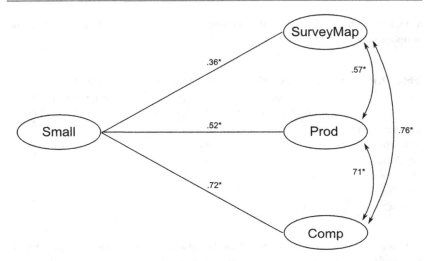

Figure 12.2 Completely standardized solution of the reduced SEM analysis relating the class of small-scale spatial abilities to three subclasses large-scale spatial abilities. *indicate highly significant path coefficients ($p < .001$)

Table 12.17 Standardized Latent Correlation Estimates for Reduced SEM II and Full SEM III

Correlation	Reduced SEM I (WLSMV, n=215)	Reduced SEM II (MLR, n=215)	Reduced SEM III (WLSMV, n=240)
Small–SurveyMap	.36	.37	.32
Small–Prod	.52	.53	.59
Small–Comp	.72	.73	.77

To test whether those results concerning the extent of overlap were even more generalizable and to prepare subsequent analyses, a second-order model involving one single factor for small-scale spatial abilities and a second-order model for large-scale spatial abilities[6] was fitted to the data. The exact-fit hypothesis was rejected, $\chi^2(87) = 113.22$, $p = .031$. The CFI was .096, the TLI was 0.95, and the RMSEA was .04 [.01, .06]. Given that the fit indices were not robust but only scaled, and given the result that not all of the 30 data sets were well explained by the model

---

[6]One general second-order latent factor, denoted as Large, and three subfactors SurveyMap, Prod, and Comp.

(results not reported here), those results indicated that this second-order model did not explain the data very well. It was therefore withdrawn from further analyses.

Finally, a group-wise reduced model was computed to test for possible sex differences. However, the high number of predictors involved in the model and the low numbers of cases left after splitting the sample in two groups resulted in estimation problems due to Heywood cases. Consequently, no results concerning the possible influences of sex differences on the relation between small-scale and large-scale spatial abilities can be reported.

In summary, results of latent correlation analyses indicated that small-scale spatial abilities are partially but not totally dissociated from large-scale spatial abilities. Those results were independent of the estimator and the sample used. The following section only reports the results of the WLSMV estimation based on the 30 MI data sets with 215 cases.

Results of the confirmatory factor analysis further allowed for determining the extent of overlap between both classes of spatial abilities and yielded moderate to strong relation. Results of the full SEM (Figure 12.1, p. 352), in partiular, indicated that not all subclasses of spatial abilities are related to the same extent. To examine this issue, latent regression analyses were computed.

12.2.2 Latent Regression Models

Since both subclasses of small-scale spatial abilities were highly related to each other ($r = .79$) and since all three subclasses of large-scale spatial abilities were related to each other ($r = .57$ to $r = .76$), some of the correlations reported in Figure 12.1 (p. 352) represent the latent correlations between small-scale and large-scale spatial abilities without taking into account the shared variance of subclasses of spatial abilities at each scale of space.

Two further structural equation models were computed to control for shared variance. A first one considered the shared variance between OB and EGO and described unique contributions of each of these two subclasses when it came to predict the three subclasses of large-scale spatial abilities. A second model considered the shared variance between SurveyMap, Prod, and Comp and described unique contributions of each of these three subclasses when it came to predict the two subclasses of small-scale spatial abilities. Since both models contributed to understand which subclass of spatial abilities was involved in predicting another subclass, those two models were denoted as latent regression models. The 'small-to-large' regression is reported first.

OB and EGO as Unique Predictors of Subclasses of Large-Scale Spatial Abilities

A first SEM taking into account the considerable amount of shared variance between OB and EGO (62%) was fitted to the data. The exact-fit hypothesis for the model was not rejected, $\chi^2(84) = 69.38$, $p = .796$, both the CFI and TLI were 1.00, and the RMSEA was .00 [.00, .03]. Consequently, the fit indices suggested good fit. Figure 12.3 shows the completely standardized solution in the first regression model. Loadings of the different measures reflecting each latent factor are omitted for the sake of simplicity of presentation. Because the model took into account the latent correlation between OB and EGO, the paths coefficients in the model are similar to multiple regression coefficients. In other words, they might be interpreted as unique contribution of each of the two subclasses OB and EGO to the prediction of variance in SurveyMap, Prod, and Comp while controlling for shared variance with the other variable.

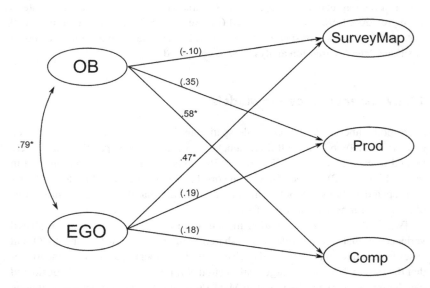

Figure 12.3 Completely standardized solution of the regression analysis with the two subclasses of small-scale spatial abilities predicting the three subclasses of large-scale spatial abilities. *indicate highly significant path coefficients ($p < .100$)

In line with what has been observed in Figure 12.1 (p. 352) and Figure 12.2 (p. 355), both OB and EGO predicted the lowest extent of variance in SurveyMap

(15.7%). They predicted about a quarter of variance in Prod (26.2%) and over half of variance in Comp (52.2%).

According to the assumptions outlined in Section 5.1.1, EGO but not OB was assumed to make a unique contribution to the prediction of SurveyMap. Both OB and EGO were assumed to make a unique prediction of Comp and Prod but OB was assumed to make a higher unique contribution. The standardized path coefficients show that the initial assumptions were only partly valid. EGO was found to make a significant contribution to SurveyMap (.47, $p = .079$). This subclass of small-scale spatial abilities, however, did not make a unique contribution to Prod and Comp. In other words, EGO was not able to contribute to the prediction of Prod and Comp above what has already been predicted by OB.

As expected, OB made no significant contribution to the prediction of Survey Map. It made a unique contribution to the prediction of Prod that was slightly not found to be significant (.35, $p = .117$).[7] It made a unique contribution to the prediction of Comp (.58, $p = .046$). In other words, in regard to those two subclasses, OB was most likely able to contribute to the prediction of Prod and Comp above what has already been predicted by EGO.

SurveyMap, Prod, and Comp as Unique Predictors of Subclasses of Small-Scale Spatial Abilities

A second SEM taking simultaneously into account the considerable amount of shared variance between SurveyMap and Prod (32%), between SurveyMap and Comp (58%) and Prod and Comp (50%) was fitted to the data. The exact-fit hypothesis for the model was not rejected, $\chi^2(81) = 77.85$, $p = .578$. In addition, both the CFI and TLI were 1.00, and the RMSEA was .00 [.00, .04]. Consequently, the fit indices suggested good fit (Table 5.1). Figure 12.4 shows the completely standardized solution in the second regression model. Again, loadings of the different measures reflecting each latent factor are omitted for a simplified presentation. Because the model took into account the latent correlation between SurveyMap, Comp, and Prod, the paths coefficients might be interpreted as unique contribution to each of the three subclasses to the prediction of variance in OB and EGO while controlling for shared variance with the other variable. Note that the path coefficient between SurveyMap and OB could not be computed due to numeric problems.

In line with what has been observed in Figure 12.1 (p. 352) and Figure 12.2 (p. 355), all three subclasses predicted about the same extent of variance in OB

---

[7]When considering an alternative model, however, in which OB was not allowed to load on EGO, this path coefficient became significant. In subsequent discussion, this path coefficient estimate will be considered as a significant one.

(38.8%) and EGO (35.6%). In other words, they were able to predict more than a third in variances in both subclasses of small-scale spatial abilities.

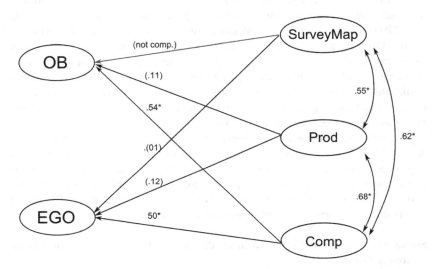

Figure 12.4 Completely standardized solution of the regression analysis with the three subclasses of large-scale spatial abilities predicting the two subclasses of small-scale spatial abilities. *indicate highly significant path coefficients ($p < .100$)

According to the predictions outlined in Section 5.1.1, SurveyMap and Comp, but probably not Prod, were assumed to make a unique contributen to the prediction of EGO. Comp and Prod were assumed to make a unique contribution to the prediction of OB. The standardized path coefficients followed the predictions only partly. SurveyMap was not found to make a significant unique prediction of EGO. Strictly speaking, SurveyMap was not able to contribute to the prediction of EGO above what has already been predicted by Prod and Comp. Similarly, Prod was not found to make a significant unique prediction of both OB and EGO, that is, it did not explain variance in those subclasses of small-scale spatial abilities above what has already been explained by SurveyMap and Comp. As expected, however, Comp made a significant contribution to explain variance in OB (.54, $p = .003$) and made also a significant contribution to explain variance in EGO (.50, $p = .012$).

12.2.3 The Role of Visuospatial Working Memory

To investigate the role of visuospatial working memory capacity (VSWM) concerning the relation between small- and large-scale spatial abilities, a latent factor representing visuospatial working memory capacity (VSWM) as reflected by the measures ATFS and ATBS was added to the reduced SEMs. This was done due to the fact that direct investigation of the question whether VSWM mediates small-scale and large-scale spatial abilities would have required to fit a second-order model to the data. A previous attempt, however, showed that this kind of model did not explain the data well. Figure 12.5 summarizes the results from the alternative approach taken to investigate the role of VSWM. In the figure, loadings of the different measures reflecting each latent factor are omitted and results that were reported above are shown in gray for a simplified presentation. The exact-fit hypothesis for the model was not rejected, $\chi^2(109) = 105.80$, $p = .569$. The CLI and TLI were 1.00, the RMSEA was .01 with a 95% confidence interval [.00, .03].

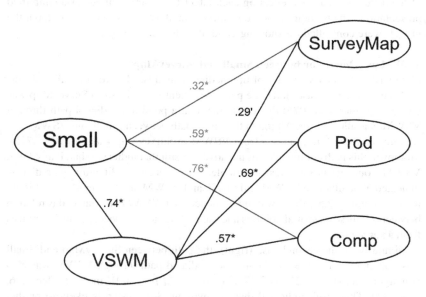

Figure 12.5 Completely standardized solution the parsimonious model including VSWM. *indicate highly significant path coefficients ($p < .001$). 'indicate significant path coefficients ($p < .05$)

Estimates of the latent correlations between VSWM and Small, but also between VSWM and SurveyMap, Prod, and Comp was in line with the results in small-scale (Figure 10.3, p. 274) and large-scale of space (Figure 11.19, p. 321). There was a high latent correlation between VSWM and small-scale spatial abilities ($r = .74$ with a 95% confidence interval [.57, .90]). There was a moderate correlation between VSWM and Survey Map ($r = .29$ with a 95% confidence interval [.08, .51]). The correlation between VSWM and Prod ($r = .69$ with a 95% confidence interval [.51, .87]) was higher than the correlation between VSWM and Comp ($r = .57$ with a 95% confidence interval [.36, .57]).

To further investigate whether VSWM capacity mediates the relationship between small-scale spatial abilities and large-scale spatial abilities, latent analyses in which both a direct path from the small-scale spatial abilities factor to each of the three subclasses of large-scale spatial abilities and an indirect path assuming that VSWM mediates the corresponding relation were considered and vice versa.

In the subsequent summarizes of the results of the mediation analyses, loadings of the different measures reflecting each latent factor are omitted for a simplified presentation. The fit indices for the mediation models are also not reported but the models were controlled for showing good fit to the data during analyses.

VSWM as a Mediator between Small and SurveyMap

Figure 12.6 shows a summary of the mediation analyses between Small and SurveyMap. In the first case (left), the path coefficient from Small to SurveyMap was non-significant ($p = .075$) due to the fact that a possible mediated path through VSWM was allowed for. Although the direct path between Small and VSWM was significant ($p < .001$), the one from VSWM to SurveyMap was not ($p = .725$). Moreover, this path coefficient was negative, a result that indicated that, if anything, VSWM would act as a suppressor variable. The result on this direction of prediction indicated, that although VSWM and Small, and VSWM and SurveyMap are highly related (Figure 12.5), there was no evidence that VSWM mediated the relation between small-scale spatial abilities and the abilities to encode and draw inferences from a survey map.

Similarly, in the second case (right), the path between SurveyMap and Small was non-significant ($p = .073$) and the mediated path through VSWM was also non-significant ($\beta_1 = .23$, p = .057, and $\beta_2 = .72$, $p < .001$, indirect effect: .16, $p = .11$). This result indicated that, if anything, SurveyMap is likely to predict Small even though VSWM is allowed to mediate this relation. Therefore, as for the other direction, VSWM was not likely to be a mediator concerning this relation.

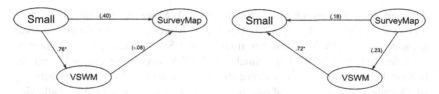

Figure 12.6 Analysis of VSWM as a mediator in the relationship between Small and SurveyMap. *indicate significant path coefficients ($p < .05$)

VSWM as a Mediator between Small and Prod

Figure 12.7 shows a summary of the mediation analyses between Small and Prod. In the first case (left), the path coefficient from Small to Prod became non-significant ($p = .075$) when a mediated path through VSWM was allowed for. Both the direct path from Small to VSWM and the one from VSWM to Prod were significant ($p < .001$ and $p = .022$, respectively.). For this first model, the total effect was significant ($p < .001$), the direct effect was close to significance ($p = .086$) and the indirect effect was non-significant ($p = .998$). Similar results were found in the second case (right). When VSWM was entered as a mediator, the direct path between Prod and Small became non-significant ($p = .998$) and the indirect path was close to significance ($p = .051$). These results suggested that VSWM was not only highly correlated to Small and Prod (Figure 12.5) but was most likely to fully mediate the relation between the class of small-scale spatial abilities and this subclass of large-scale spatial abilities allowing children to solve production tasks.

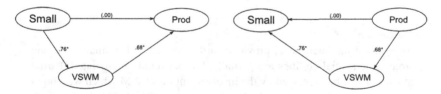

Figure 12.7 Analysis of VSWM as a mediator in the relationship between Small and Prod. *indicate significant path coefficients ($p < .05$)

VSWM as a Mediator between Small and Comp

Figure 12.8 shows a summary of the mediation analyses between Small and Comp. In the first case (left), the path coefficient from Small to Comp remained significant ($p = .006$) although a possible mediated path through VSWM was allowed for.

Although the direct path between Small and VSWM was significant ($p < .001$), the one from VSWM to Comp was not ($p = .89$). Moreover, this path coefficient was negative, a result that indicated that, if anything, VSWM would act as a suppressor variable. Although VSWM and Small and VSWM and Comp are highly related (Figure 12.5), there was no evidence that VSWM mediated the relation between small-scale spatial abilities and this subclass of large-scale spatial abilities allowing children to solve comprehension tasks.

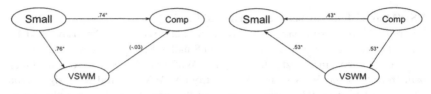

Figure 12.8 Analysis of VSWM as a mediator in the relationship between Small and Comp. *indicate highly significant path coefficients ($p < .05$)

In the second case (right), the path between Comp and Small remained significant ($p = .005$) but also the indirect, mediated path through VSWM was significant ($\beta_1 = .53$, p = .001 and $\beta_2 = .53$, $p < .006$, indirect effect: .43, $p = .0505$). This result indicated that Comp is likely to predict Small even though VSWM might, if anything, partially mediate this relationship.

Conclusion

Analyses at the latent level provided further evidence that small-scale and large-scale spatial abilities are partially, but not totally dissociated. Partial relation might be explained by the involvement of VSWM which was, however, not found to mediate both classes of spatial abilities. There were moderate to strong correlations at the measurement error-free level. There were especially high correlations between subclasses of spatial abilities involving reasoning with the help of a depiction (paper-and-pencil task, map). The empirical relations between OB and the two subclasses of map use were so substantial that OB remained a significant predictor even when controlling for shared variance with EGO. EGO, in turn, emerged as a unique predictor of SurveyMap. Vice versa, the relations between Comp and both classes of small-scale spatial abilities were so substantial that Comp remained a significant predictor when controlling for shared variance with the two other subclasses of large-scale spatial tasks.

12.3   Discussion of the Empirical Findings

This study related performances in the paper-and-pencil tests to the ones in the map-based tests to investigate the relation between children's small-scale and large-scale spatial abilities, that is, to study whether scale matters when it comes to children's spatial thought in two different settings in geometry education. Empirical relations were studied at the manifest and latent level. In addition, the role of individual spatial tasks and visuospatial working memory were addressed. This final section discusses the findings concerning the empirical relations found with respect to the theoretical framework outlined in Part I.

The Relation between Small-Scale and Large-Scale Spatial Abilities
In line with general psychological and environmental theories (e. g., ACREDOLO 1981; ITTELSON 1973; MANDLER 1983; MONTELLO 1993; WEATHERFORD 1982) and in line with previous studies in adults (e. g., ALLEN ET AL. 1996; HEGARTY ET AL. 2006; KOZHEVNIKOV, MOTES, RASCH, & BLAJENKOVA 2006) and children (FENNER ET AL 2000; LIBEN ET AL. 2013), this study found a partial but not total dissociation between small-scale and large-scale spatial abilities. Results from correlation, regression, and structural equation model analyses showed that there were moderate to strong empirical relations between both classes of spatial abilities. Results showing scale dependent differences in individual performances suggested, however, that both classes of spatial abilities were also dissociated. More specifically, the findings of this study demonstrated that an individual who performs well in paper-and-pencil tasks that require him or her to use small-scale spatial abilities, does not automatically perform well in map-based orientation tasks that ask him or her to use large-scale spatial abilities, and vice versa.

There are several explanations for the dissociation found. This study argued that spatial tasks that reflect both classes of spatial abilities involved different types of spatial information which, in turn, required the children to engage in different classes of spatial abilities. Whereas pieces of information in paper-and-pencil tasks can be perceived from one single vantage point, pieces of information in map-based orientation tasks require the individual to move in space and to perceive them from multiple vantage points. The relative size of this information, scale, becomes a determinant of spatial cognition in a way that influences the processing of the perceived information. The processing of small-scale spatial information required the children to encode an exact, probably metric image of the stimulus and to mentally transform it (e. g., WIMMER ET AL. 2016). To process large-scale spatial information they needed to integrate the distorted information from multiple vantage points in a survey map (e. g., DOWNS & STEA 1973; MEILINGER 2007; MONTELLO 1992). Consequently, although this study did not address the underlying spatial representation in detail, it is safe to assume that two qualitatively different spatial representations were involved in spatial abilities at both scales of space (e. g., ALLEN 1999b). This is also in line with results showing that children use different types of frames of reference to encode spatial information at different scales of space (e. g., ALLEN 1999a; BELL 2002).

An additional explanation might be that the small-scale spatial tasks in this study were developed in such a manner that they involved one type of mental transformation only. Some tasks reflected the abilities to perform mental transformations of objects whereas others required the children to transform the imagined self. Although some of the tasks, for example PFT, required the children to engage in multistep solution processes (LINN & PETERSEN 1985) and to hold spatial information in visuospatial working memory, those involved all the same type of mental transformation, namely manipulations of the imagined piece of paper. Similarly, some other tasks, such as Emil, required children to perform multistep mental transformations but they were all imagined transformations of the self in a labyrinth. In contrast, all the large-scale spatial tasks required children to perform different types of cognitive processes that were, most likely, highly intertwined such as mentally aligning the map to perform a dimensional transformation of spatial relations on the environment presented on the map to the environment (e. g., MACEACHREN 1995; OTTOSSON 1987; SEILER 1996). Consequently, the complexity of the cognitive processes involved might have led to the dissociation (LOCKMAN & PICK 1984).

A third explanation for the dissociation found might be that small-scale spatial tasks did not require the children to relate the depicted spatial information to the referent space. In other words, they had to imagine being on a playground rather than actually experiencing this space. Consequently, they had to engage in abstract problems of spatial thought that might have required them to use abstract spatial concepts (LIBEN 1981). When solving map-based orientation tasks, children had to constantly refer to the depicted real space while experiencing it. Consequently, they had to engage in rather specific problems of spatial thought. This would be in line with LIBEN's (1981) conceptual distinction between different types of spatial thought.

Another explanation for the dissociation among the tasks involving visual representation (paper-and-pencil tests, map-based tests) might be that solving written tasks require the children to perform a mental transformation of 2D abstract spatial stimuli into 3D mental representations, whereas when solving map-based tasks, individuals relate 3D realistic stimuli to a 2D representation. Those dimensional conversions have been empirically shown to be partially distinct, at least for mental rotation tasks (IWANOWSKA & VOYER 2013)

Finally, the test format itself might have accounted for the individual differences found at the different scales of space. Indeed, small-scale spatial abilities were measured using a test format that is well known for the children. In addition, solving paper-and-pencil tasks does not involve motoric engagement. In contrast, the test format of the map-based orientation test was unknown to the children and involved the interaction with an experimenter and demanded motoric engagement when answering the individual items. Given that previous studies have shown an impact of the test format on individual performances in measures of small-scale spatial abilities, for example, when comparing computer-based tasks with paper-and-pencil tasks (e.g., LOWRIE & LOGAN 2015; LOGAN 2015), it is likely to assume that the different test formats may have contributed, at least to some extent, to differences in performances at both scales of space.

The Role of Subclasses of Spatial Abilities at Both Scales of Space
The high latent correlation between both classes of spatial abilities indicated that they shared a considerable amount of variance. This finding suggests that although they rely on specific cognitive processes, both classes of spatial abilities seem to rely also on a set of common cognitive processes. These involve the maintenance of simultaneously perceived spatial information in working memory (e.g., ALLEN ET AL. 1996; COLUCCIA ET AL. 2007; HEGARTY ET AL. 2006). Moreover, the results of this study showed that a specific set of cognitive processes is most likely to be involved in spatial abilities at both scales of space.

Similar to HEGARTY ET AL. (2006), this study found low shared variance between small-scale spatial abilities and survey mapping abilities. This study identified an even smaller latent correlation (.36) concerning the extent of overlap than HEGARTY ET AL. (2006) did (.50), which can be explained by the fact that the subjects of their study were focused on learning environmental features, whereas the children in this study completed a range of tasks and, besides, needed to learn locations of landmarks on the campus which may have biased their actual performances in this task. Moreover, HEGARTY ET AL. (2006) used a broader set of measures than this study did, involving a measure of visuospatial working memory. Since visuospatial working memory has been shown to be closely related to small-scale spatial abilities in this study, including the measures in the small-scale spatial abilities test battery might have led to similar findings concerning the extent of overlap as in HEGARTY ET AL. (2006). The fact that small-scale spatial abilities and the abilities to encode a survey map are only related to a weak extent was further shown by results of multiple regression analyses. Indeed, performances in spatial tasks predicted under 10% of variances in the two tasks that required children to point to memorized landmarks.

Additionally, this study was able to show that this shared variance is particularly governed by the abilities to perform egocentric perspective transformations. Similar to previous findings in adults, this study found, both at the manifest and latent level, that the abilities to perform mental transformations of the self predicted unique variance for survey mapping abilities (e. g., ALLEN ET AL. 1996; FIELDS & SHELTON 2006; HEGARTY ET AL. 2006; KOZHEVNIKOV ET AL. 2006). In contrast, the abilities to perform mental transformations of objects were not only weakly related to survey mapping abilities but were also not able to predict unique variance for survey mapping abilities. This finding is in line with previous studies in children that showed that object manipulation abilities seem to be dissociated from the abilities to learn about an environment (JANSEN 2009; QUAISER-POHL ET AL. 2004). Consequently, cognitive processes that are unique for this subclass of small-scale spatial abilities, that is, updating the relation of the egocentric frame of references with the environmental and intrinsic one, seem to be involved in both subclasses of spatial thought, independently of scale. Egocentric perspective transformation abilities allow children to keep track of varying spatial relations between the self and landmarks during mental (as in the case of paper-and-pencil tasks) or actual movement (as in the case of map-based tasks) in an environment (that is implicitly given by the paper-and-pencil tasks or the directly surrounding real world). Indeed, to solve perspective tasks, children have to constantly manipulate their egocentric frame of reference (YU & ZACKS 2017) which is also the case during cognitive mapping in real spaces (FIELDS & SHELTON 2006; WOLBERS & WIENER 2014).

Similar to studies that investigated the issue at the manifest level (LIBEN ET AL. 2008; 2013), this study found strong relations between both subclasses of small-scale spatial abilities and both subclasses of large-scale spatial abilities requiring the children to keep track of the map-environment-self relations. Multiple regression analyses revealed that up to a quarter of variances in performances in solving production and comprehension tasks (Dots and Disks) could be explained by performances in spatial tasks. This degree of relations found is, at least for Dots, in line with what another study in children found in a similar setting (LIBEN ET AL. 2013). Results from structural equation modeling analyses revealed that there were strong relations between both subclasses of small-scale spatial abilities and both subclasses of large-scale spatial abilities involving map use. Additionally, this study was able to show that this shared variance was particularly governed by object manipulation abilities. Semipartial correlation analyses and results of a structural equation model analysis that took into account the shared variance between the two subclasses of small-scale spatial abilities, revealed that cognitive resources that are unique to object manipulation abilities appear to have affected the performances in tasks that require children to maintain the map-environment-self relation while moving in the environment. These involve the updating of relations with the intrinsic, object-based frame of reference with both the egocentric and environmental one. This might be the case during mental rotations of objects or mental scanning of objects (KOSSLYN ET AL. 1978; SHEPARD & METZLER 1971). The fact that mental rotations are involved when mentally aligning the map, has not only been shown in chronometric studies (ARETZ & WICKENS 1992; SHEPARD & HURWITZ 1984) but has also been shown to be an important strategy when it comes to using maps in real space (CHRISTENSEN 2011).

Results of regression analyses have shown that performances in small-scale spatial tasks are not only predictors of performances in large-scale ones, but also performances in large-scale spatial tasks that account for up to a third of variance in performances in small-scale spatial tasks. Consequently, large-scale spatial abilities are considerably related to small-scale ones. This study was the first one to investigate the role of individual subclasses of large-scale spatial abilities when it comes to explaining relations between both subclasses of spatial abilities. Results from semipartial correlation analyses and results from a structural equation model analysis revealed that the subclass of abilities that enable children to solve comprehension tasks accounted for shared variance with both subclasses of small-scale spatial abilities. Consequently, cognitive processes that are unique to solving these tasks, developing a hypothesis on how two-dimensional abstract information might look like in three-dimensions using visualization, seem to be involved in all these subclasses of spatial abilities, independent of scale. Indeed, visualization has been described as the mental operation of seeing in the mind's, egocentric eye what has been depicted in a spatial stimuli which is also perceived from an egocentric perspective (LIBEN 1999; LOBBEN 2004; OTTOSSON 1987). Consequently, the spatial relations depicted in a map might be similar to what has been described as decoding of spatial information in paper-and-pencil tasks (LOWRIE & LOGAN 2018).

The Role of Individual Spatial Tasks
Results of hierarchical multiple regression analyses revealed a set of tasks that were most likely to play a particular role when it came to understanding the relations

between small-scale and large-scale spatial abilities at the level of tasks reflecting those latent constructs.

Similar to the findings in adults, this study found PFT to be an important predictor when it comes to locating oneself on a map (LIBEN ET AL. 2008, 2010) but also in comprehension tasks such as placing disks and pointing to unseen landmarks with the help of the map (LIBEN ET AL. 2010). This contrasts the findings of LIBEN ET AL. (2013) who did not identify PFT but the WLT as an important predictor of self-locating on a map. One possible explanation might be that their study used the original speeded PFT by EKSTROM, FRENCH, HARMAN, AND DERMEN (1976) with an instructional item, whereas this study used a set of age-specific items that could be solved unlimited in time and was introduced by a demonstration of the paper folding procedure. One interpretation for PFT being such an important predictor might lie in the task's nature to require multistep solution processes involving a sequence of transformations of complex stimuli (LINN & PETERSEN 1985). Consequently, the individual is required to maintain different pieces of information in working memory. This is also important when performing tasks that mimick the everyday use of maps in real space (LIBEN ET AL. 2010).

LR emerged as an important predictor for the tasks Dots and Disks which are complex and highly dynamic tasks of map use in real space. In line with previous findings in adults, the abilities to mentally rotate a map is an important predictor (e. g., LOBBEN 2007). In addition, this study did not find any predictive relations between LR and MFlag or MDisk, a result that is in line with previous studies on layout learning using maps on adults (FIELDS & SHELTON 2006). Indeed, active map use in real space requires the subject to successfully relate the map to the environment, both when identifying the own location and when navigating real space with the help of the map. Given the fact that the children in this study were not allowed to turn the map during usage and given that the spatial layout of the campus is rather perpendicular, this result is not surprising. Indeed, solving the LR task demands children to subsequently mentally turn the map to determine whether a turning is left or right. When children determined their location on the map or used the map to navigate to a landmark, they were also required to mentally turn the map and to track left-right turning of their body in real space to the corresponding left-right turnings on the map. This strong relation found further empirical support by the fact that both tasks, Dots and Disks, were significant predictors of performances in LR in regression analyses. Thus, cognitive processes similar to mental rotation seem to be involved in all of these tasks (STEINKE & LLOYD 1983; SHEPARD & HURWITZ 1984; ARETZ & WICKENS 1992).

Similar to findings with children in real space (LIBEN ET AL. 2013) and in the classroom (LOGAN ET AL. 2017), this study found that the two perspective taking tasks Boxes and Meadow were important predictors when it came to solving tasks with the help of a map. Both tasks require children to align possible depicted viewpoints with a certain location. In the task Boxes, this location is situated on a plan, and in the task Meadow it has to be imagined on a cognitive map. While solving these tasks, one might argue that children visualize how two-dimensional information would appear in three-dimensions (Boxes) or one might argue that they are required to transform spatial information perceived from their egocentric perspective in some kind of bird's-eye perspective that, in turn, allows them to specify which way they have taken (Meadow). Thus, results support the idea that being able to imagine how configurations look from different perspectives is an important skill when it comes to determining the own location on a map while solving production but also comprehension tasks.

This study was the first one to investigate the role of individual large-scale spatial tasks when it came to the prediction of performances in small-scale spatial tasks. Results showed that the tasks Dots, Disks, and ActiveHP that require the children to constantly transform information from the map to the environment and vice versa while being aware of the map-environment-self relation during movement in real space, were good predictors of performances in paper-and-pencil spatial tasks. In other words, large-scale spatial tasks that required the children to engage in multistep solution processes based on complex spatial information and which involved a range of dimensional transformation seemed to be particularly related to small-scale spatial tasks.

The Role of Visuospatial Working Memory
In line with the literature (e. g., SHOLL & FRAONE 2004), this study found strong latent correlations between visuospatial working memory capacity and both classes of spatial abilities. Visuospatial working memory capacity was found to be especially related to both subclasses of small-scale spatial abilities but also to survey mapping and to the abilities to solve production tasks and comprehension tasks.

Results of hierarchical multiple regression analyses provided further insights into the results found at the latent level. When it came to predicting performances in large-scale spatial tasks using multiple regression models, measures of visuospatial working memory emerged as important predictors in the production tasks Dots and Dir. Results of hierarchical multiple regression analyses for those two tasks revealed that when measures of visuospatial working memory were added to models predicting performances in Dots and Dir by means of small-scale spatial tasks, there was a significant improvement of the total variance explained. In other words, when entering measures of visuospatial working memory as predictors in the models, they significantly contributed and beyond what has already been explained by performances in small-scale spatial tasks to explain variance in both production tasks. This was not the case when it came to predictions of the other map-based orientation tasks. Consequently, in addition to the maintenance of spatial information as required when solving smallscale spatial tasks, visuospatial working memory capacity played an important role when it came to solve production tasks (see Section 11.4, for further discussion).

When it came to predicting performances in large-scale spatial tasks using multiple regression models, measures of visuospatial working memory emerged as important predictors in all small-scale spatial tasks. Results of hierarchical multiple regression analyses for those tasks revealed that in case measures of visuospatial working memory were added in models predicting performances by means of large-scale spatial tasks, there was a significant improvement of the total variance explained. In other words, when entering measures of visuospatial working memory as predictors in the models, they significantly contributed and beyond what has already been explained by performances in large-scale spatial tasks to explain variance in all paper-and-pencil tasks. Consequently, in addition to the maintenance of spatial information as required when integrating information from multiple vantage points and when maintaining information perceived on a map, visuospatial working memory capacity played an important role when it came to solve small-scale spatial tasks (see Section 10.5, for further discussion).

Results of mediation analyses at the latent level generalized these findings. Visuospatial working memory was found to fully mediate the relationship between small-scale spatial abilities and the subclass of solving production tasks. In contrast to findings in adults (e. g., ALLEN ET AL. 1996), visuospatial working was not found to mediate anything when it came to the relation with the abilities to encode a survey map. Moreover, it partially mediated the relation between the abilities to solve comprehension tasks and small-scale spatial abilities.

A tentative interpretation of these results would be that this demonstrates how challenging production tasks are for children. They require them to integrate multiple spatial cues to develop a hypothesis concerning their self-location on the map which needs to be tested and refined to finally knowing where they are located on the map. To do so, children have to carefully select important features in the environment which is an effortful cognitive task that requires them to control attention. Consequently, solving production tasks can be considered as a particular problem solving tasks (see also LOBBEN 2004). The fact that problem solving is related to visuospatial working memory capacity has been emphasized in the literature (e. g., HAMBRICK & ENGLE 2003).

Discussion, Limitations, and Implications 13

Children need to process spatial information perceived in spaces of different sizes in their everyday life. Although geometry education aims to address phenomena of children's day-to-day experiences, there has been little research on the impact of scale on children's spatial thought. This study addressed this gap in the literature and examined the effect of scale when it comes to children's spatial cognition in geometry education and considered to which extent small-scale spatial abilities reflect the same or a different class of spatial abilities than large-scale spatial abilities. In particular, this study investigated the relation between the former class, which allow children to solve paper-and-pencil tasks, for example in a classroom setting, and the latter, which allow children to solve map-based orientation tasks that mimic map use in real space in everyday life.

Part I of this study addressed the cognitive dimension of scale. Chapter 2 described, from a theoretical point of view, how scale affects spatial behavior in an indirect manner. The theoretical framework of this study described how scale, when it is not only considered as a spatial but also as a psychological construct, differentiates space into different classes with respect to the type of interaction that is afforded by the corresponding spatial information available. It further described how scale might lead to differences in the type of information processing that could contribute to differences in spatial behavior and, consequently, in differences in performances in spatial tasks. Chapter 3 provided theoretical insights concerning the modes of processing using small-scale and large-scale spatial information and showed that scale, indeed, indirectly implies different processing demands on individuals not only during encoding and maintenance of those pieces of information, but also when it comes to making inferences from the corresponding representations. The analyses of cognitive processes allowed for development of hypotheses concerning the question of how small-scale and large-scale spatial abilities could be related and how those abilities could be modeled. Chapter 4, then, reviewed

C. Heil, *The Impact of Scale on Children's Spatial Thought*, Studien zur theoretischen und empirischen Forschung in der Mathematikdidaktik, https://doi.org/10.1007/978-3-658-32648-7_13

empirical evidences for the main assumptions of this study and showed that scale indeed may influence individual performances in spatial tasks, an issue that has not sufficiently been addressed when it comes to investigating children's spatial thought.

Part II addressed the empirical dimension of this study. Chapter 5 stated the research question and described the design that was chosen to answer the research questions and introduced the methodological background by addressing the literature on psychometric measuring, multivariate analyses methods involving semipartial correlations, multiple regression, confirmatory factor analyses, structural equation modeling, and multiple imputation treatment of missing data. Moreover, this part involved a presentation of the set of different paper-and-pencil tasks that were developed to reflect the two-component model describing the construct of small-scale spatial abilities (Ch. 6). It also presented a set of seven map-based orientation tasks that were developed to reflect the multidimensional model describing the construct of large-scale spatial abilities (Ch. 7). In Chapter 8, additional material of the empirical study was presented, and Chapter 9 specified the testing procedure.

Part III of this study reported the empirical results concerning small-scale spatial abilities (Ch. 10), large-scale spatial abilities (Ch. 11), and the relation between both classes (Ch. 12). Results of the analyses were reported and discussed in detail with respect to the theoretical framework established in Part I.

This chapter discusses the findings. After addressing limitations of this study, the practical implications for geometry education and implications for future research are specified. Chapter 14 concludes.

13.1   Discussion of the Empirical Findings

This section summarizes the findings of this study and discusses them with respect to the three hypotheses outlined in Section 5.1.1. Before doing so, this section discusses the individual differences as measured by the two different test instruments. It also discusses to what extent sources of individual differences can be explained by the hypothesized theoretical models representing small-scale and large-scale spatial abilities.

13.1.1   Children's Individual Differences in Spatial Abilities at Different Scales of Space

The results of this study demonstrate that many children are able not only to solve a variety of spatial tasks presented in a written setting but also to keep track of the changing relations between the self and features of the environment during map-based navigation in real space at the end of primary school. At that time, spatial abilities allow children to engage in spatial activities that require them to perform multiple mental transformation of complex spatial information not only in rather limited settings of geometry education but also in large and unknown spaces that surround them. This is not to say that these tasks do not pose a challenge. But the fact that even many adults still struggle to mentally engage with complex spatial stimuli or to use a map in real space, in particular whenever it is not aligned with an unknown environment, demonstrates that children have developed their spatial abilities to a considerable extent by the end of primary school.

The findings of this study showed that paper-and-pencil spatial tasks were rich sources for cognitive reasoning. They required the children to engage with maps, labyrinths, and configuration plans, to mentally fold papers, to rotate complex two- and three-dimensional objects, to take imagined perspectives at discrete locations in space, and to imagine perspective shifts along routes. The descriptive analyses showed that some children scored high in almost all tasks, a result that indicated that the small-scale spatial tasks were suitable for age-specific testing. Because some children scored very low in almost all tasks, it can be assumed that the tasks were not too easy for that age group. Those findings suggest that the underlying spatial abilities that are reflected by those measures, small-scale spatial abilities, are well but not yet fully developed by the end of primary school.

This study proposed the use of maps as cognitive tools and as a means to observe how children reason about the varying spatial relations in real space. The results demonstrated that maps can be rich sources for spatial thought. In tasks that mimic everyday uses, children were asked to locate themselves on the map, to identify their viewing direction, to mentally rotate the map for indicating directions toward unseen landmarks, and to navigate toward pre-defined goals. Strictly speaking, they had to solve a range of comprehension and production tasks, and, in a subsequent step, to make inferences on directions toward encountered places from knowledge stored in the survey map. The descriptive results of this study suggested that almost all of map-based orientation tasks represented a challenge. Even without movement in space, children struggled to establish geometrical correspondences between spatial relations in real space and the map as well as to indicate directions toward unseen landmarks. Moreover, it was surprising to see that some children had a hard time to locate their position on the map even after walking a short distance to the first flag, whereas others found their location on the map with ease, even after walking a long route involving traverses of buildings. This might have been due to the particular demand of this study that children were not allowed to physically turn the map. Those findings suggest that the underlying spatial abilities that are reflected by those measures and the chosen experimental conditions, large-scale spatial abilities, seem not to be fully developed by the end of primary school.

One important finding of this study is that spatial abilities should be considered as a differentiated, multidimensional construct at both scales of space. Results from confirmatory factor analyses demonstrated that measures of small-scale spatial abilities defined two separable factors. Those were characterized by the underlying mental transformations that were intended to be used when solving the corresponding tasks. Although all measures required the children to encode the visual stimuli and to maintain a high quality depictive representation in visuospatial working memory for further transformation, measures reflecting the abilities to perform object-based transformations (OB) did not reflect the same subclass of abilities as measures reflecting the abilities to perform imagined transformations of the egocentric perspective (EGO). These mental transformations emerged as cognitive processes that seemed to be unique for each subclass of small-scale spatial abilities. Consequently, although both subclasses of small-scale spatial abilities were highly correlated ($r = .80$) and shared a considerable amount of variance, good performances of individuals in OB tasks may not automatically lead to good performances of the same individuals in EGO tasks.

Moreover, the findings of this study suggested that sources of individual differences in measures of large-scale spatial abilities reflected three distinct subclasses of cognitive abilities that are characterized by the involvement of a map and the type of the map-based task performed. Although all measures required the children to integrate and process visuospatial information from multiple viewpoints and to maintain the resulting fragmented, distorted and non-metric information in working memory, measures that reflected the abilities to mentally represent, integrate, and to make inferences from spatial relations that have been perceived and experienced in real space and on the map (SurveyMap) did not reflect the same subclass of abilities as measures that required the children to reason about the constantly changing map-environment-self relation (MapEnvSelf). Furthermore, measures that reflected the abilities to using their environment-to-map transformation abilities to solve production tasks by making inferences from multiple positions perceived in space on the own location on the map (Prod) did not reflect the same subclass of abilities as measures requiring the children to use their map-to-environment transformation abilities to solve comprehension tasks by reasoning about spatial relations in real space based on spatial information found on the map (Comp). Consequently, although all three subclasses of large-scale spatial abilities were highly correlated ($r = .59$ to $r = .77$) and shared a considerable amount of variance, good performances of individuals in one subset of tasks, for example production tasks, may not necessarily imply that they will achieve similar results in another subset of tasks, for example comprehension tasks.

13.1.2 The Relation Between Small-Scale and Large-Scale Spatial Abilities

The major goal of this study was to assess whether and to which extent small-scale spatial abilities and large-scale spatial abilities are related. The first hypothesis H_1 stated that there is a partial relationship between the class of small-scale spatial abilities and the class of large-scale spatial abilities, that is, that the partial dissociation model describing spatial abilities at different scales of space holds true.

The results indicated that spatial abilities in primary children should not be treated as an undifferentiated construct. The empirical findings provided evidence for a partial rather than a total dissociation between the class of small-scale spatial abilities that were reflected by a set of paper-and-pencil measures and the class of large-scale spatial abilities that were reflected by a set of map-based orientation measures. H_1 may therefore be considered to be true.

These findings offer a strong support for the assumption that scale has—at least partially—an impact on children's spatial thought. Indeed, it requires them to cognitively engage with different pieces of spatial information generated by different types of interaction with space that children are needed to master. These qualitative differences in the children's behavior are afforded by different types of information processing resulting from the scale of the space they engage with and, consequently, different types of cognitive engagement with the changing frames of reference when reasoning about the relations between a spatial product, the space that surrounds them, and themselves.

In the case of small-scale spatial abilities, the spatial product involved in the tasks is typically an abstract depiction of some kind of object or spatial situation of everyday life. While solving the spatial tasks, children only have to ascribe meaning to the space that surrounds them to the extent that it might lead to conflicting frames of reference. Children have to be aware that their initial egocentric perspective on the representation might be needed to be suppressed to solve some of the tasks, in particular the tasks asking for perspective changes. In other words, when solving small-scale spatial tasks, children have to reason about a time-invariant physical relation between themselves, the depicted spatial information, and the surrounding environment. They do not physically have to interact with the space they are mentally engaged in since they can perceive all information needed from a single vantage point. Consequently, children's small-scale spatial thought is characterized by abstract reasoning about space, that is, understanding it in terms of spatial concepts.

In the case of large-scale spatial abilities used by individuals to stay oriented in real space with the help of spatial product, the map, especially during movement, children have to constantly be aware of the changing, time-variant relations between the map, the environment, and the self that result from movement in space. By doing so, children do need to ascribe a meaning to the spatial relations depicted on the map by considering their egocentric perspective on the map and the environment as well as by aligning the map, at least mentally, with the environment. During movement in space, they constantly refer to both the map and the depicted space, that is, the spatial information that they perceive in the surrounding environment and that they constantly use for the development of spatial hypotheses needed for locating themselves on the map or to test the ones established for identifying the location of landmarks. In other words, when solving large-scale spatial tasks, children have to reason both about the map and the space represented on the map in an intertwined way, thereby processing and assimilating different types of spatial information from multiple vantage points. Consequently, children's large-scale spatial thought is characterized by specific reasoning about space, that is, to understand it in terms of locations, directions, and distances.

The empirical results, however, also indicated that both classes of spatial abilities are not completely dissociated. Indeed, the extent to which both classes of spatial abilities were related was high, with latent correlations ranging from .27 to .71, indicating that the amount of individual variances in the two latent variables tended to overlap up to 50%. These results showed that despite the fact that scale affects the information processing when engaging in different classes of spatial cognition, there seem to be some shared cognitive processes. Understanding those was also a goal of this study. Empirical findings that yield assumptions about possible underlying cognitive processes based on the theoretical framework will be discussed in a next step.

13.1.3 The Role of Subclasses of Spatial Abilities

The second major goal was to examine the patterns of correlations found to disentangle the empirically observed partial relation between small-scale and large-scale spatial abilities. The second hypothesis H_2 stated that not all subclasses of small-scale spatial abilities are related to the same extent to all subclasses of large-scale spatial abilities, a fact which might be explained by a subset of spatial tasks at both scales of space.

The correlations found among the two subclasses of small-scale spatial abilities, OB and EGO, and the three subclasses of large-scale spatial abilities, SurveyMap, Prod, and Comp, in the full correlation model (Figure 12.1, p. 352) and the reduced model (Figure 12, p. 355) showed that not all of these subclasses were related to the same extent. This plausible observation was confirmed in a latent regression analysis that indicated that OB and EGO explained about almost a fifth of the variance in SurveyMap, more than a quarter of the variance in Prod, and more than half of variance in Comp. The three subclasses of large-scale spatial abilities were able to explain more than a third of variance in each subclass OB and EGO. H_2 may therefore be considered to be true.

These findings suggest that the relation between small-scale and large-scale spatial abilities might be higher among subclasses involving the spatial reasoning with a spatial product. The relation seems to be less specific with the subclass involving spatial reasoning based on the information encoded from direct experience in space. Consequently, both classes of spatial abilities seem to rely particularly on cognitive processes that allow children to engage with the varying frames of reference induced by the use of a spatial product. Although this finding suggested that shared underlying cognitive processes might rely on those cognitive processes that involve perceiving and inference making from visual spatial information, this result was rather unspecific.

In addition, the results were biased due to the high latent correlations between factors of small-scale and large-scale spatial abilities. In other words, although results suggested that the subclasses OB and EGO, for example, were related to a considerable amount with Prod and Comp, this might have been due to the shared variance between the subclasses of spatial abilities at both scales of space only. Latent regression analyses that considered the shared variance between subclasses confirmed these observations for the full correlation model. When partialling out the shared variance, only a few subclasses remained significantly related to each other. The following sections present results of these analyses and interpret them with respect to the theoretical framework.

The Role of Object-Based Transformations
This study found relations between OB and all three subclasses of large-scale spatial abilities, with particular high ones to Prod and Comp. A subsequent goal was to identify whether those were so substantial that their involvement in those subclasses of large-scale spatial abilities were still significant when controlling for shared variance with EGO. An initial assumption derived from the literature stated that this could be expected for the involvement of OB in Prod and Comp, but not for SurveyMap.

Results of a latent regression analysis showed that the involvement of OB in both Prod and Comp, but not on SurveyMap, was so substantial that OB emerged as a significant predictor of both factors Prod and Comp when controlling for the shared variance with EGO. Consequently, OB did most likely contribute to explain variances in Prod and Comp above what can be explained by EGO. In other words, the results suggested that object-based mental transformations, a unique characteristic of OB, were substantially used by the children for making inferences when solving production and comprehension tasks. Indeed, children were not allowed to physically turn the map. Consequently, they had to mentally align the map; a cognitive operation that requires the child to update the relations of the environmental and egocentric frame of reference with the intrinsic frame of reference of the map, that is, to perform an object-based transformation. In addition, they constantly had to scan the map for spatial information, another important object-based transformation. The initial assumption therefore turned out to be true.

The findings suggested that those latent relations emerged from relations in performances of the tasks requiring children to decide on left-right turns in maps (LR) and to imagine folding processes, a multistep spatial reasoning task with complex spatial stimuli (PFT). Those tasks did not only allow for differentiation between children with lower and higher performances in small-scale spatial tasks, but performances in these tasks were found to be important predictors for performances in the set of large-scale spatial tasks. This finding was even robust when controlling for possible influences of sex differences. It is therefore most likely to assume that these tasks contributed to explain the empirical relation between OB and Prod, and OB and Comp in an important manner.

The Role of Imagined Transformations of the Egocentric Perspective

This study found relations between EGO and all three subclasses of large-scale spatial abilities, with particular high ones to the subclass Comp, but also with SurveyMap and Prod. A subsequent goal was to identify whether those were so substantial that their involvement in subclasses of large-scale spatial abilities were still significant when controlling for shared variance with OB. An initial assumption derived from the literature stated that this could be expected for the involvement of EGO in all three subclasses.

Results of a latent regression analysis showed that the involvement of EGO in SurveyMap, but not in Prod and Comp, was so substantial that EGO emerged as a significant predictor for SurveyMap when controlling for the shared variance with OB. The initial assumption therefore turned out to be partially true. EGO did most likely contribute to explain variances in SurveyMap above what has been explained by OB. In other words, the results suggested that egocentric perspective transformations, a unique characteristic of EGO, might have been used by the children for making inferences from the encoded survey map. Indeed, accurate performance in pointing to landmark tasks seems to be based on the organized encoding of distorted spatial information in one coherent mental representation and to accurately manipulate it to make inferences on possible directions from an egocentric point of view. Imagined transformations of the imagined self allow individuals to mentally organize and integrate pieces of information from spatial stimuli from different perspectives. By doing so, they are able to make inferences on all the spatial relations involved, not only with respect to the own viewpoint but also for imagined ones. During reasoning, they mentally update the relations of the environmental and intrinsic frame of references with the egocentric one. Consequently, abilities to perform transformation of the imagined self allow an individual to literally gain insight into a situation as is also the case when perceiving and processing information from multiple vantage point and when being aware of the own locations in space with respect to landmarks.

The results suggested that empirical relations emerged from relations in performances of the tasks requiring children to reason about views from different vantage points (Boxes), imagined perspective shifts (Meadow), and imagined views along a route (Emil). Performances in those tasks predicted performances in pointing tasks without a map. They were, however, not found to be robust against possible influences of sex differences. It is therefore likely to assume that those tasks contributed to explain the empirical relation between EGO and SurveyMap, but only for the boys.

Latent regression analysis showed that performances in EGO did most likely not contribute to explain Prod and Comp above what has already been explained by OB. That is, when it came to solving production and comprehension tasks, not the egocentric transformations per se might have led to the relations found in the correlation model. Those were most likely due to the shared cognitive processes with OB, such as maintenance of spatial information in working memory. In other words, the results suggested that it remains unclear whether egocentric transformations are used by children for making inferences when establishing configurational correspondences and developing spatial hypotheses during solving production and comprehension tasks. It is not clear whether visualizing how spatial relations on the map could look like in the real environment or vice versa is qualitatively similar to performing a transformation of the imagined self.

Indeed, the results did not clearly indicate that egocentric perspective transformations were not involved at all during solving of production and comprehension tasks. Increased uncertainty arose from the fact that tasks requiring children to reason about views from different vantage points (Boxes) and imagine perspective shifts (Meadow) were important predictors when it came to the prediction of map-based performances. This held true when considering possible effects of sex differences. Consequently, although EGO did not predict Prod and Comp above the extent which was already explained by OB at the latent level, findings at the manifest level suggested that they might be involved, at least to a small extent.

The Role of Self-to-Space Experiences During Survey Mapping
This study found small relations between SurveyMap with EGO and OB. A subsequent goal was to identify whether they were substantial or resulted from shared variance with Prod and Comp. An initial assumption stated that this was to be expected for the involvement of SurveyMap in the subclass, EGO, but not for OB.

When controlling for shared variance with Prod and Comp, SurveyMap did not contribute to explain EGO and OB above what has been explained by the other two subclasses. The latter findings were rather surprising since analyses of semipartial correlations at the manifest level suggested that SurveyMap was an important, yet not significant predictor for EGO. When additionally considering those findings, one could argue that when individuals keep track of the varying relations between the self and landmarks during movement, they enrich spatial experiences with the coordination of different perspectives. These perspectives can then be emulated when engaging in tasks requiring mental forms of taking different perspectives. When controlling for the influence of production and comprehension tasks, neither MFlags nor MDisks significantly predicted performances for EGO tasks. Consequently, solving those tasks required cognitive processes that were involved in solving production or comprehension tasks such as maintenance of visual information in working memory when using the arrow-and-circle device. The initial assumption therefore turned out to be false.

The Role of Environment-To-Map Transformations
This study found high relations between Prod with both subclasses of small-scale spatial abilities. A subsequent goal was to identify whether those were so substantial that their involvement were still significant when controlling for shared variance with SurveyMap and Comp. An initial assumption derived from the literature stated that this was probably the case for the involvement of Prod in OB, but not for EGO.

When controlling for shared variance with each of the two other subclasses of large-scale spatial abilities, Prod predicted neither OB nor EGO. Consequently, Prod did most likely not contribute to explain variances in OB and EGO above what has been explained by SurveyMap and Comp. That is, when it came to solve small-scale spatial tasks, the dimensional transformations of three-dimensional information to map two-dimensional ones might have probably not explained the initial correlations that were found. Those seemed to be rather due to shared cognitive processes with Comp and SurveyMap, such as mental map alignment or map-to-environment transformations. In addition, locating oneself on the map, a typical production tasks, has been found to be a prerequisite for solving comprehension tasks. Partialling out shared variance with Comp out, in turn, might have obscured possible unique involvement of Prod.

In other words, the results suggested that it remained unclear whether cognitive processes, that were unique for solving production tasks, were used by the children for solving small-scale spatial tasks, in particular because the semipartial correlations at the manifest level suggested that Prod had some of important involvement in these tasks. This assumption was further supported by the fact that the task Dots significantly predicted six out of eight small-scale spatial tasks. This finding was also often robust against influences of sex differences. However, performances in Dir were not found to have significant influence on performances in small-scale spatial tasks. Taken together, these findings suggested that the initial assumption was most likely found to be false.

The Role of Map-To-Environment Transformations
This study found high relations between Comp with both subclasses of small-scale spatial abilities. A subsequent goal was to identify whether those were so substantial that their involvement was still significant when controlling for shared variance with SurveyMap and Prod. An initial assumption derived from the literature stated that this was the case for the involvement in both subclasses of small-scale spatial abilities.

Indeed, the involvement of Comp was so substantial that when controlling for shared variance with SurveyMap and Prod, Comp emerged as a significant predictor for both OB and EGO. Consequently, Comp did most likely contribute to explain variances in OB and EGO above what has been explained by SurveyMap and Prod. In other words, the results suggested that spatial hypotheses development based on map-to-environment transformations, unique characteristics of Comp, were substantially used by the children for solving small-scale spatial tasks. Although solving comprehension tasks also involves mental map alignment, performing environment-to-map transformations, and being able to indicate directions with respect to the egocentric frame of reference, individual differences due to these cognitive processes were most likely partialled out when controlling for shared variance with Prod and SurveyMap.

As a consequence, map-to-environment transformations, that is, visualization of information from a map to the environment, and spatial hypothesis development may have contributed to solutions processes in small-scale spatial tasks. Indeed, children were given two-dimensional material of mostly three-dimensional objects or real world-situations. When solving OB tasks, for example, children needed to perform a dimensional transformation of the stimulus to transform two-dimensional information in some form of three-dimensional mental image, that can then manipulated and updated. When solving EGO tasks, children needed to formulate a spatial hypothesis of how flat, two-dimensional views of objects might appear to be three-dimensional, for example for reasoning about objects in fore and background. In other words, they had to visualize objects and situations in their inner eye, probably by using knowledge from long-term memory such as how to read areal plans and perspective drawings or to draw upon schemata to understand the depicted real world situations based on what they had experienced before.

The findings suggest that indicating directions of important landmarks (ActiveHP) and navigating towards a pre-defined goal with the help of the map (Disks) was not only found to be important measures to differentiate between children with lower and higher performances in large-scale spatial tasks, but performances in these tasks were also found to be important predictors of performances in all small-scale spatial tasks, a finding that was even robust when controlling for possible influences of sex differences. It is most likely to assume that they explained the empirical results concerning the relations outlined above.

Taken together, these findings allowed for explaining underlying cognitive processes involved in the use of a spatial product at both scales of space, that is, when perceiving and drawing inferences from visual spatial information. The findings outlined above suggest that at both scales of space, children have to complete some kind of dimensional conversions whenever spatial products in paper-and-pencil format are involved. Since these are abstractions of the real space, using them requires the children to perform a transformation from 2D stimuli to 3D mental representations. This type of dimensional transformation, visualizing two-dimensional information in the inner eye, might explain the stronger relations between both subclasses of small-scale spatial abilities and both classes of large-scale spatial abilities involving map-use. It is most likely that the unique characteristic of large-scale spatial abilities, transformation of 3D stimuli to 2D representations, might explain the dissociation to some extent.

13.1.4 The Role of Visuospatial Working Memory

The third goal of this study was to understand the role of visuospatial working memory capacity (VSWM) in greater detail. The third hypothesis H_3 stated that VSWM mediates the relationship between small-scale and large-scale spatial abilities. Similar to the two classes of spatial abilities, VSWM was reflected by two psychometric Corsi Block Test measures.

The results of this study indicate that although VSWM was considerably related to all subclasses of small-scale and large-scale spatial abilities, there was not sufficient evidence that VSWM mediated all of the relations between spatial abilities at both scales of space. H_3 may therefore be considered to be false.

These findings, however, should be considered in a differentiated manner. Because the sample size did not allow for computing second-order models, this study was not able to directly address the research question outlined in Section 5.1.3. The conclusion that H_3 may not be true was based on a series of mediation analyses for three relations between the class of small-scale spatial abilities and the three subclasses of large-scale spatial abilities. Findings of these analyses indicated that VSWM mediates, at least partly, some of the relations.

VSWM did not mediate the relation between small-scale spatial abilities and the abilities to encode and draw inferences from a survey map. This finding was not surprising, given the fact that SurveyMap represents the subclass of all those large-scale spatial abilities that allow children to acquire survey knowledge, to encode it in working memory, and to transform this information into long-term knowledge.

One surprising finding was, however, that VSWM partially mediated the relationship when it came to Comp predicting both small-scale spatial abilities. No mediation was found on the other way round. In other words, when considering VSWM as a mediator, Comp remained a significant predictor for small-scale spatial abilities, but the indirect paths involving VSWM was also significant. In particular with respect to the fact that VSWM was not found to be a mediator when it came to small-scale spatial abilities predicting Comp, this finding could be explained by the fact that both Comp and both subclasses of small-scale spatial abilities seemed to involve VSWM, but the latter one involved it to a much higher extent. This finding was further supported by evidence found at the manifest level, with measures of VSWM emerging as one important predictor in addition to performances in spatial tasks.

VSWM fully mediated the relationship between small-scale spatial abilities and Prod. In other words, when VSWM was considered as a mediator between the relationship of small-scale spatial abilities and Prod, direct prediction paths became non-significant and only the indirect paths significant. This finding indicated that the VSWM could fully explain the relations found between OB and Prod which was possibly caused by an extra cognitive load that resulted from the need to mentally align the map. If this had been the sole explanation, VSWM would have also mediated the relation between small-scale spatial abilities and Comp because solving comprehension tasks also involves aligning the map.

Consequently, these findings rather indicated that VSWM is involved in spatial tasks that require the individual to maintain correctly encoded spatial information in working memory when performing multistep mental transformations of these to find the own location on a map. Indeed, production tasks require the children to perceive and process spatial information from multiple vantage points. Until the child is ask to place a sticker, pieces of information have to be maintained and assimilated in working memory. Doing so, represents a high load on working memory capacity. This is in line with what has been observed for most small-scale spatial tasks, in particular when those are designed as power rather than speed tasks. This does not mean that solving comprehension tasks does not involve maintaining information in working memory, but children can rely on information from the map by constantly consulting it, which might decrease the cognitive load.

13.1.5 An Extended Model of Spatial Abilities at Different Scales of Space

In line with the literature, this study argued that a major goal of geometry education should be to help children to understand their surrounding space well enough to be able to interact successfully with it. To do so, children have to be able to grasp the changing relations and their orientations with respect to the self and other objects. This can be particularly challenging when these objects or children are moving in space. The findings of this study showed that to solve spatial problems in everyday life, children need to engage two distinct classes of spatial abilities. In their broadest sense, the findings of this study suggested that when conceptualizing the underlying spatial abilities that should be promoted to help children to 'grasp space' in geometry education, this should involve at least two major spatial behaviors: (a) to interact with objects in graphical space that might serve as tools to solve geometric problems by manipulating them or the perspective on them using small-scale spatial abilities; and (b) to interact with real space by successfully using a map as a referent in graphical space, by keeping track of the changing relations of landmarks and the self during movement in real space, such as during wayfinding enabled by large-scale spatial abilities. Since both latent constructs seem, as suggested by the empirical findings, to be only partially related, a conceptualization of spatial abilities in small-scale space only might be too limited for geometry education.

An extended model conceptualizing spatial abilities in geometry education suggested by the theoretical and empirical findings is shown in Figure 13.1. It combines a theoretical model from the cognitive psychology literature proposed by LIBEN (1981) with the one from the mathematics education literature proposed by PERRIN-

GLORIAN ET AL. (2013). In line with the literature on spatial cognition, this model describes spatial abilities as the cognitive abilities that enable children to understand the complex relationship between physical space, the space they interact with in their everyday life, and the corresponding psychological space that represents the space of their subjective experience of the surrounding environment. In line with PERRIN-GLORIAN ET AL. (2013), the model describes spatial abilities also as those cognitive abilities that allow children to successfully master the interaction between, on the one hand, physical space, the space of real geometric problems and day-to-day spatial experiences, and, on the other hand, graphical space, the space of visual representations that maps those problems or whole spaces by means of abstraction.

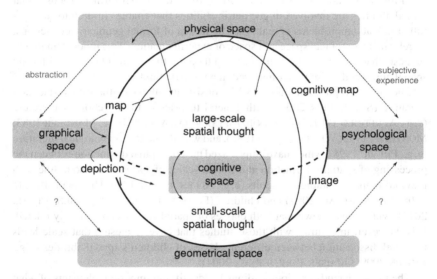

Figure 13.1 Small-scale and large-scale spatial abilities in geometry education

Elements of graphical space, that is, maps and depictions of real world situations, provide sources of different types of spatial information. Small-scale spatial thought requires the processing of depictive spatial information only, which are abstractions of the real space. In contrast, large scale spatial thought requires the processing of depictive information from a map and the subsequent map-mediated integration of spatial information perceived during interaction with real space during movement. This results in different types of spatial storage, that is, elements of psychological space, and consequently, qualitatively different types of spatial thought at both scales of space.

In line with LIBEN (1981), this study argued that large-scale spatial thought is rather specific, and often linked with direct experiences of and in real space, that is, experience of locations and places. Small-scale spatial thought is rather abstract, requiring the children to conceptualize space by means of understanding topological, projective, and Euclidean concepts. In other words, at both scales of space, children have to reason about the varying spatial relations between objects and the self, but large-scale spatial thought is based on specific concepts such as relations between locations, time-variant directions, and distances, whereas small-scale spatial thought is based on abstractions of these.

In line with PERRIN-GLORIAN ET AL. (2013), the model further proposes that spatial abilities are involved in geometric abilities that enable children to success-fully think about problems of real space by means of formal geometric deductions, thereby using graphical space as a space of experimentation. This model tentatively suggests that it is probably abstract spatial thought, that is, the class of small-scale spatial abilities that are rather involved in geometric thinking.

The model proposed in Figure 13.1 is consistent with the mathematics education literature (SOUVIGNIER 2000), with general theories in the psychology literature (e. g., ACREDOLO 1981; ITTELSON 1973; LOCKMAN & PICK 1984; MANDLER 1983; MONTELLO 1993; WEATHERFORD 1982), and with the neuropsychological literature (e. g., PREVIC 1998), which have emphasized the conceptual distinction between the processing of spatial information at different scales of space. It is also in line with previous empirical studies on adults (e. g., ALLEN ET AL. 1996; HEGARTY ET AL. 2006; LIBEN ET AL. 2008) and children (FENNER ET AL. 2000; LIBEN ET AL. 2013), which have shown that both classes of spatial abilities are partially related. It is, however, in contrast with those studies that have suggested that scale leads to a total dissociation between different classes of children's spatial abilities (e. g., JANSEN 2009; QUAISER-POHL ET AL. 2004).

The model extends previous findings by identifying important elements of what is here referred to as cognitive space, that is, the space of cognitive processes and spatial concepts involved in both classes of spatial thought. In other words, cognitive space constitutes spatial thought that is invariant at different scales of space. In line with HEGARTY ET AL. (2006), this study showed that different processes attributed to visuospatial working memory, such as the individuals' abilities to maintain spatial information in working memory for further processing, to use purposeful knowledge stored in long-term memory, or to activate schemata that might support to perceive the given spatial information are important elements of cognitive space.

The model also extends previous findings of studies in the individual differences literature that have emphasized the importance of several spatial tasks when it comes to explaining the relation between small-scale and large-scale spatial abilities of children (LIBEN ET AL. 2013, 2008). This study identified important subclasses of small-scale and large-scale spatial abilities that were found to be unique predictors for other subclasses of spatial abilities. Using the theoretical insights from the spatial cognition literature, this study concludes that the underlying cognitive processes that are characteristic for those predictors, are likely to be important elements of cognitive space. In line with the literature, these are mental transformations of objects (K. J. BRYANT 1982; LIBEN ET AL. 2008; MALINOWSKI 2001; SHEPARD & HURWITZ 1984), mental transformations of the imagined self (FIELDS & SHELTON 2006; KOZHEVNIKOV ET AL. 2006), and visualization as defined as transformation of three-dimensional spatial information from two-dimensional abstractions (IWANOWSKA & VOYER 2013; LOCKMAN & PICK 1984). All three mental transformations require the child to be aware of three different frames of reference, the egocentric, environmental, and intrinsic one, and to consciously change the relations between those when mentally moving objects or the imagined self in small-scale space, or when maintaining the map-environment-self relation during movement in real space.

13.2 Limitations

This study has several limitations. First, there were limitations related to basic assumptions and the design of the study. Second, the experimental design may have led to some artefacts. Third, it is possible that the results were biased due to certain methodological choices. Fourth, a set of variables that were not controlled for may have also affected the results of this study.

13.2.1 Basic Assumptions and Design of the Study

Perhaps the most important limitation of this study was its cross-sectional research design. Observations of spatial behavior, defined here as the individual test results of both paper-and-pencil and map-based orientation test, were conducted at a single point of time, which provided a 'snapshot' of the abilities considering the organisational costs to be managed by an individual in a research project. Although the correlational patterns identified suggested relations between small-scale and large-scale spatial abilities, a cross-sectional research design usually does not allow researchers to draw any inferences regarding (causal) relations (e. g., LEVIN 2006). To provide further evidence for relations identified here, it would be important to confirm the findings of this study either in a longitudinal or an intervention-based study.

A second limitation was that this study assumed strategy homogenity in the solution processes of the spatial tasks used to measure both classes of spatial abilities. It also did not consider related metacognitive concepts such as individual preferences in the processing of spatial information. These might have had an important influence in the solution process, but it was assumed that all children would solve tasks consistently using the same, pre-defined, and intended strategy derived from the literature. In other words, although individual differences in both small-scale and large-scale spatial abilities were documented in this study, the underlying strategies were only assumed and not measured directly. Since strategy choice might have contributed to the partial relations that have been identified in the literature (e. g., GRÜSSING 2002; KASTENS & LIBEN 2010), future studies should address strategy use when measuring spatial abilities at different scales of space.

Third, it is important to bear in mind that in addition to not measuring the individual strategies, this study did neither attempt to measure the underlying (neural) representations nor the corresponding cognitive processes. By using an approach that measured the outcome of these representations and processes, that is, the corresponding spatial behavior, it was not possible to measure individual spatial abilities as such. Measurement was also affected by a range of non-spatial abilities mandatorily required for solving spatial tasks, such as encoding or manipulating of spatial stimuli. Although the theoretical framework pointed to several cognitive processes that might explain the empirical findings of this study, these explanations remain tentative. These underlying cognitive processes could be examined in future studies.

Given the constitution of the sample, caution must be applied when interpreting and generalizing the empirical findings. First, the choice of the sample was closely related to the organizational resources which resulted in a sample that consisted of children from the same town, and, in some cases, from the same school. Second, the sample size was also limited, which might have affected the statistical analyses. Although this study controlled for intraclass effects and that a reasonable number of cases per estimation parameter were given, the reduced sample size might have led to biases. Reproducing the findings, ideally using a larger and more diverse sample, would strengthen the explanatory power of this study.

13.2.2 Experimental Setup

Although one major goal of this study was to take place in real space, the complex environmental setting of the map-based orientation test limited the amount of control over several factors that may have influenced individual performances of the children. For example, because solving spatial tasks in real space involves interaction, they are inextrieably linked to the test environment. Indeed, all measures of large-scale spatial abilities are dependent on the environment chosen and are, for example, sensitive to its complexity (SAS & MOHD NOOR 2009) or to the characteristics of the routes walked (WRENGER 2015, p. 180). Layout and architecture of the university campus may have affected individual performances, but it is not clear to which extent. This study demonstrated that measuring individual differences in spatial thinking is even possible in environmental settings, but to fully understand the relative importance of individual variables it will be necessary to examine them in more restricted experimental settings.

It cannot be ruled out that the behavior of individual experimenters had an effect on children's performances. Although all experimenters were trained, they may have behaved in a way that instruction-based variables such as completeness and correctness of instructions or motivational-affective variables such as creating a trustful test atmosphere or motivating the children for doing the treasure hunt may have been influenced. It could be argued that the former could have been controlled for by videotaping in each of the test situations and deriving further co-variates to compensate for incorrect and incomplete instructions. Doing so would, however, have resulted in largely incomplete data. Introducing these variables, then, would have further reduced the number of available cases and impeded statistical analyses. Future studies could control for the influence of the experimenter by videotaping or recording to understand to what extent these issues may influence tests in environmental settings such as the one used here.

Children's individual responses to the test situations might have likewise influenced the findings. The tasks in the tests at both scales of space differed in many respects, and this might have contributed to the dissociations found. The small-scale spatial abilities test was more closely related to test situations that most of the children have probably already encountered in their school life, whereas the large-scale spatial abilities test was completely new for them and involved interacting with the experimenter. The present data cannot explain whether and to what extent the individual responses to these test situations contributed to the differences observed. Future studies could benefit from controlling for this variable.

Finally, the decision to collect data just before the children's summer holiday may have affected the outcome of this study. Although this implied sunny weather and allowed more children to participate in the study because the overall school year was finished, this might have decreased motivation among the children. In particular when solving the small-scale spatial tests, some children did so very quickly, and one might wonder whether they consciously completed the task.

13.2.3 Methodology

An important methodological limitation of this study was that the children were not allowed to physically turn the map, but to put their finger on the map. Although the experimenters did not recommend this strategy, more than half of the sample did so. Using the finger might have probably decreased the cognitive load and probably affected results concerning the involvement of specific mental transformations in map-based orientation such as the ones allowing children to transform their imagined self. Due to the limited sample size, this variable could not be included into the models to control for it in post-analysis. The present data can therefore not explain whether and to what extent using the finger on the map as a strategy to maintain orientation contributed to the individual differences observed. Future studies should either allow or prohibit all types of strategies such as physically turning the map and using the finger.

Another important limitation was that test validity for both the paper-and-pencil and map-based orientation test was only considered by verifying construct validity. Having related the tests to instruments that intend to measure similar constructs, having related them to other test measuring related constructs (e. g., mathematics achievement), or having asked experts whether the tests reflect what they claim to reflect, would have provided further insights of whether the tasks really measure what they claim to measure. Consequently, results such as a low extent of variance explained in the tests could not be explained. Ideally, future studies would consider validity not in such a restrictive way.

The study might be somehow limited by the fact that the large-scale spatial abilities test was piloted with a few cases only. Consequently, the map-based orientation test included some measures that were characterized by bad properties from a psychometric point of view. This study, however, included some of these measures in the subsequent statistical analyses to make sure to keep an acceptable number of measures that reflected the subclasses of spatial abilities assumed in the hypothesized models. Including measures with these types of characteristics might have, for example, biased the estimated amount of shared variance. Ideally, future studies would avoid this potential problem by using larger samples during pilot testing.

Another limitation might be due to a trade-off between an ideal construction of the test instruments and the fact that it were children who were tested. Although both tests involved a wide range of measures that reflected the underlying theoretical constructs, some individual measures were limited in terms of the number of items. As indicated in Part II, using at least 3 or 4 items per measure was important to treat the variables as metric and not categorical, at least for the corresponding statistical analyses. This issue could have been addressed by using polychoric correleations rather that Pearson ones (HOLGADO-TELLO ET AL. 2010). In the study, an WLSMV estimator was chosen to control for effects resulting from almost categorical variables. Small biases in the estimates due to using Pearson rather than polychoric correlations were neglected during correlation and regression analysis. Future studies should avoid those biases using measures with sufficient high numbers of categories.

Although all measures of small-scale spatial abilities have been piloted, some suffered from a limited range of items, which led to ceiling and floor effects. In addition to excluding data of two of the measures from the analysis and not considering the cases which scored highest or lowest, the WLSMV estimator was used to provide distribution-free estimates with robust standard errors. Involving measures with more items that can be used to differentiate between children with high and low scores could increase the explanatory power of the study and allow for the use of more effective and less sample-size sensitive estimators.

Although structural equation modeling (SEM) has been proven to be a powerful tool to examine the relations among different cognitive abilities, this method usually requires a large sample size during parameter estimation with a high number of indicators. The full model Figure 12.1 (p. 352) required the estimation of 40 parameters (with 120 covarianes available, thus implying a $df = 80$). The model was computed using 215 cases, thus providing a N:p-ratio of 5.375 which is at the lower limit of being acceptable for robust estimations (e. g., R. B. KLINE 2015). A further limitation of the methodological choices was that this study was not designed using a prior sample size estimation using Monte Carlo simulation of power analysis as provided in MPLUS (L. K. MUTHÉN & MUTHÉN 2004). As a consequence, no results on statistical power can be provided for the given estimations. Future studies should invest in a careful pre-analysis concerning the number of cases required to achieve a certain level of robustness in parameter estimations and to determine the statistical power of the results.

Finally, the treatment of missing values might be another limitation of this study. Treatment of them involved a multiple imputation procedure. Although the robustness of computing those imputations has been shown using parameter studies, estimates of performances tend to be biased since they are computed based on data available for every case and do not reflect the actual performance of children in individual tasks (which could, in turn, be better or worse than performances in the other tasks). Subsequent statistical analyses using a set of 30 imputations were also found to be restrictive in a way that it was not possible to report the necessary fit indices in the multiple regression analyses and that it led to further biases regarding SEM-related fit indices. If possible, future studies should try to limit the number of missing values by carefully planning the organizational framework.

13.2.4 Neglected Covariates

This study did not control for a range of variables to limit the extent of diagnostic effort for the children. These variables may, however, have affected results regarding the individual differences found in the study. As a consequence, it is not possible to conclude that the measured individual differences are only due to individual differences in the underlying spatial abilities, but they might have been affected by these covariates.

Not controlling for cognitive variables, such as general intelligence, might be a limitation of this study. More specifically, given that spatial abilities have been shown to be one dimension of human intelligence (e. g., HEGARTY & WALLER 2005), controlling for this variable could have yielded more detailed results concerning the extent of overlap between both classes of spatial abilities. In particular, it remains an open question if the overlap found was due to a relation between small-scale and large-scale spatial abilities or due to a general relation with intelligence.

Although a measure of visuospatial working memory, the Corsi Block Test (CBT), was included in this study, it is not clear whether this measure reflects VSWM capacity. Given the fact that executive resources have been shown to be involved when primary school children were subjected to the CBT test (ANG & LEE 2008), it is likely that the results rather reflected VSWM capacity as related to attentional control and central executive (CE) functions. Since subclasses of spatial abilities have been found to involve different CE capacities (MIYAKE ET AL. 2001), including an additional measure reflecting CE would have shed light on the relations and dissociations addressed in this study.

Two variables that would have been beneficial concerning measuring large-scale spatial abilities but were not included would have covered children's pre-requisites in understanding and applying the concept of scale and angle. The children had to reason correctly about scale, and consequently, distances to understand spatial relations between the map and the referent space, since they had to determine locations of the self or landmarks in space and on the map. Pointing to landmarks involved understanding the concepts of angle and shortest distance (UTTAL 1996). Both variables were not sufficiently considered, but it is safe to assume that they had an impact on individual performances regarding large-scale spatial tasks. As a consequence, the large-scale measures used here did not measure large-scale spatial abilities as such but also implicitly measured concept of scale and angle. This conflation of constructs implies that at the level of individual differences, it was not possible to determine whether errors in the tasks occurred due to a low level of spatial abilities or missing or misapplied concepts of scale and angle.

Another limitation was that this study only inadequately controlled for individual spatial activity in real space or experience and prior knowledge using a map (DAVIES & UTTAL 2007). Individual differences in spatial activity have been shown to affect pointing accuracy in tasks requiring path integration by children in kindergarden (NEIDHARDT & POPP 2010). In addition, primary school children's previous knowledge in map-reading has been shown to be a strong predictor for various spatial tasks in real space involving map-based navigation (HEMMER ET AL. 2012, 2015). Controlling for the children's previous formal (i.e., school training) and informal (i.e., exposure in informal, familiar contexts) experience with maps would have been beneficial. Although three items were included to test for previous knowledge concerning prior experience with maps, the children struggled to provide answers on a five-point Likert scale. Therefore, these variables were not useful and could not be integrated into the analyses. Future studies should consider age-specific questionnaires and teacher's self-evaluations in the research design.

Finally, this study neither controlled for the role of spatial thinking tasks in each of the classes nor the text books used. Given that teachers decide how and to which extent they foster the development of spatial abilities in the classroom and the fact that different text books treat this issue differently, these variables could have provided more prfound insights how to interpret individual differences in the children's performances.

13.3 Implications

Despite the limitations outlined above, this study has important implications both for research in mathematics education and psychology and for practical geometry education. When considering the implication for geometry education at school, one has to be aware that this study was neither a longitudinal nor an intervention evaluation study.

13.3.1 Implications for Research in Mathematics Education

This study had two primary motivations for examining the relationship between small-scale and large-scale spatial abilities. First, it intended to contribute to the literature of spatial abilities in the context of geometry education. Second, it aimed to contribute to the literature by establishing a theoretical and empirical basis of how to design future interventions to leverage spatial abilities more effectively. Consequently, this study has major implications for research addressing the understanding of spatial abilities and research addressing the development of interventions. Last but not least, it has implications for future research on the role of spatial abilities.

On the Understanding of Spatial Abilities
This study has two major implications concerning the understanding of spatial abilities and gives possible directions for further research which are outlined in the following sections:

Understanding the Impact of Scale on Spatial Thought   This is the first study to examine whether previous theoretical constructs of spatial abilities in the context of geometry education were in line with the intended conceptualization of geometry education that should enable children to 'grasp space.' The findings of this study indicated that conceptualizations that referred to small-scale spatial abilities may have had a restricted focus. Indeed, this class of spatial abilities represents children's abilities to grasp space only to a limited extent: It represents them when it comes to solving tasks that require the mental manipulation of spatial information perceivable from one single vantage point. This class of abilities does not, however, fully explain individual differences engaging in spatial thought in real spaces because doing so requires the integration of spatial information from multiple vantage points. The first major implication for mathematics education therefore is that scale affects spatial thought in children. Future studies should not only carefully consider which class of spatial abilities they want to address. This will also need to further investigate how scale affects spatial thought in children. Given that scale, considered as a psychological construct, describes a broad spectrum of qualitatively different ways to interact with spatial information, future research might address more than two classes of spatial abilities. This might, for example, involve studying spatial thought in spaces sized as large as classrooms or gyms.

   This study also has an indirect impact on research on spatial abilities in secondary school. The results of this study indicated that children do not fully master those cognitive abilities by the end of primary school. Although at least within the German mathematics education system, spatial abilities are not explicitly addressed in the curriculum any more, they continue to play an important role in geometry education, for example when using geometry-related software. Dynamic Geometry Software (DGS) has recently become increasingly popular (e. g., SOURY-LAVERGNE & MASCHIETTO 2015). Understanding which class of spatial abilities is involved in the case of DGS remains an open question. Examining whether those abilities should be rather understood as small-scale or large-scale ones or whether they should be conceptualized as a qualitatively separate class of spatial abilities (e. g., dynamic ones, see HEGARTY & WALLER 2005) could not only help to develop a more nuanced conception of spatial abilities in the context of mathematics education, but also be the basis for the development of intervention programs that may be applied from the early grades of secondary school to university mathematics programs.

Understanding Cognitive Space The findings of this study demonstrated that small-scale and large-scale spatial abilities, although being conceptually different, are not completely dissociated. This study aimed to examine this relationship from an individual differences perspective and beyond a cognitive psychology theoretical framework. By assuming that individual differences in solving spatial tasks at both scales of space result from individual differences regarding the cognitive processes involved in those tasks and by using a structural equation modeling approach that not only allowed for specifying correlations but also unique contribution of subclasses of spatial abilities, this study was able to describe a set of cognitive processes that are elements of cognitive space, that is, the space of shared cognitive and metacognitive abilities that constitute spatial thought invariant at different scales of space.

Measuring spatial abilities by means of individual differences is, of course, only one means of approaching cognitive space. Observation from this study showed that children solved map-based tasks using different strategies by verbalizing ideas and co-thought gesturing. Indeed, metacognitive abilities such as adaptive strategy use have been shown to be an important determinant of spatial thought, both when solving small-scale and large-scale spatial tasks. Concerning the latter, much research has been done on the verbal and written clues that children use in large-scale space when solving tasks requiring them to mark their location in space (KASTENS & LIBEN 2007, 2010) or during wayfinding (GERBER & KWAN 1994; WRENGER 2015). In particular, it had been stressed that individual strategy use predicted scores for locating oneself on a map both in adults (LIBEN ET AL. 2008) and children (KASTENS & LIBEN 2010; LIBEN ET AL. 2013). Future research should address the question of strategy use to gain a deeper understanding of the relation between both classes of spatial abilities. One direction to take would be to study similarities and differences of strategy use at different scales of space. Another approach could be to combine insights from strategy use and to measure individual differences in performances in spatial tasks. Since map use strategies, for example, have been shown to have the effect that spatial task lose their predictive power due to interaction effects (LIBEN ET AL. 2013), combining both perspectives might lead to a more nuanced understanding of cognitive space, and, it turn, of the findings of this study. Initial research with adults has been conducted by HEGARTY (2010) and could inspire future studies on children.

Another way to study individual approaches to small-scale and large-scale spatial tasks could be to study differences in the emergence of spontaneous co-thought gesturing (e. g., NEWCOMBE ET AL. 2013), which has been shown to be an indicator of cognitive challenging situations. Studies that examine the emergence of co-thought gesturing during solving spatial tasks at both scales might help to sensitize teachers to identify students that have difficulties in solving spatial tasks (e. g., LOGAN, LOWRIE, & DIEZMANN 2014). Moreover, a comparison of gestures at both scales of space could be one way to qualitatively approach cognitive space.

Analyses on strategy or gesture use at various scales of space could be completed by taking into account different individual differences in acquiring and processing spatial information (see KOZHEVNIKOV 2007). For example, planning and following a route could be done by relying on geometrical correspondences extracted from the map (relying on spatial imagery) or by recalling a sequence of landmarks extracted from the map (relying on object imagery) (KOZHEVNIKOV ET AL. 2005). Investigating those concepts when solving small- and large-scale spatial tasks could be one starting point to investigate similarities and differences between both spatial abilities at different scales of space.

Taking a linguistic approach to spatial abilities, one could investigate the vocabulary of spatial descriptions in written and map-based settings and compare language aspects such as, for example, the use of PIAGET AND INHELDER's (1948/1956) topological, projective and Euclidean expressions in clue-reports at both scales of space. Although there have been initial linguistic studies in the case of small-scale spatial abilities (e. g., MIZZI 2017) and in the case of map use in real space (KASTENS & LIBEN 2010), a systematic comparison could provide new insights concerning the question how children perceive and make use of spatial information in small-scale and large-scale space.

Last, but definitely not least, the results of this study showed that there seem to be sex differences in spatial abilities at both scales of space, a result that is in line with studies in adults (e. g., HEGARTY ET AL. 2006). Future studies should re-analyze these findings and discuss them with respect of how sex might affect elements such as possible shared cognitive processes involved in cognitive space.

On the Design of Interventions

Although this study did not investigate an intervention design, it has important implications in the design of studies that will adress intervention programs. Spatial abilities can be trained, and this is, as outlined in many national curricula, an important goal of geometry education. This can be done by increasing the amount of instruction and practice in the respective tasks in class such as mental rotation (BRUCE & HAWES 2015). The empirical findings of this study concerning the relation of spatial abilities might help researchers to design intervention programs that do not only foster one class of spatial abilities alone but several ones simultaneously. Consequently, another direction of future research in mathematics education might address the design and empirical validation of pedagogical interventions.

Some important spatial tasks that could be used in the design of pedagogical interventions in tabletop environments or real space have been identified above. Since this study only examined relations between performances in tasks and the corresponding latent abilities, those suggestions are tentative and deserve further research concerning their pedagogical effectiveness. Beyond the background of the findings of this study, one hypothesis would be that solving tasks in small-scale spatial settings might not only have an impact on small-scale spatial abilities, but also, at least to some extent, on large-scale spatial abilities, in particular those involving maps, and vice versa.

Two important research questions to answer in future research are

- What is the impact of small-scale learning environments on the improvement of small-scale spatial abilities? How do they improve large-scale spatial abilities?
- To which extent do map-based learning environments in real-spaces foster the improvement of not only large-scale spatial abilities, but also small-scale ones?

Maps as Cognitive Tools to Improve Spatial Abilities Maps have previously been shown to be important cognitive tools that can be used to develop spatial abilities at school. DAVIES AND UTTAL (2007), LIBEN (2006), UTTAL (2000a), and WIEGAND (2006) suggested that children should be exposed to maps as early as in the first grade to give them the possibility to acquire experience with maps. Training of related small-scale spatial abilities such as mental rotation and support children to interpret maps as representations of a referent space involving a double-spatial feature could not only foster their abilities to use maps, but also other forms of spatial knowledge that will become more and more important during the curriculum (e. g., DAVIES & UTTAL 2007).

Because maps are spatial themselves, many of the ideas of abstract spatial thought might find their analagon in large-scale spatial settings. To solve the tasks proposed in this study, for example, children had to address not only topological (i.e., reason about the proximity of certain landmarks), projective (i.e., reason about the direction of a landmark in the case of unaligned maps), but also Euclidean (i.e., reason about distances, scale and angle) ones. All these concepts have been shown to improve children's conception of space.

One tentative implication of the model proposed here ((Figure 13.1, p. 387) might be that pedagogical interventions that intend to foster the use of these spatial concepts or subclasses of small-scale spatial abilities such as mental rotation should be designed in such a way that children first directly experience those in map-based settings in real space. Mental rotation, for example, could be first experienced in real space, thereby learning on the enactive level. Then, it could be addressed in paper-and-pencil tasks, thereby involving learning at the iconic, and later being transferred to the symbolic level, for example using spatial descriptions and rather abstract tasks of mental rotation (BRUNER 1964). Spatial reasoning with maps in real space might therefore allow teachers to design experience-based learning environments for spatial thinking that can be transferred to the classroom, and vice versa. One way to achieve this kind of transfer could be to include, in addition to production and comprehension tasks, the representational correspondence and meta-representational methods suggested by LIBEN (1997). In this manner, teachers could address Euclidean concepts such as understanding scale, calculating distances, or determining directions in experienced-based learning environments.

Indirectly improving certain small-scale spatial abilities as a result of map-based interventions could be a means to develop those abilities that have been shown to be difficult to foster in small-scale settings, such as perspective taking abilities (LOWRIE ET AL. 2017). Indeed, it has been argued that by the age of 10 to 12, children have developed their perspective taking abilities well enough to solve a range of paper-and-pencil based tasks. To reach the next level, they might need different sets of tasks that require more complex, multistep perspective taking processes such as those involved during map interpretation in real space (see also LOWRIE ET AL. 2017). Fostering those abilities could not only help children to better understand the space they interact with, but also to improve in problem solving, since these abilities have been shown to involve organization processes that relate to some steps of mathematical problem solving (TARTRE 1990).

'Spatializing the curriculum' Many researchers have stressed the need not only to integrate spatial education as a particular isolated issue of mathematics or science education, but also to take a more integrative approach in which "spatially based education become[s] a natural part or even a focus on ongoing activities in mathematics and science" (NEWCOMBE ET AL. 2013, p. 583). This approach has also been referred to as the need to 'spatialize the curriculum' (see also LIBEN 2006).

To achieve this goal, future research could start by identifying the implicit impact of a regular geometry class, but also classes in other disciplines dealing with, in one way or the other, with spatial abilities at both scales of space. For example, one might investigate the extent to which specific geometrical activities, such as estimation or measuring, but also mathematical ones, such as interpreting graphs or geographical ones, indirectly affect the improvement of spatial abilities at both scales of space. One approach to do so would be to identify the underlying mental transformations that are necessary when solving a particular tasks, such as when assembling figures during measurement problems on the geoboard, reflecting them to compare areas, or disassembling them to derive the area from another figure. Another approach could be to study correlational patterns between performances in solving a set of spatial tasks and another set of task of interest such as tasks requiring the children to estimate sizes, for example weights, lengths, areas, or volumes.

On the Role of Spatial Abilities
Concerning the role of spatial abilities, this study has one implication. It found empirical evidence for a set of cognitive processes involved in spatial thought at both scales of space. These were object-based transformations, transformations of the imagined self, dimensional transformation of spatial information, and maintaining information in working memory. Future research investigating the role of spatial abilities in mathematics could focus on these processes to understand their influence on learning in the context of mathematics. By doing so, these studies could extend the literature on the role of egocentric perspective transformation abilities (e. g., TARTRE 1990; VAN DEN HEUVEL-PANHUIZEN ET AL. 2015), object-based transformations with mental rotation in particular (e. g., BRUCE & HAWES 2015; BÜCHTER 2011; YOUNG, LEVINE, & MIX (2018)), and dimensional transformations when working with spatial products in graphical space (e. g., GUTIÉRREZ 1996), or all of these (see LOWRIE & LOGAN 2018, for further elaboration).

Future research should not only consider VSWM in research of spatial abilities at both scales of space, but also individual differences in children's VSWM capacity, also when it comes to investigating performances in mathematics. Female adults, for example, have been shown to process information at a slower speed of processing when solving tasks that tap VSWM (LORING-MEIER & HALPERN 1999), and sex differences in spatial abilities in adults have been explained by sex differences in VSWM (KAUFMAN 2007). For children and younger adults, these relations have not been investigated, and doing so might not only shed light on individual differences when solving spatial tasks but also on related constructs such as mathematics performance. Extending the study of BÜCHTER (2011), and searching to deeper understand why sex differences in mathematics performances can be accounted for by sex differences in spatial abilities by addressing VSWM involvement could be one step in this direction.

13.3.2 Implications for Research in Psychology

This study is fundamental research from the perspective of mathematics education. Drawing on insights of the cognitive and developmental psychology, this study might also have implications in psychology.

Implications for the Cognitive Psychology Literature

This study has two major implications for the cognitive psychology literature. First, it proposed a whole comprehensive set of map-based orientation tasks for children. Previous studies in the cognitive psychology literature have examined large-scale spatial abilities in children either by addressing survey knowledge acquisition or by analyzing children's map use abilities in real space. These studies commonly used only a few individual tasks. The relations between individual differences in a set of large-scale spatial tasks have not been examined in these studies. Although the set of tasks presented in this study is far from being complete, the approach chosen here led to a classification of individual differences in map-based orientation tasks in a three-component model. Each of the three components could be described with a profile of the main cognitive demands, and it was shown that learning the layout of an environment requires separable abilities from the ones used for making inferences from information in real space on relations presented on a map, or vice versa.

Second, this study addressed the relation between small-scale and large-scale spatial abilities in a differentiated manner both at the manifest and the latent level. Previous studies on children and adults have examined this relation using either analyses at the manifest level (e. g. multiple regression analyses) or at the latent level using structural equation modeling, thereby considering small-scale spatial abilities as an undifferentiated construct. Using analyses at the manifest level, it was possible to identify specific measures that predicted individual performances in other tasks. Results of the latent approach, structural equation modeling, did not only lead to an estimation of the extent of overlap between small-scale and large-scale spatial abilities, but also addressed the relations of subclasses of spatial abilities at both scales of space. Relating these results of the latent correlation analyses to the theoretical framework, this study identified cognitive processes that may underlie particular subclasses of small or large-scale spatial abilities. It also showed which of these cognitive processes might be shared when it comes to explaining the relation of both classes of spatial abilities.

These conclusions, however, were theory-driven and require further investigation. Future empirical studies could lead to better understanding of why small-scale and large-scale spatial abilities are related, but also why they are dissociated. This, in turn, could result in a more refined taxonomy of spatial abilities at various scales of space and other theoretical advancements. Thus, future research should analyze the complex small-scale and large-scale spatial tasks and related predictors. In particular the complex map-reading tasks involve spatial but also non-spatial concepts. Analyses could yield a decomposition of these complex tasks and an identification of basic cognitive processes such as the underlying mental transformations or different processes involving the control and manipulation of spatial frames of reference on children. Diagnosing those issues, at least partially, by traditional means of investigations such as analyses of reaction times, neuroscientific approaches such as neural imaging using fRMI, or recent methods such as eye-tracking, would lead to a more comprehensive model of spatial cognition.

Implications for the Developmental Psychology Literature
This study has two major implications for the developmental psychology literature. First, it provided empirical support for the assumption that a dissociation between object manipulation abilities and egocentric perspective transformations abilities can be found in children at the end of primary school. Previous developmental research has produced mixed results concerning the relation between these two abilities. Investigating this issue has been very difficult since it required researchers to operationalize each subclass in a set of age-specific measures without floor or ceiling effects. For this study, a range of age-specific paper-and-pencil measures were developed which were shown to have psychometric properties that were, despite minor floor and ceiling effects, acceptable. In particular, this study was able to propose different measures reflecting egocentric perspective transformation abilities. Moreover, results of the statistical analyses revealed whether and to which extent object transformation abilities are dissociable from egocentric perspective transformation abilities.

Second, it provided empirical evidence that the construct of spatial abilities is indirectly dependent on the influence of spatial scale. Previous developmental research has investigated spatial abilities in children either in small-scale spatial settings, examined children's cognitive mapping abilities, or the children's abilities to use maps. The relationship between those abilities has, however, not been established. This study found empirical evidence that spatial abilities in children are a differentiated construct that is dependent on scale to the extent that the underlying cognitive processes seem to be different which might explain differences in individual performances in spatial tasks posed at different scales of space.

It is important to note that these conclusions were based on the findings of a cross-sectional research design. Future research should examine both dissociations in greater detail by proposing research designs that are better suited to provide further empirical evidence concerning different classes of spatial abilities of children. Focusing on small-scale spatial abilities or the relation of spatial abilities at different scales of space, researchers could investigate possible dissociations using invention-based designs. They could examine whether the same pedagogical intervention leads to different effects in both classes of spatial abilities, which would, in turn, provide further evidence that there is a dissociation in the underlying cognitive abilities. Another important direction in further research would be to analyze spatial abilities using longitudinal study designs involving children aged from pre-school to the end of primary school and to examine whether there are different developmental trajectories for both classes of spatial abilities. In particular, the findings of longitudinal studies could provide a theoretical explanation of why there seem to be two developmental trajectories, how they are characterized, and which factors may influence these developments.

13.3.3 Implications for Geometry Education in Primary Schools

Although this study addressed a range of future directions that could be taken to gain a better understanding of the impact of scale on children's spatial thought and resulting directions for the design of pedagogical interventions, it also has some direct implications on practical geometry education.

Although children aged from 9 to 12 have made an important cognitive development, the findings demonstrate that solving spatial tasks should be an important activity in geometry classes at least until the end of primary school. The individual differences that were found in both small and large-scale spatial abilities show that there is a need to constantly foster spatial abilities in primary school. Moreover, the established theoretical framework and the empirical findings concerning the relation between both classes of spatial abilities could help teachers to design pedagogical interventions.

The abilities to represent and mentally manipulate the varying relations of objects and the self are important cognitive abilities that allow children to draw inferences how their actions might change relations in the future and to reason about what past events might have contributed to current spatial configurations that they perceive. To help children to engage in spatial thought in situations that mimic their everyday life, ideally involving different scales of spaces, should be the goal of geometry education. The findings demonstrate that even at the end of primary school, teachers should anticipate substantial student-to-student variations in the accuracy in which children solve spatial tasks both in classroom settings and those in real space.

Maybe the most important implication of this study is that spatial abilities cannot be treated as an unidimensional construct in geometry education when children engage in spatial thought at different scales of space. In other words, performances of children do not only differ within a learning group, but might differ when the same child completes paper-and-pencil and real-space based spatial tasks. Teachers that support the process of spatial learning by scaffolding should be aware that changes of relative size of the spatial information provided have an impact on the way children interact with space, and should, for example, not assume that solving spatial tasks in the classroom will prepare the children to use maps in real space. Consequently, addressing spatial abilities in the context of geometry eduction should not only happen in limited settings such as the classroom but also in real space.

Visuospatial working memory capacity is generally not addressed in geometry education, but the findings of this study emphasize its particular role when it comes to solving spatial tasks at both scales of space. Teachers should anticipate the involvement of working memory capacity by analyzing given tasks with respect to the demand of controlling attention, for example during sorting, or concerning the amount of spatial information that have to be kept in mind. Individual differences in visuospatial working memory capacity might account, at least to some extent, to individual differences in solving spatial tasks at both scales of space.

The findings of this study suggest that by the end of primary school, children in classroom should be given complex spatial tasks that involve multistep solution processes, that demand the children to sort pictures, or to make inferences from spatial information presented in maps, labyrinths, and plans. This study emphasized the important role of two subclasses of small-scale spatial abilities that children use to engage with these tasks, object-based transformation abilities and the abilities to transform the imagined self. Practicing the former subclass of abilities is of essential value, not only for the sake of 'grasping space' in classroom, but also because they are essential when it comes to map-reading in real space. Educational interventions that focus on this subclass of small-scale spatial abilities may therefore be expected not to improve the cognitive abilities themselves but also enhance map-based spatial thinking in real space. Tasks that relate to the mental rotation of maps, such as LR, or tasks that require children to mentally engage with a complex stimuli that is transformed in multiple subsequent steps, such as the task PFT, could stimulate spatial thought concerning the changing relations that the movement of an object might have at both scales of space.

Moreover, this study emphasized the important role of egocentric perspective transformation abilities in geometry classroom. It showed that they can be addressed even in paper-and-pencil format, and that they are important not only in classroom but also when it comes to acquiring survey knowledge in an unknown environment. Moreover, they are also involved in map use in real space. Hereby, tasks that require the child to mentally engage with abstract spatial configurations that require them to control egocentric thinking, such as Boxes, or tasks that simulate the imagined or pre-defined navigation in space, such as Meadow and Emil, could stimulate spatial thought concerning the changing relations between the self and the environment with movement in space, both at the mental and physical level.

Given the fact that this study suggested that children's abilities to mentally transform objects in tasks, such as mental rotation and paper folding, might not reflect the children's abilities to mentally transform the imagined self in tasks involving reasoning about different perspectives on configurations of objects or mental movement in a labyrinth-like situation, teachers that plan learning environments should recognize the importance of both subclasses of spatial abilities and reflect whether they designed tasks that require the children to use both object-based mental transformations and perspective transformations.

The findings of this study suggested that children in primary school should be given spatial tasks in real space. This study emphasized the particular role of maps as a cognitive tool when it comes to spatial reasoning in real space. The use of maps might help children to think about the space that they perceive and experience in a more abstract way since they can be interpreted as a suggestion of a particular spatial conception of the surrounding space. Given that a landmark might not only be a single object but also a topographical object such as an intersection of routes, a map offers the opportunity to perceive the whole topographical organization of the surrounding space from one single viewpoint. Using maps may therefore enrich direct experience that requires the integration of multiple, separate viewpoints on these objects. The information on the map may therefore complete the distorted and fragmented information that a subject acquires from real space by merely experiencing it in a metric and ordered way.

This study proposed a range of map-based tasks in real space that required the children to record the location of themselves on a map, to point to unseen landmarks, or to navigate to pre-defined goals while keeping the map aligned with the environment. Although all of these tasks proposed in this study mimic everyday behavior in real space, with or without using maps, teachers should design the corresponding learning environments with care. The findings of this study suggest that different subclasses of large-scale spatial abilities are involved when it comes to reasoning about memorized landmarks and using maps in real space. Moreover, it turned out that another subclass of abilities is required to solve production tasks than it is the case for comprehension tasks. In other words, performances of children did not only differ with respect to the involvement of a map when engaging in spatial cognition in real space. Children may also perform differently when they complete production or comprehension tasks. Teachers who plan a pedagogical intervention in real space should be aware of this, analyze the tasks that they plan to use carefully, and adopt their scaffolding during the learning process. Consequently, when addressing spatial abilities in real space, it should involve both production and comprehension tasks on the map, but also making inferences from knowledge encoded, addressing different concepts of space.

One remarkable finding of this study is the important role of solving comprehension tasks during map-based navigation in space. Those tasks do not only require the child to act as an active navigator who extracts directions to unseen landmarks, plans a possible route towards them, and more, reasons about perceived spatial relations in the environment based on hypotheses developed from what can be seen on the map. The abilities to solve comprehension tasks have moreover been shown to be related to both object-based transformation abilities and egocentric perspective transformation abilities. Given the fact that perspective taking abilities for example, have been shown to be difficult to be practiced in small-scale spatial settings, solving comprehension tasks such as Disks and ActiveHP in real space could not only improve the children's abilities to use maps but represent one way to foster related small-scale spatial abilities indirectly.

The findings of this study emphasized that children at the end of primary school face substantial problems when being asked to locate themselves on a map during production tasks. To do so, they have to maintain spatial information perceived from multiple vantage points in the working memory, and subsequently assimilate them to make inferences on their own location on the map. Mastering this challenge is not only important for everyday life, but is a prerequisite for mastering comprehension tasks since they require children to constantly track the own location on the map. Consequently, spatial activities in the real space should explicitly address tasks involving the location of the self.

Conclusion

14

> *We are not lost. We are locationally challenged.*
> —*John M. Ford*

So here I am in the very last chapter of this study. During the last five years, I have been given the chance to explore a set of cognitive abilities, which are both a fascinating and challenging research subject. While conducting research on spatial thought in primary school children, I came to realize how important these abilities are for my own life, for example when I struggled to navigate unknown territory or tried to squeeze furniture into rooms. In light of these experiences, I was even more impressed by the performances of the children who participated in my study. One of the most striking findings is that even at the age of 9, children seem to have developed the abilities to solve a wide range of spatial tasks mostly unknown to them. This involved not only reasoning about spatial relations in two-dimensional representations of realistic settings but also navigating in real space using an unaligned map. Conducting research on spatial abilities and trying to grasp them both at the theoretical and empirical level felt like trying to grasp smoke at some points during the project. My findings provide further evidence for the highly complex and intertwined nature of those abilities, which are, as I showed, closely related to a wide range of other cognitive abilities and capacities. For this reason, conducting research on those abilities was both challenging and rewarding. Although the act of trying to grasp them seemed already to disperse their purely mental nature, I was not only able to find solutions for these challenges but also to identify new perspectives for future research. Another truly rewarding dimension of my research project was to see how children experience space.

Indeed, starting from toddlerhood, children explore the varying spatial relations between themselves and objects in the environment, which change either due

C. Heil, *The Impact of Scale on Children's Spatial Thought*, Studien zur theoretischen und empirischen Forschung in der Mathematikdidaktik, https://doi.org/10.1007/978-3-658-32648-7_14

to manipulation of those objects or their movement in space. Geometry education aims to organize and structure those experiences to allow children to develop an abstract conception of space. Fostering spatial abilities and spatial thought therefore is a central activity of geometry education because it enables children to success-fully interact with their environment, but it also helps them to develop geometrical knowledge.

Following Freudenthal's ideas on geometry education, I argued that spatial abil-ities should be considered not only when it comes to small-scale spatial settings involving written or manipulable material, but also in the case of tasks that mimic the complex spatial challenges in real space. I wondered whether the previous con-cepts of spatial abilities in mathematics education were too limited since they have only been defined for the mental manipulation of small-scale spatial information. Although these stimuli typically depict situations from real space, they can also be solved without drawing on information from the depicted space. Addressing these unresolved issues, I proposed a broader concept of spatial abilities, one that would, for example, also involve solving map-based spatial tasks in real space. In these tasks, spatial information only becomes meaningful when the represented real space is also considered, ideally from multiple vantage points. I assumed that the abilities to solve small-scale spatial tasks reflect the abilities to solve map-based orientation tasks in real space only to a limited extent. Gaining a deeper understand-ing of the relation between small- and large-scale spatial abilities by analyzing their similarities and differences became the primary goal of this study.

Since large-scale spatial abilities are highly complex, their relationship to small-scale spatial abilities is particularly hard to distangle. In this study, I proposed a setting of spatial activities for both small-scale and large-scale of space that are meaningful in the context of mathematics education and empirically related individ-ual performances in these tasks using a quantitative study design. Two key findings of this study are that there are large individual differences in spatial abilities at both scales of space and that children between the age of 9 and 12 who have usually developed their small-scale spatial abilities well enough to solve complex paper-based tasks, may struggle when being asked to solve map-based orientation tasks. Conversely, children who have developed the abilities to locate themselves in real space with the help of a map may struggle when it comes to solving spatial tasks in the classroom that are based on abstractions of real situations. The empirical results of this study show that both classes of spatial abilities are not completely dissociated, that is, that there might be some cognitive processes that are required for solving spatial problems at both scales of space. It will be necessary to investi-gate those in greater detail, not only because large-scale spatial abilities are closely related to solving spatial problems in everyday life, but also because they can offer new theoretical insights regarding small-scale spatial abilities. This might lead to a better understanding of children's spatial thought, which, in turn, could serve as a basis for developing interventions that are effective for fostering spatial abilities at different scales of space.

Investigating how scale of space relates to children's spatial thought when it comes to 'grasping space', this study offers support for the assumption that scale is indirectly linked to children's spatial abilities used in different settings in geometry education. The findings presented here are, however, merely a starting point, and a range of further research questions can be derived from my study. I addressed some questions in previous sections and the critical reader might probably find more ideas in the limitations section of this study. One of the most important implications of this study for researches in mathematics education is that it will be necessary to develop and test pedagogical interventions involving spatial settings at various scales of space. The question is how these interventions, both those restricted to tabletop settings in the classroom and those involving real space, need to be combined to improve children's spatial abilities. After all, we do not want our students to end up like poor Pooh and Piglet who struggle to navigate in real space, running around in circles and becoming scared of more and more footprints of whatever-it-is. Researchers and teachers could include spatial activities outside the classroom, perhaps using maps, and thereby help students to make the connection between abstract spatial thought and their day-to-day lives.

Small-Scale Spatial Abilities Test

A

Im Labyrinth der Bilder

Das kann ich schon alles.

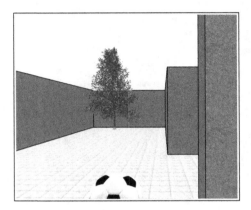

Fantasiename: ....

☐ Ich bin ein Mädchen.   ☐ Ich bin ein Junge.

Geburtsdatum: ....

Mein Alter: ...... Jahre

AUF DEM SPIELPLATZ

Hier siehst du 4 Kinder auf einem Spielplatz. Male jeweils **aus der Sicht der Kinder** alle rechten Schuhe rot aus.

LINKS ODER RECHTS?

Hier siehst du eine Schatzkarte mit einem Weg von oben.

Flo läuft nun den Weg vom Start bis zum Schatz entlang. Schreibe an jede Ecke in den Kreis, ob Flo dabei nach **rechts (R)** oder **links (L)** abbiegt.

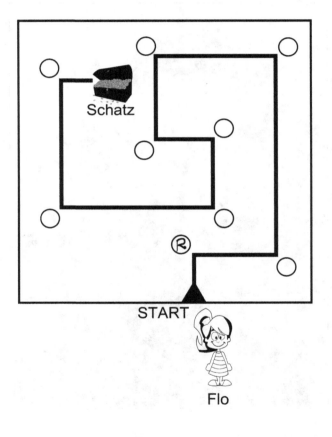

GEDREHTE FIGUREN

Hier siehst du
eine Figur.

Kann man jede Figur auf dieser Seite so drehen,
dass die Figur links entsteht?

☒ ja ☐ nein   ☐ ja ☒ nein

1.

☐ ja ☐ nein   ☐ ja ☐ nein   ☐ ja ☐ nein

2.

☐ ja ☐ nein   ☐ ja ☐ nein   ☐ ja ☐ nein

3.

☐ ja ☐ nein   ☐ ja ☐ nein   ☐ ja ☐ nein

4.

☐ ja ☐ nein   ☐ ja ☐ nein   ☐ ja ☐ nein

DURCH DAS LABYRINTH

Claudia läuft durch ein Labyrinth und macht dabei 6 Fotos.

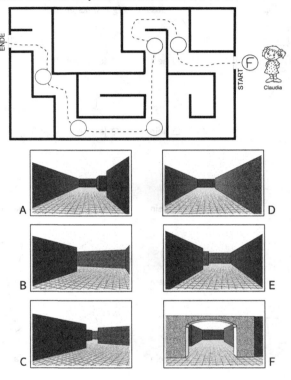

Welches Foto wurde wo im Labyrinth gemacht?

Trage jeweils den richtigen Buchstaben in den Kreis ein.

WILDE SCHLANGEN

Hier siehst du | **Zwei** der vier Bilder zeigen die gleiche Schlange.
eine Schlange. | Welche **zwei** Schlangen musst du ankreuzen?

| | ☐ | ☒ | ☐ | ☒ |

1.

| | ☐ | ☐ | ☐ | ☐ |

2.

| | ☐ | ☐ | ☐ | ☐ |

3.

| | ☐ | ☐ | ☐ | ☐ |

4.

| | ☐ | ☐ | ☐ | ☐ |

5.

| | ☐ | ☐ | ☐ | ☐ |

Kontrolle: Hast du wirklich überall **zwei** Schlangen angekreuzt?

VERSCHACHTELT

Bei dieser Aufgabe siehst du gleich große
Schachteln von oben auf einem Spielbrett.

Welche Seitenansicht gehört zu welchem Buchstaben?

Trage die richtigen Buchstaben in die Kästchen ein.
Achtung, ein Bild gehört nicht dazu (dort lässt du das Kästchen frei).

(1)   von oben   von der Seite

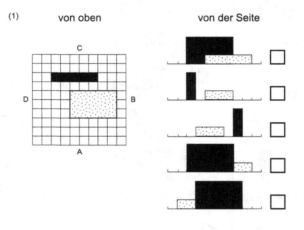

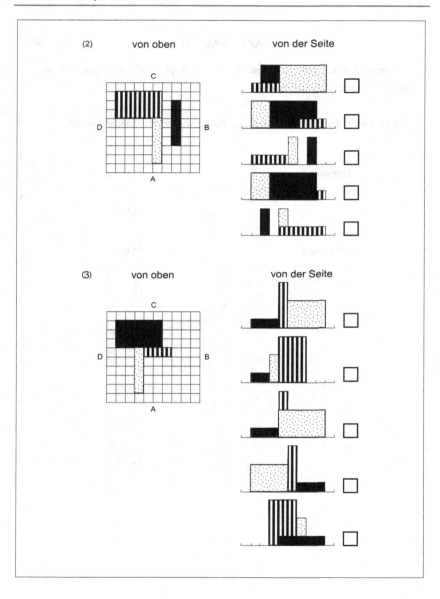

PAPIERFALTEN

Bei dieser Aufgabe geht es darum, dass du im Kopf ein Papier faltest und mit
dem Stift ein Loch hinein machst.

Wo sind dann die Löcher, wenn du das Papier wieder auseinander faltest?

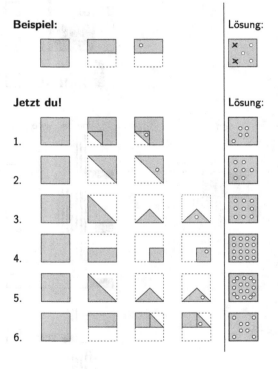

NOCH EIN LABYRINTH

Emil läuft durch ein Labyrinth und macht dabei 6 Fotos.

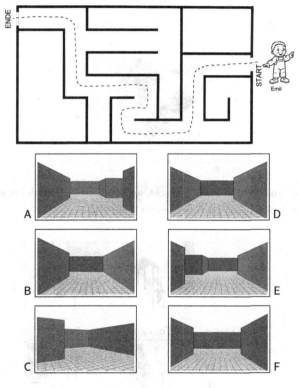

Die Fotos von Emil sind durcheinander geraten. Sortiere sie.

1. _____   4. _____

2. _____   5. _____

3. _____   6. _____

WO WURDEN DIE FOTOS GEMACHT?

Ben sieht sich einen Spielplatz an. Der sieht von oben so aus:

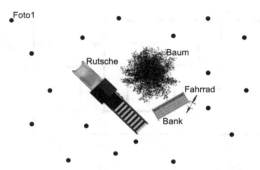

Ben hat ein Foto an dem Punkt gemacht, wo Foto 1 steht. Danach hat er an 5 weiteren Punkten auch ein Foto gemacht (Foto 2 bis Foto 6).

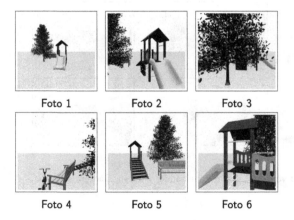

Wie ist Ben gelaufen um seine Fotos zu machen?
Verbinde die richtigen <u>6 Punkte</u> oben in der Karte.

DIE SCHIFFFAHRT

Der Kapitän eines Schiffes hat auf seiner Fahrt an dieser Küste Fotos gemacht.

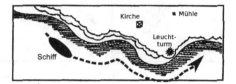

Kannst du die Fotos in die richtige Reihenfolge bringen?

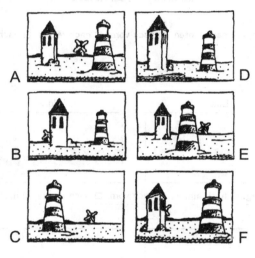

1. __B__   4. _____

2. _____   5. _____

3. _____   6. _____

AUF DER WIESE

Bild 1	Bild 2

Wie bist du von Bild 1 zu Bild 2 gegangen?

Kreuze die richtigen **zwei** Kästchen an.

Mein Weg: ☒ nach hinten ☐ nach vorn ☒ nach rechts ☐ nach links

Jetzt du!

Bild 1	Bild 2

Mein Weg: ☐ nach hinten ☐ nach vorn ☐ nach rechts ☐ nach links

- -

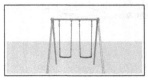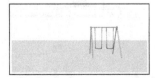

Mein Weg: ☐ nach hinten ☐ nach vorn ☐ nach rechts ☐ nach links

- -

Bild 1   Bild 2

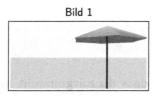   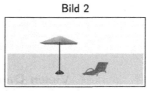

Mein Weg: ☐ nach hinten   ☐ nach vorn   ☐ nach rechts   ☐ nach links

Mein Weg: ☐ nach hinten   ☐ nach vorn   ☐ nach rechts   ☐ nach links

Mein Weg: ☐ nach hinten   ☐ nach vorn   ☐ nach rechts   ☐ nach links

Mein Weg: ☐ nach hinten   ☐ nach vorn   ☐ nach rechts   ☐ nach links

Kontrolle: Hast du bei jeder Aufgabe **zwei** Mal angekreuzt?

Vielen Dank, dass du mitgemacht hast!

Extra Tables

B

C. Heil, *The Impact of Scale on Children's Spatial Thought*, Studien zur
theoretischen und empirischen Forschung in der Mathematikdidaktik,
https://doi.org/10.1007/978-3-658-32648-7

Table B.1 Revisions After Pilot Study in the Paper-and-Pencil Test

Measure	Observation	Revision
Shoes	Nine children constantly confound yellow and blue and score with zero.	Modify instruction to "color all right shoes red".
LR	More than 5% of the children score with zero, probably because they did not understand the task instruction. Questions on task instruction. Loads on OB rather than on EGO.	Introduction of a protagonist (Flo) who walks along the way.
2DMR	Item 2 has a bad discrimination value and does not scale according to the Rasch model. The example is hard to understand since it figures a complex object.	Eliminate item 2 and insert it slightly modified as example. Adopt scoring to polytomous one.
Claudia	Problem with sorting picture D, probably because of distance estimation problems.	Complete revision, see below
3DMR	Item 4 and 5 feature complex objects with more than 7 cubes. Item 1 and 5 feature bad discrimination values, all other items show moderate discrimination values. The distractors show snakes of complete different form.	Complete revision, see below.
Meadow	Bimodality and ceiling effect, but good item characteristics.	no revisions
PFT	Item 1 has a low discrimination value (.23) and is very easy (.93). There is another easy item 2. Picture in the task instruction is confusing.	Delete item 1 and conceptualize another item with a non-symmetric folding as item 6. Put task instruction and example into the general moderation block.
Anna, Ben	Anna task demands for mental transformations in a non-human, horizontal movement way. Ceiling effects in the Ben-task.	Remove task Anna. Keep task Ben, but do not take task instruction into the general moderation block.
Cruise	Low reliability in the two-factor model (+low portion of explained variance).	Keep for reasons of comparability to other samples. Adopt scoring (Table C.12 in ESM).
Boxes	53% of the children score with 1 in the first item, probably because the letters "A" and "B" suggest a subjective simplicity of the first item. Distractor in item 3 does picture only two boxes. Ceiling effects in all items.	Remove letters "A" and "B" in the first item. Put a third box into the distractor of item 3 such that it corresponds to a reflection of a good view. Adopt scoring of the measure (Table C.17 in ESM).
The Way	High guessing probability (1/3), long task instruction.	Remove and keep Meadow as route identification task.
Maps	37 cases are missing due to time problems; limited confidence in the statistical values, two identical figures in item 2.	Promising approach to test 2D mental rotation. 2DMR was kept to compare the results to the previous studies in the literature.
Dirk	Solution is not unique.	Replace by Emil-task, see below.

Table B.2 Small-Scale Spatial Abilities Tests for the Main Study

Nb.	Task name	Construct	Nb. of items	Item scoring	Task scoring	Estimated time
1	Shoes	starter	4	–	none	3 min
2	LR	OB	8	dichotomous (CMC)	sumscore if Rasch scalable	3 min
3	2DMR	OB	4	polytomous	sumscore if Rasch scalable	3 min
4	Claudia	EGO	5	dichotomous	sumscore if Rasch scalable	5 min
5	3DMR	OB	5	dichotomous (CMC)	sumscore if Rasch scalable	3,5 min
6	Boxes	EGO	3	polytomous sum score	sumscore if Rasch scalable	7 min
7	PFT	OB	6	dichotomous (CMC)	sumscore if Rasch scalable	5 min
8	Emil	EGO	1	–	sumscore over the sorting	4 min
9	Ben	EGO	5	dichotomous	sumscore if Rasch scalable	3,5 min
10	Cruise	EGO/OB	1	–	sumscore over the sorting	4 min
11	Meadow	EGO	6	dichotomous (CMC)	sumscore if Rasch scalable	4 min

CMC = Complex Multiple Choice

Table B.3 Large-Scale Spatial Abilities Tests for the Main Study

Nb.	Task name	Construct	Nb. of items	Item scoring	Task scoring
1	MapRot	Comp	6	dichotomous	sumscore if Rasch scalable
2	Dots	Prod	3	polytomous	sumscore if Rasch scalable
3	Dir	Prod	3	dichotomous	sumscore if Rasch scalable
4	ActiveHP	Comp	6	dichotomous	sumscore if Rasch scalable
5	Disks	Comp	3	polytomous	sumscore if Rasch scalable
6	MFlag	SurveyMap	3	dichotomous	sumscore if Rasch scalable
7	MDisk	SurveyMap	3	dichotomous	sumscore if Rasch scalable

Table B.4 Different Methods to Study Unaided Large Scale Navigation and Learning

behavior	task	dependent variable(s)	example
distance estimation	pairwise distance estimation (self-landmark, landmark-landmark)	accuracy (mean error)	HEGARTY ET AL. (2006)
	route distance estimation	mean error	ALLEN ET AL. (1996)
	distance ranking	rank order correlations	KIRASIC (2000)
orientation judgement	pointing to landmarks	time	HOLDING AND HOLDING (1989)
	pointing to landmarks	mean error	HEGARTY ET AL. (2006)
map drawing	sketch map	number of qualitative errors	HEGARTY ET AL. (2006)
	landmark and route locating	number of correct landmarks	PEARSON AND IALONGO (1986)
navigation	find a pre-defined way	efficiancy in terms of shortest distance	KIRASIC (2000)
	find a shortcut	number of correct segments of the route	KOZHEVNIKOV ET AL. (2006)
	find a pre-defined way	number of wrong directions, turns	MOFFAT, HAMPSON, AND HATZIPANTELIS (1998)
	Route reversal	time	LAWTON (1996)
	Route reversal	number of correct turns	KOZHEVNIKOV ET AL. (2006)
	landmark sequencing task	number of correct landmarks	ALLEN ET AL. (1996)

Table B.5 A Summary of Tasks That Measured Map-Based Navigation in Adults and Children. All Studies Have Been Done Using Depictive, Opaque Maps and Not Areal Photographs

method	study	subjects	task	role of the experimenter	environmental context
comprehension method	GERBER AND KWAN (1994)	children (12 years)	walk along a 2 km pre-defined route + qualitative behavior observation	shadowing	familiar suburban area
comprehension method	MALINOWSKI AND GILLESPIE (2001)	adults	find eight locations with the help of a map (results reported on score card)	–	(unfamiliar) forest
production method	LIBEN ET AL. (2008)	adults	mark five locations on a map with an arrow sticker (guided walk)	lead participants along a pre-defined way; observe strategies	unfamiliar campus
production method	KASTENS and LIBEN (2010)	children (fourth-graders)	mark locations on a map with an sticker (free walk) + write explanation	–	unfamiliar campus
production method	LIBEN ET AL. (2010)	adults	mark 5 locations and directions on a map with an arrow sticker (guided walk); give 5 directions to unseen buildings with a map	lead participants along a pre-defined way; observe strategies	unfamiliar campus
production method	LIBEN ET AL. (2013)	children (9 to 10 years)	mark locations on a map with an sticker (free walk)	shadow children and observe strategies	unfamiliar campus
comprehension method	WRENGER (2015)	children(5th grade)	walk along a pre-defined route on the map	shadowing	city

Table B.6 Summary of Literature Concerning the Relation Between Performances in Small-Scale Spatial Tasks and Tasks of Unaided Navigation (extended version of the analysis by HEGARTY AND WALLER (2005), p.151)

authors(year)	sample size	significant correlations reported	measures of small-scale abilities	measures of large-scale abilities	learning environment
WALSH, KRAUSS, AND REGNIER (1981)	31	2 (.47 to .57)	ETS building memory test; Figure rotation test (PMA)	Accuracy of locating neighborhood landmarks	video
K. J. BRYANT (1982)	85	2 (.26 to .39)	Vandenberg Mental Rotations Test	Sense of direction self-report; Worrying of becoming lost self-report; Pointing accuracy	familiar environment
GOLDIN AND THORNDYKE (1982)	94	6 (of 18) (?? to .30)	Building Memory; Form Board Test; Hidden Figures; Tests of vividness of imagery	location recognition test; serial order of locations; route distance estimation; pointing to landmarks; euclidean distance estimation; landmark placement task	video, bus tour narration
PEARSON AND IALONGO (1986)	353	7 (of 8) (.09 to .30)	Group Embedded Figures Test (EFT); Vandenberg's Mental Rotation Test; Differential Aptitude Test (DAT); ETS building memory test	Route and landmark knowledge	slide-based
SHOLL (1988) (study 1)	28	4 (.07 to .19)	GZ-Spatial Orientation Test; Paper folding; Card rotations; Vandenberg Mental Rotation	Sense of Direction Scale (9-point-scale, 1 question)	—
ROVINE AND WEISMAN (1989)	45	6 (of 18) (.30 to .40)	Sense of Direction Scale (20 items); Group Embedded Figures Test; Spatial Relation Test (PMA)	Sketch Map Measures (number of landmarks and paths, complexity, topological accuracy); Wayfinding Tasks (direct route taking, distance traveled, number of turns)	university building

(Continued)

Table B.6 (Continued)

authors(year)	sample size	significant correlations reported	measures of small-scale abilities	measures of large-scale abilities	learning environment
ALLEN ET AL. (1996)(study 1/2)	100/103	42 (.02 to .37/.00 to .41)	Surface development , Cube comparison, Hidden Figures, Gestalt completion, Map memory, Map planning (Kit of Factor-Referenced Cognitive tasks)	Route reversal, Euclidean distance and direction judgement, Scene recognition, Scene sequencing, Intraroute distance judgement, Map placement test	unfamiliar environment (outdoors)
JUAN-ESPINOSA, ABAD, COLOM, AND FERN'ANDEZ-TRUCHAUD (2000)	111	4+2 (.20 to .31)	Displacement test; Solid Figures Test	Reference Points-Updating (building); Starting-Point-Updating (building); Destination-point-updating (VE)	university building + VE
KIRASIC (2000)	240	30 (.05 to .39)	Surface Development Test, Card rotation, Hidden Figures, Gestalt Completion, Map planning	Scene recognition, distance ranking, map placement (environmental learning); Route execution, Walking observation, standing observation (wayfinding behavior)	supermarket
MALINOWSKI (2001)	211	2 (.27 to .32)	Vandenberg Mental Rotation test	map compass classical orienteering task (find 10 locations)	military territory
KOZHEVNIKOV ET AL. (2006) (study2)	54	4 (of 8) (.29 to .43)	computerized PTSOT; Shepard-Metzeler Mental Rotations Test	Route retracing, Floor plan route drawing, shortcut finding, direction pointing	unfamiliar university building
EGARTY ET AL. (2006)	221	12+12+12 (.11 to .28; .13 to .46; .09 to .45)	Group Embedded Figures test, Vandenberg Mental Rotations test, Perspective taking, SBSOD, Arrow Span	Direction estimation, Distance estimation, map sketching	unknown campus building/ video/VE

PMA = Primary Mental Ability Test; ETS = Educational Testing Service; VE = Virtual environment

References

ACREDOLO, L. P. (1981). Small- and large-scale spatial concepts in infancy and childhood. In L. S. Liben, A. H. Patterson, & N. S. Newcombe (Eds.), *Spatial representation and behavior across the life span* (pp. 63–81). New York, NY: Academic Press.

ALLEN, G. L. (1999a). Children's control of reference systems in spatial tasks: Foundations of spatial cognitive skill? *Spatial Cognition and Computation, 1*(4), 413–429.

ALLEN, G. L. (1999b). Spatial abilities, cognitive maps, and wayfinding. In R. G. Golledge (Ed.), *Wayfinding behavior: Cognitive mapping and other spatial processes* (pp. 46–80). Baltimore, MD: John Hopkins University Press.

ALLEN, G. L., KIRASIC, K. C., DOBSON, S. H., LONG, R. G., & BECK, S. (1996). Predicting environmental learning from spatial abilities: An indirect route. *Intelligence, 22*(3), 327–355.

AMORIM, M.-A., ISABLEU, B., & JARRAYA, M. (2006). Embodied spatial transformations: "body analogy" for the mental rotation of objects. *Journal of Experimental Psychology: General, 135*(3), 327–347.

AMORIM, M.-A., & STUCCHI, N. (1997). Viewer- and object-centered mental explorations of an imagined environment are not equivalent. *Cognitive Brain Research, 5*(3), 229–239.

ANDERSON, J. R. (1978). Arguments concerning representations for mental imagery. *Psychological Review, 85*(4), 249–277.

ANDERSON, J. R. (1987). Methodologies for studying human knowledge. *Behavioral and Brain Sciences, 10*(3), 467–477.

ANDERSON, J. R., & FUNKE, J. (Eds.). (2013). *Kognitive Psychologie* (7th ed.). Berlin, Heidelberg: Springer.

ANDERSON, R. E. (2008). Implications of the information and knowledge society for education. In J. Voogt & G. Knezek (Eds.), *International handbook of information technology in primary and secondary education* (pp. 5–22). New York, NY: Springer.

ANG, S. Y., & LEE, K. (2008). Central executive involvement in children's spatial memory. *Memory, 16*(8), 918–933.

APPLEYARD, D. (1970). Styles and methods of structuring a city. *Environment and Behavior, 2*(1), 100–117.

ARETZ, A. J., & WICKENS, C. D. (1992). The mental rotation of map displays. *Human performance, 5*(4), 303–328.

ARNOLD, A. E., BURLES, F., KRIVORUCHKO, T., LIU, I., REY, C. D., LEVY, R. M., & IARIA, G. (2013). Cognitive mapping in humans and its relationship to other orientation skills. *Experimental Brain Research, 224*(3), 359–372.

ATIT, K., SHIPLEY, T. F., & TIKOFF, B. (2013). Twisting space: Are rigid and nonrigid mental transformations separate spatial skills? *Cognitive Processing, 14*(2), 163–173.

BAGOZZI, R. P., & PHILLIPS, L. W. (1982). Representing and testing organizational theories: A holistic construal. *Administrative Science Quarterly, 27*(3), 459–489.

BARRATT, E. S. (1953). An analysis of verbal reports of solving spatial problems as an aid in defining spatial factors. *The Journal of Psychology, 36*(1), 17–25.

BARRETT, P. (2007). Structural equation modelling: Adjudging model fit. *Personality and Individual Differences, 42*(5), 815–824.

BARTOLOMEO, P. (2002). The relationship between visual perception and visual mental imagery: A reappraisal of the neuropsychological evidence. *Cortex, 38*(3), 357–378.

BEHRMANN, M. (2000). The mind's eye mapped onto the brain's matter. *Current Directions in Psychological Science, 9*(2), 50–54.

BELL, S. (2002). Spatial cognition and scale: A child's perspective. *Journal of Environmental Psychology, 22*(1–2), 9–27.

BENTLER, P. M., & CHOU, C. P. (1987). Practical issues in structural modeling. *Sociological Methods & Research, 16*(1), 78–117.

BERCH, D. B., KRIKORIAN, R., & HUHA, E. M. (1998). The corsi block-tapping task: Methodological and theoretical considerations. *Brain and Cognition, 38*(3), 317–338.

BERLINGER, N. (2015). *Die Bedeutung des räumlichen Vorstellungsvermögens für mathematische Begabungen bei Grundschulkindern: theoretische Grundlegung und empirische Untersuchungen.* Münster: WTM.

BERTHELOT, R., & SALIN, M.-H. (1993). L'enseignement de la géométrie à lécole primaire. *Grand N, 53*, 39–56.

BESUDEN, H. (1999a). Raumvorstellung und Geometrieverständnis. *Mathematische Unterrichtspraxis, 20*(3), 1–10.

BESUDEN, H. (1999b). *Raumvorstellung und Geometrieverständnis–Unterrichtsbeispiele.* Zentrum für pädagogische Berufspraxis.

BIKNER-AHSBAHS, A., & VOHNS, A. (2019). Theories of and in mathematics education. In H. N. Jahnke & L. Hefendehl-Hebeker (Eds.), *Traditions in German-speaking mathematics education research* (pp. 171–200). Cham: Springer International Publishing.

BISHOP, A. J. (1980). Spatial abilities and mathematics education-a review. *Educational Studies in Mathematics, 11*(3), 257–269.

BLADES, M. (1989). Children's ability to learn about the environment from direct experience and from spatial representations. *Children's Environments Quarterly, 6*(2/3), 4–14.

BLADES, M., & SPENCER, C. (1986). Map use in the environment and educating children to use maps. *Environmental Education and Information, 5*(3), 187–204.

BLADES, M., & SPENCER, C. (1987). The use of maps by 4-6-year-old children in a large-scale maze. *British Journal of Developmental Psychology, 5*(1), 19–24.

BLUESTEIN, N., & ACREDOLO, L. P. (1979). Developmental changes in map-reading skills. *Child Development, 50*(3),691–697.

BOND, T., & FOX, C. M. (2015). *Applying the Rasch model: Fundamental measurement in the human sciences* (3rd ed.). New York, London: Routledge.

BORICH, G. D., & BAUMAN, P. M. (1972). Convergent and discriminant validation of the French and Guilford-Zimmerman spatial orientation and spatial visualization factors. *Educational and Psychological Measurement, 32*(4), 1029–1033.

BORTZ, J., & DÖRING, N. (2016). *Forschungsmethoden und Evaluation für Human- und Sozialwissenschaftler* (5th ed.). Heidelberg: Springer.

BRUCE, C. D., DAVIS, B., SINCLAIR, N., MCGARVEY, L., HALLOWELL, D., DREFS, M., . . . MULLIGAN, J. (2017). Understanding gaps in research networks: Using "spatial reasoning" as a window into the importance of networked educational research. *Educational Studies in Mathematics, 95*(2), 143–161.

BRUCE, C. D., & HAWES, Z. (2015). The role of 2D and 3D mental rotation in mathematics for young children: What is it? Why does it matter? And what can we do about it? *ZDM, 47*(3), 331–343.

BRUNER, J. S. (1964). The course of cognitive growth. *American Psychologist, 19*(1), 1–15.

BRUNER, J. S. (1973). *Beyond the information given: Studies in the psychology of knowing.* New York, Norton.

BRUNER, J. S., OLVER, R. R., & GREENFIELD, P. M. (1966). *Studies in cognitive growth.* Oxford: Wiley.

BRYANT, D. J., & TVERSKY, B. (1999). Mental representations of perspective and spatial relations from diagrams and models. *Journal of Experimental Psychology: Learning, Memory, and Cognition, 25*(1), 137–156.

BRYANT, K. J. (1982). Personality correlates of sense of direction and geographic orientation. *Journal of Personality and Social Psychology, 43*(6), 1318–1324.

BRYANT, P. (2009). Understanding space and its representation in mathematics. In T. Nunes, P. Bryant, & A. Watson (Eds.), *Key understandings in mathematics learning* (pp. 1–40). London: Nuffield Foundation.

BÜCHTER, A. (2011). *Zur Erforschung von Mathematikleistung: theoretische Studie und empirische Untersuchung des Einflussfaktors Raumvorstellung.* Doctoral dissertation. Retrieved from http://hdl.handle.net/2003/27660 (17.06.2019)

BUNDESEN, C., & LARSEN, A. (1975). Visual transformation of size. *Journal of Experimental Psychology: Human Perception and Performance, 1*(3), 214–220.

BYRNE, B. M., (2013). *Structural equation modeling with Mplus: Basic concepts, applications, and programming.* New York, NY: Routledge.

CAEYENBERGHS, K., TSOUPAS, J., WILSON, P. H., & SMITS-ENGELSMAN, B. C. M. (2009). Motor imagery development in primary school children. *Developmental Neuropsychology, 34*(1), 103–121.

CARPENTER, P. A., & JUST, M. A. (1982). Spatial ability: An information processing approach to psychometrics. In R. J. Sternberg (Ed.), *Advances in the psychology of human intelligence* (Vol. 3, pp. 221–252). Hillsdale, NJ: Erlbaum.

CARR, M., ALEXEEV, N., WANG, L., BARNED, N., HORAN, E., & REED, A. (2018). The development of spatial skills in elementary school students. *Child Development, 89*(2), 446–460.

CARROLL, J. B. (1993). *Human cognitive abilities: A survey of factor-analytic studies.* Cambridge, UK: Cambridge University Press.

CARROLL, J. B. (1997). The three-stratum theory of cognitive abilities. In D. P. Flanagan, J. L. Genshaft, & P. L. Harrison (Eds.), *Contemporary intellectual assessment: Theories, tests, and issues* (pp. 122–130). New York, NY: The Guilford Press.

CHENG, Y.-L., & MIX, K. S. (2014). Spatial training improves children's mathematics ability. *Journal of Cognition and Development, 15*(1), 2–11.

CHRISTENSEN, A. E. (2011). *Not all map use is created equal: Influence of mapping tasks on children's performance, strategy use, and acquisition of survey knowledge*. Doctoral dissertation. Retrieved from https://pdfs.semanticscholar.org/bd27/fe571982886361035f719953c57b2e389749.pdf (17.06.2019)

CLEMENTS, D. H. (1999). Geometric and spatial thinking in young children. In J. V. Copley (Ed.), *Mathematics in the early years* (pp. 66–79). Reston, VA: National Council of Teachers of Mathematics.

CLEMENTS, D. H., & BATTISTA, M. T. (1992). Geometry and spatial reasoning. In D. A. Grouws (Ed.), *Handbook of research on mathematics teaching and learning: A project of the National Council of Teachers of Mathematics* (pp. 420–463). New York, NY: Macmillan.

CLEMENTS, K. (1982). Visual imagery and school mathematics. *For the Learning of Mathematics, 2*(3), 33–39.

COHEN, J. (1988). *Statistical power analysis for the behavioral sciences* (2nd ed.). Hillsdale, NJ: Erlbaum Associates.

COHEN, S. L. & COHEN, R. (1985). The role of activity in spatial cognition. In R. Cohen (Ed.), *The development of spatial cognition* (pp. 199–224). Hillsdale, NJ: Erlbaum.

COLUCCIA, E., BOSCO, A. & B RANDIMONTE, M. A. (2007). The role of visuo-spatial working memory in map learning: New findings from a map drawing paradigm. *Psychological Research, 71*(3), 359–372.

COLUCCIA, E. & LOUSE, G. (2004). Gender differences in spatial orientation: A review. *Journal of Environmental Psychology, 24*(3), 329–340.

COMREY, A. L. & LEE, H. B. (1992). *A first course in factor analysis* (2nd ed.). Hillsdale, NJ: Psychology Press.

CORP, I. B. (2017). IBM SPSS statistics for windows, version 25.0. *Armonk.*

CRAMPTON, J. (1992). A cognitive analysis of wayfinding expertise. *Cartographica: The International Journal for Geographic Information and Geovisualization, 29*(3-4), 46–65.

CREEM, S. H., WRAGA, M. & PROFFITT, D. R. (2001). Imagining physically impossible self-rotations: Geometry is more important than gravity. *Cognition, 81*(1), 41–64.

CRESCENTINI, C., FABBRO, F. & URGESI, C. (2014). Mental spatial transformations of objects and bodies: Different developmental trajectories in children from 7 to 11 years of age. *Developmental Psychology, 50*(2), 370–383.

CRONBACH, L. J. (1951). Coefficient alpha and the internal structure of tests. *Psychometrika, 16*(3), 297–334.

CUTTING, J. E., VISHTON, P. M. & BRAREN, P. A. (1995). How we avoid collisions with stationary and moving objects. *Psychological Review, 102*(4), 627–651.

DAVIES, C. & UTTAL, D. H. (2007). Map use and the development of spatial cognition. In J. M. Plumert & J. P. Spencer (Eds.), *The emerging spatial mind* (pp. 219–247). New York, NY: Oxford University Press.

DAVIS, B. & SPATIAL REASONING STUDY GROUP. (2015). *Spatial reasoning in the early years: Principles, assertions, and speculations*. New York and London: Routledge.

DE LANGE, J. (1984). Geometry for all or: No geometry at all. *ZDM, 3*, 90–97.

DELOACHE, J. S. (1987). Rapid change in the symbolic functioning of very young children. *Science, 238*(4833), 1556–1557.

DELOACHE, J. S. (1989). Young children's understanding of the correspondence between a scale model and a larger space. *Cognitive Development*, *4*(2), 121–139.

DENHIÈRE, G., LEGROS, D. & TAPIERO, I. (1993). Representation in memory and acquisition of knowledge from text and picture: Theoretical, methodological, and practical outcomes. *Educational Psychology Review*, *5*(3), 311–324.

DEPARTMENT FOR EDUCATION. (2014). *The national curriculum in England: key stages 3 and 4 framework document.* Retrieved from www.gov.uk/government/publications/national-curriculum-in-england-secondary-curriculum

DIEZMANN, C. M. & LOWRIE, T. (2008). Assessing primary students' knowledge of maps. *Proceedings of PME32 and PME-NA 30*, *2*, 414–421.

DIEZMANN, C. M. & LOWRIE, T. (2009). Primary students' spatial visualization and spatial orientation: An evidence base for instruction. *Proceedings of PME33*, *2*, 417–424.

DOGU, U. & ERKIP, F. (2000). Spatial factors affecting wayfinding and orientation: A case study in a shopping mall. *Environment and Behavior*, *32*(6), 731–755.

DOMNICK, I. (2005). *Probleme sehen – Ansichtssache.* Doctoral dissertation. Retrieved from https://refubium.fu-berlin.de/handle/fub188/1187 (17.06.2019)

DOWNS, R. M., & STEA, D. (Eds.). (1973). *Image and environment: Cognitive mapping and spatial behavior.* Chicago, IL: Aldine.

DOWNS, R. M. & STEA, D. (1977). *Maps in minds: Reflections on cognitive mapping.* New York, NY: Harper & Row.

DOWNS, R. M. & STEA, D. (1981). Maps and mappings as metaphors for spatial representation. In L. S. Liben, A. H. Patterson, & N. S. Newcombe (Eds.), *Spatial representation and behavior across the life span* (pp. 143–166). New York, NY: Academic Press.

EASTON, R. D. & SHOLL, M. J. (1995). Object-array structure, frames of reference, and retrieval of spatial knowledge. *Journal of Experimental Psychology: Learning, Memory, and Cognition*, *21*(2), 483–500.

EDWARDS, J. R. & BAGOZZI, R. P. (2000). On the nature and direction of relationships between constructs and measures. *Psychological Methods*, *5*(2), 155–174.

EID, M. & SCHMIDT, K. (2014). *Testtheorie und Testkonstruktion.* Göttingen: Hogrefe Verlag.

EKSTROM, R. B. (1976). *Kit of factor-referenced cognitive tests.* Princeton, NJ: Educational Testing Service.

EKSTROM, R. B., FRENCH, J. W., HARMAN, H. H., & DERMEN, D. (Eds.). (1976). *Manual for kit of factor-referenced cognitive tests.* Princeton, NJ: Educational Testing Service.

ELIOT, J. (1987). *Models of psychological space: psychometric, developmental, and experimental approaches.* New York, NY: Springer.

ELIOT, J. & HAUPTMAN, A. (1981). Different dimensions of spatial ability. *Studies in Science Education*, *8*(1), 45–66.

ELIOT, J. & SMITH, I. M. (1983). *An international directory of spatial tests.* Windsor: NFER-Nelson.

ENDERS, C. K. (2013). Analyzing structural equation models with missing data. In G. R. Hancock & R. O. Mueller (Eds.), *Structural equation modeling: A second course* (pp. 493–519). Charlotte, NC: Information Age Publishing.

ENGELKAMP, J. & PECHMANN, T. (1993). Kritische Anmerkungen zum Begriff der mentalen Repräsentation. In J. Engelkamp & T. Pechmann (Eds.), *Mentale Repräsentation* (pp. 7–16). Bern, Göttingen, Toronto, Seattle: Huber.

EPLEY, N., KEYSAR, B., VAN BOVEN, L. & GILOVICH, T. (2004). Perspective taking as egocentric anchoring and adjustment. *Journal of Personality and Social Psychology, 87*(3), 327–339.

EPLEY, N., MOREWEDGE, C. K. & KEYSAR, B. (2004). Perspective taking in children and adults: Equivalent egocentrism but differential correction. *Journal of Experimental Social Psychology, 40*(6), 760–768.

ESTES, D. (1998). Young children's awareness of their mental activity: The case of mental rotation. *Child Development, 69*(5), 1345–1360.

EVANS, G. W. & PEZDEK, K. (1980). Cognitive mapping: knowledge of real-world distance and location information. *Journal of Experimental Psychology: Human Learning and Memory, 6*(1), 13–24.

EYSENCK, M., PAYNE, S. & DERAKSHAN, N. (2005). Trait anxiety, visuospatial processing, and working memory. *Cognition & Emotion, 19*(8), 1214–1228.

FARRAN, D. C., LIPSEY, M. W. & WILSON, S. (2011). *Experimental evaluation of the tools of the mind pre-k curriculum.* Washington, DC.

FARRELL PAGULAYAN, K., BUSCH, R. M., MEDINA, K. L., BARTOK, J. A. & KRIKORIAN, R. (2006). Developmental normative data for the Corsi Block-Tapping task. *Journal of Clinical and Experimental Neuropsychology, 28*(6), 1043–1052.

FENNER, J., HEATHCOTE, D. & JERRAMS-SMITH, J. (2000). The development of wayfinding competency: Asymmetrical effects of visuo-spatial and verbal ability. *Journal of Environmental Psychology, 20*(2), 165–175.

FIELDS, A. W. & SHELTON, A. L. (2006). Individual skill differences and large-scale environmental learning. *Journal of Experimental Psychology: Learning, Memory, and Cognition, 32*(3), 506–515.

FILIMON, F. (2015). Are all spatial reference frames egocentric? Reinterpreting evidence for allocentric, object-centered, or world-centered reference frames. *Frontiers in Human Neuroscience, 9,* 648.

FINNEY, S. J., & DISTEFANO, C. (2013). Non-normal and categorical data in structural equation modeling. In G. R. Hancock & R. O. Mueller (Eds.), *Structural equation modeling: A second course* (Vol. 10, pp. 439–492). Charlotte, NC: Information Age Publishing.

FLAVELL, J. H., EVERETT, B. A., CROFT, K., & FLAVELL, E. R. (1981). Young children's knowledge about visual perception: Further evidence for the Level 1–Level 2 distinction. *Developmental Psychology, 17*(1), 99–103.

FLORA, D. B. & CURRAN, P. J. (2004). An empirical evaluation of alternative methods of estimation for confirmatory factor analysis with ordinal data. *Psychological Methods, 9*(4), 466–491.

FRANKLIN, N. & TVERSKY, B. (1990). Searching imagined environments. *Journal of Experimental Psychology: General, 119*(1), 63–76.

FRENCH, J. W. (1951). *The description of aptitude and achievement tests in terms of rotated factors.* Chicago, IL: University of Chicago Press.

FRENCH, JOHN W., EKSTROM, RUTH B., & PRICE, LEIGHTON A. (1963). *Manual for kit of reference tests for cognitive factors: (revised 1963).* DTIC Document

FREUD, E., PLAUT, D. C. & BEHRMANN, M. (2016). 'What' is happening in the dorsal visual pathway. *Trends in Cognitive Sciences, 20*(10), 773–784.

FREUDENTHAL, H. (1971). Geometry between the devil and the deep sea. *Educational Studies in Mathematics, 3*(3–4), 413–435.

FREUDENTHAL, H. (1973). The case of geometry. In H. Freudenthal (Ed.), *Mathematics as an educational task* (pp. 401–511). Dordrecht: Springer.

FREUNDSCHUH, S. M. & EGENHOFER, M. J. (1997). Human conceptions of spaces: Implications for geographic information systems. *Transactions in GIS, 2*(4), 361–375.

FRICK, A. (2019). Spatial transformation abilities and their relation to later mathematics performance. *Psychological Research, 83*, 1465–1484.

FRICK, A., DAUM, M. M., WALSER, S. & MAST, F. W. (2009). Motor processes in children's mental rotation. *Journal of Cognition and Development, 10*(1–2), 18–40.

FRICK, A., MÖHRING, W. & NEWCOMBE, N. S. (2014). Picturing perspectives: Development of perspective-taking abilities in 4- to 8-year-olds. *Frontiers in Psychology, 5*, 386.

GAL, H. & LINCHEVSKI, L. (2010). To see or not to see: Analyzing difficulties in geometry from the perspective of visual perception. *Educational Studies in Mathematics, 74*(2), 163–183.

GALLISTEL, C. R. (1990). *Learning, development, and conceptual change. (t]he organization of learning.* Cambridge, MA: The MIT Press.

GÄRLING, T., BÖÖK, A. & LINDBERG, E. (1986). Spatial orientation and wayfinding in the designed environment: A conceptual analysis and some suggestions for postoccupancy evaluation. *Journal of Architectural and Planning Research, 3*(1), 55–64.

GÄRLING, T. & GOLLEDGE, R. G. (1989). Environmental perception and cognition. In E. H. Zube & G. T. Moore (Eds.), *Advance in environment, behavior, and design* (pp. 203–236). Boston, MA: Springer.

GEISER, C. (2012). *Data analysis with Mplus.* New York: Guilford Press.

GERBER, R. & KWAN, T. (1994). A phenomenographical approach to the study of pre-adolescents' use of maps in a wayfinding exercise in a suburban environment. *Journal of Environmental Psychology, 14*(4), 265–280.

GOLDIN, S. E. & THORNDYKE, P. W. (1982). Simulating navigation for spatial knowledge acquisition. *Human Factors: The Journal of the Human Factors and Ergonomics Society, 24*(4), 457–471.

GOLLEDGE, R. G. (1999). Human wayfinding and cognitive maps. In R. G. Golledge (Ed.), *Wayfinding behavior* (pp. 5–45). Baltimore, MD: John Hopkins University Press.

GRASS, K. H. & KRAMMER, G. (2018). Direkte und indirekte Einflüsse der Raumvorstellung auf die Rechenleistungen am Ende der Grundschulzeit. *Journal für Mathematik-Didaktik, 39*(1), 43–67.

GRÖMPING, U. (2006). Relative importance for linear regression in R: The package relaimpo. *Journal of Statistical Software, 17*(1), 1–27.

GRÜSSING, M. (2002). Wieviel Raumvorstellung braucht man für Raumvorstellungsaufgaben? Children's spatial problem solving strategies. *ZDM, 34*(2), 237–45.

GRÜSSING, M. (2012). *Räumliche Fähigkeiten und Mathematikleistung. Eine empirische Studie mit Kindern im 4. Schuljahr.* Münster: Waxmann.

GUILFORD, J. P. (1956). The Guilford-Zimmerman aptitude survey. *Journal of Counseling & Development, 35*(4), 219–223.

GUILFORD, J. P. & ZIMMERMAN, W. S. (1948). The Guilford-Zimmerman aptitude survey. *Journal of Applied Psychology, 32*(1), 24–34.

GUNALP, P., MOOSSAIAN, T. & HEGARTY, M. (2019). Spatial perspective taking: Effects of social, directional, and interactive cues. *Memory & Cognition, 47*(5), 1031–1043.

GUTIÉRREZ, Á. (1996). Visualization in 3-dimensional geometry: In search of a framework. In L. Puig & Á. Gutiérrez (Eds.), *Proceedings of the PME 7* (pp. 3–19). Valencia.

GUTTMAN, R., EPSTEIN, E. E., AMIR, M. & GUTTMAN, L. (1990). A structural theory of spatial abilities. *Applied Psychological Measurement, 14*(3), 217–236.

HAMBRICK, D. Z. & ENGLE, R. W. (2003). The role of working memory in problem solving. In J. E. Davidson & R. J. Stemberg (Eds.), *The psychology of problem solving* (pp. 176–206). New York, NY: Cambridge University Press.

HARDWICK, D. A., MCINTYRE, C. W. & PICK, H. L. (1976). The content and manipulation of cognitive maps in children and adults. *Monographs of the Society for Research in Child Development, 41*(3), 1–55.

HARRIS, J., HIRSH-PASEK, K. & NEWCOMBE, N. S. (2013). Understanding spatial transformations: Similarities and differences between mental rotation and mental folding. *Cognitive Processing, 14*(2), 105–115.

HARRIS, J., NEWCOMBE, N. S. & HIRSH-PASEK, K. (2013). A new twist on studying the development of dynamic spatial transformations: Mental paper folding in young children. *Mind, Brain, and Education, 7*(1), 49–55.

HART, R. A., & MOORE, G. T. (1973). The development of spatial cognition: A review. In R. M. Downs & D. Stea (Eds.), *Image and environment: Cognitive mapping and spatial behavior* (pp. 246–288). Chicago, IL: Aldine.

HARTLEY, T., & BURGESS, N. (2002). Models of spatial cognition. In L. Nadel (Ed.), *Encyclopaedia of cognitive science*. London: Palgrave MacMillan.

HATTERMANN, M., KADUNZ, G., & REZAT, S. (2015). Geometrie: Leitidee Raum und Form. In R. Bruder, L. Hefendehl-Hebecker, B. Schmidt-Thieme, & H.- G. Weigand (Eds.), *Handbuch der Mathematikdidaktik* (pp. 185–219). Berlin, Heidelberg: Springer.

HEGARTY, M. (2010). Components of spatial intelligence. In B. H. Ross (Ed.), *Psychology of learning and motivation* (pp. 265–297). San Diego, CA: Elsevier Academic Press.

HEGARTY, M. (2018). Ability and sex differences in spatial thinking: What does the mental rotation test really measure? *Psychonomic Bulletin & Review, 25*(3), 1212–1219.

HEGARTY, M., MONTELLO, D. R., RICHARDSON, A. E., ISHIKAWA, T. & LOVELACE, K. (2006). Spatial abilities at different scales: Individual differences in aptitude-test performance and spatial-layout learning. *Intelligence, 34*(2), 151–176.

HEGARTY, M., RICHARDSON, A. E., MONTELLO, D. R., LOVELACE, K. & SUBBIAH, I. (2002). Development of a self-report measure of environmental spatial ability. *Intelligence, 30*(5), 425–447.

HEGARTY, M.,& WALLER, D. (2004). A dissociation between mental rotation and perspective-taking spatial abilities. *Intelligence, 32*(2), 175–191.

HEGARTY, M.,& WALLER, D. (2005). Individual differences in spatial abilities. In P. Shah & A. Miyake (Eds.), *The cambridge handbook of visuospatial thinking* (pp. 121–169). New York, NY: Cambridge University Press.

HEIL, C. (2017). Double perspective taking processes of primary children–adoption and application of a psychological instrument. In T. Dooley & G. Gueudet. (Ed.), *Proceedings of CERME10* (pp. 613–620). Dublin: DCU Institute of Education and ERME.

HELLMICH, F., & HARTMANN, J. (2002). Aspekte einer Förderung räumlicher Kompetenzen im Geometrieunterricht. *ZDM, 34*(2), 56–61.

HEMMER, I., HEMMER, M., KRUSCHEL, K., NEIDHARDT, E., OBERMAIER, G., & UPHUES, R. (2012) Zur Relevanz ausgewählter personenbezogener Einflussfaktoren auf die kartengestützte Orientierungskompetenz. In A. Hüttermann, P. Kirchner, St. Schuler, & K. Drieling (Eds.), *Räumliche Orientierung.* (pp. 64–73). Braunschweig: Westermann.

HEMMER, I., HEMMER, M., NEIDHARDT, E., OBERMAIER, G., UPHUES, R., & WRENGER, K. (2015). The influence of children's prior knowledge and previous experience on their spatial orientation skills in an urban environment. *Education 3-13, 43*(2), 184–196.

HERRMANN, T. (1993). Mentale Repräsentation-ein erläuterungswürdiger Begriff. In J. Engelkamp & T. Pechmann (Eds.), *Mentale Repräsentation* (pp. 17–30). Bern, Göttingen, Toronto, Seattle: Huber.

HERSHKOWITZ, R., PARZYSZ, B., & VAN DORMOLEN, J. (1996). Space and shape. In A. J. Bishop, M. K. Clements, K. Clements, C. Keitel, J. Kilpatrick, & C. Laborde (Eds.), *International handbook of mathematics education* (pp. 161–204). Dordrecht: Kluwer Academic Publishers.

HOLDING, C. S., & HOLDING, D. H. (1989). Acquisition of route network knowledge by males and females. *The Journal of General Psychology, 116*(1), 29–41.

HOLGADO-TELLO, F. P., CHACÓN-MOSCOSO, S., BARBERO-GARCÍA, I., & VILA-ABAD, E. (2010). Polychoric versus pearson correlations in exploratory and confirmatory factor analysis of ordinal variables. *Quality & Quantity, 44*(1), 153–166.

HOLM, S. (1979). A simple sequentially rejective multiple test procedure. *Scandinavian Journal of Statistics, 6*(2), 65–70.

HOOGLAND, J. J., & BOOMSMA, A. (1998). Robustness studies in covariance structure modeling: An overview and a meta-analysis. *Sociological Methods & Research, 26*(3), 329–367.

HU, L., & BENTLER, P. M. (1999). Cutoff criteria for fit indexes in covariance structure analysis: Conventional criteria versus new alternatives. *Structural Equation Modeling: A Multidisciplinary Journal, 6*(1), 1–55.

HUNT, E., & WALLER, D. A. (1999). *Orientation and wayfinding: A review*. Doctoral dissertation. Retrieved from http://citeseerx.ist.psu.edu/viewdoc/summary?doi=10.1.1.46.5608 (17.06.2019)

HUTTENLOCHER, J., & PRESSON, C. C. (1973). Mental rotation and the perspective problem. *Cognitive Psychology, 4*(2), 277–299.

HUTTENLOCHER, J., & PRESSON, C. C. (1979). The coding and transformation of spatial information. *Cognitive Psychology, 11*(3), 375–394.

ISHIKAWA, T., & MONTELLO, D. R. (2006). Spatial knowledge acquisition from direct experience in the environment: Individual differences in the development of metric knowledge and the integration of separately learned places. *Cognitive Psychology, 52*(2), 93–129.

ITTELSON, W. H. (1973). Environment perception and contemporary perceptual theory. In W. H. Ittelson (Ed.), *Environment and cognition* (pp. 1–19). New York: Seminar.

IWANOWSKA, K., & VOYER, D. (2013). Dimensional transformation in tests of spatial and environmental cognition. *Memory & Cognition, 41*(8), 1122–1131.

JANSEN, P. (2009). The dissociation of small- and large-scale spatial abilities in school-age children. *Perceptual and Motor Skills, 109*(2), 357–361.

JANSEN, P., WIEDENBAUER, G., & HAHN, N. (2010). Manual rotation training improves direction-estimations in a virtual environmental space. *European Journal of Cognitive Psychology, 22*(1), 6–17.

JANZEN, G. (2013). *Organisation räumlichen Wissens: Untersuchungen zur Orts- und Richtungsrepräsentation*. Wiesbaden: Deutscher Universitätsverlag.

JOHNSON, E. S., & MEADE, A. C. (1987). Developmental patterns of spatial ability: An early sex difference. *Child Development, 58*(3), 725–740.

JONES, K. (2003). Issues in the teaching and learning of geometry. In L. Haggarty (Ed.), *Aspects of teaching secondary mathematics* (pp. 121–139). London: Routledge.

JONES, K., & TZEKAKI, M. (2016). Research on the teaching and learning of geometry. In Á. Gutiérrez, G. C. Leder,& P. Boero (Eds.), *The second handbook of research on the psychology of mathematics education* (pp. 109–149). Rotterdam: SensePublishers.

JONES, N., ROSS, H., LYNAM, T., PEREZ, P., & LEITCH, A. (2011). Mental models: An interdisciplinary synthesis of theory and methods. *Ecology and Society, 16*(1), 46.

JORGENSEN, T. D., PORNPRASERTMANIT, S., SCHOEMANN, A., ROSSEEL, Y., MILLER, P., QUICK, C., . . . PREACHER, K. (2018). Package 'semTools'.

JUAN- ESPINOSA, M., ABAD, F. J., COLOM, R., & FERNÁNDEZ-TRUCHAUD, M. (2000). Individual differences in large-spaces orientation: g and beyond? *Personality and Individual Differences, 29*(1), 85–98.

JUST, M. A., & CARPENTER, P. A. (1985). Cognitive coordinate systems: Accounts of mental rotation and individual differences in spatial ability. *Psychological Review, 92*(2), 137–172.

KASTENS, K. A., KAPLAN, D., & CHRISTIE-BLICK, K. (2001). Development and evaluation of "Where Are We?" map-skills software and curriculum. *Journal of Geoscience Education, 49*(3), 249–266.

KASTENS, K. A., & LIBEN, L. S. (2007). Eliciting self-explanations improves children's performance on a field-based map skills task. *Cognition and Instruction, 25*(1), 45–74.

KASTENS, K. A., & LIBEN, L. S. (2010). Children's strategies and difficulties while using a map to record locations in an outdoor environment. *International Research in Geographical and Environmental Education, 19*(4), 315–340.

KAUFMAN, S. B. (2007). Sex differences in mental rotation and spatial visualization ability: Can they be accounted for by differences in working memory capacity? *Intelligence, 35*(3), 211–223.

KEEHNER, M., GUERIN, S. A., MILLER, M. B., TURK, D. J., & HEGARTY, M. (2006). Modulation of neural activity by angle of rotation during imagined spatial transformations. *Neuroimage, 33*(1), 391–398.

KERNS, K. A., & BERENBAUM, S. A. (1991). Sex differences in spatial ability in children. *Behavior Genetics, 21*(4), 383–396.

KESSELS, R. P. C., VAN ZANDVOORT, M. J. E., POSTMA, A., KAPPELLE, L. J., & DE HAAN, E. H. F. (2000). The Corsi block-tapping task: Standardization and normative data. *Applied Neuropsychology, 7*(4), 252–258.

KESSLER, K., & THOMSON, L. A. (2010). The embodied nature of spatial perspective taking: Embodied transformation versus sensorimotor interference. *Cognition, 114*(1), 72–88.

KHOOSHABEH, P., & HEGARTY, M. (2008). How visual information affects a spatial task. *Proceedings of Cognitive Science Society*, 2041–2046.

KIM, K. H. (2005). The relation among fit indexes, power, and sample size in structural equation modeling. *Structural Equation Modeling, 12*(3), 368–390.

KIRASIC, K. C. (2000). Age differences in adults' spatial abilities, learning environmental layout, and wayfinding behavior. *Spatial Cognition and Computation, 2*(2), 117–134.

KLATZKY, R. L. (1998). Allocentric and egocentric spatial representations: Definitions, distinctions, and interconnections. In C. Freksa, C. Habel, & K. Wender (Eds.), *Spatial Cognition. Lecture Notes in Computer Science* (pp. 1–17). Berlin, Heidelberg: Springer.

KLINE, P. (2013). *Handbook of psychological testing* (2nd ed.). New York, NY: Routledge.

KLINE, R. B. (2015). *Principles and practice of structural equation modeling.* New York, NY: Guilford Press.

KMK. (2004). *Bildungsstandards im Fach Mathematik für den Primarbereich.* München: Luchterhand.

KNAUFF, M. (1997). *Räumliches Wissen und Gedächtnis.* Wiesbaden: Deutscher Universitätsverlag.

KOLLER, I., ALEXANDROWICZ, R., & HATZINGER, R. (2012). *Das Rasch Modell in der Praxis: Eine Einführung in eRm.* Wien: UTB.

KOSSLYN, S. M., BALL, T. M., & REISER, B. J. (1978). Visual images preserve metric spatial information: Evidence from studies of image scanning. *Journal of Experimental Psychology: Human Perception and Performance, 4*(1), 47–60.

KOZHEVNIKOV, M. (2007). Cognitive styles in the context of modern psychology: Toward an integrated framework of cognitive style. *Psychological Bulletin, 133*(3), 464–481.

KOZHEVNIKOV, M., & HEGARTY, M. (2001). A dissociation between object manipulation spatial ability and spatial orientation ability. *Memory & Cognition, 29*(5), 745–756.

KOZHEVNIKOV, M., KOSSLYN, S. M., & SHEPHARD, J. M. (2005). Spatial versus object visualizers: A new characterization of visual cognitive style. *Memory & Cognition, 33*(4), 710–726.

KOZHEVNIKOV, M., MOTES, M. A., RASCH, B., & BLAJENKOVA, O. (2006). Perspective-taking vs. mental rotation transformations and how they predict spatial navigation performance. *Applied Cognitive Psychology 20*(3), 397–417.

KRAHA, A., TURNER, H., NIMON, K., ZIENTEK, L., & HENSON, R. (2012). Tools to support interpreting multiple regression in the face of multicollinearity. *Frontiers in Psychology 3,* 44.

KYRITSIS, M., GULLIVER, S. R., MORAR, S., & MACREDIE, R. (2009). Impact of cognitive style on spatial knowledge acquisition. In I. '09 (Ed.), *IEEE international conference on multimedia 2009* (pp. 966–969).

LABATE, E., PAZZAGLIA, F., & HEGARTY, M. (2014). What working memory subcomponents are needed in the acquisition of survey knowledge? Evidence from direction estimation and shortcut tasks. *Journal of Environmental Psychology, 37,* 73–79.

LAWTON, C. A. (1996). Strategies for indoor wayfinding: The role of orientation. *Journal of Environmental Psychology, 16*(2), 137–145.

LEAN, G., & CLEMENTS, M. K. (1981). Spatial ability, visual imagery, and mathematical performance. *Educational Studies in Mathematics, 12*(3), 267–299.

LEHNUNG, M. M. (2000). *Die Entwicklung räumlicher Repräsentationen bei Kindern im Vorschul- und Schulalter und ihre Beeinträchtigung durch Schädel-Hirn-Traumata.* München: utzverlag Wissenschaft.

LEHNUNG, M. M., HAALAND, V. Ø., POHL, J., & LEPLOW, B. (2001). Compass- versus finger-pointing tasks: The influence of different methods of assessment on age-related orientation performance in children. *Journal of Environmental Psychology, 21*(3), 283–289.

LEHNUNG, M. M., LEPLOW, B., HAALAND, V. Ø., MEHDORN, M., & FERSTL, R. (2003). Pointing accuracy in children is dependent on age, sex and experience. *Journal of Environmental Psychology, 23*(4), 419–425.

LEVIN, K. A. (2006). Cross-sectional studies. *Evidence-Based Dentistry, 7*(1), 24–25.

LEVINE, M. (1982). You-are-here maps: Psychological considerations. *Environment and Behavior, 14*(2), 221–237.

LEVINE, M., MARCHON, I., & HANLEY, G. (1984). The placement and misplacement of you-are-here maps. *Environment and Behavior, 16*(2), 139–157.

LEVINE, S. C., HUTTENLOCHER, J., TAYLOR, A., & LANGROCK, A. (1999). Early sex differences in spatial skill. *Developmental Psychology, 35*(4), 940–949.

LEVINSON, S. C. (1996). Frames of reference and molyneux's question: Crosslinguistic evidence. In P. Bloom & M. Peterson (Eds.), *Language and space* (pp. 109–169). Cambridge, MA: MIT Press.

LI, C. H. (2016). Confirmatory factor analysis with ordinal data: Comparing robust maximum likelihood and diagonally weighted least squares. *Behavior Research Methods, 48*(3), 936–949.

LIBEN, L. S. (1981). Spatial representation and behavior: Multiple perspectives. In L. S. Liben, A. H. Patterson,& N. S. Newcombe (Eds.), *Spatial representation and behavior across the life span* (Vol. 79, pp. 3–36). New York, NY: Academic Press.

LIBEN, L. S. (1982). Children's large-scale spatial cognition: Is the measure the message? *New Directions for Child and Adolescent Development, 1982*(15), 51–64.

LIBEN, L. S. (1997). Children's understanding of spatial representations of place: Mapping the methodological landscape. In N. Foreman (Ed.), *A handbook of spatial research paradigms and methodologies* (pp. 41–83). Hove: Psychology Press.

LIBEN, L. S. (1999). Developing an understanding of external spatial representations. In I. E. Sigel (Ed.), *Development of mental representation* (pp. 297–321). Mahwah, NJ: Erlbaum Associates.

LIBEN, L. S. (2003). Thinking through maps. In M. Gattis (Ed.), *Spatial schemas and abstract thought* (pp. 45–77). Cambridge, MA: MIT Press.

LIBEN, L. S. (2006). Education for spatial thinking. In K. A. Renninger & I. E. Sigel (Eds.), *Child psychology in practice* (pp. 197–247). Hoboken, NJ: John Wiley & Sons Inc.

LIBEN, L. S., & DOWNS, R. M. (1989). Understanding maps as symbols: The development of map concepts in children. In H. W. Reese (Ed.), *Advances in child development and behavior* (Vol. 22, pp. 145–201). New York, NY: Academic Press.

LIBEN, L. S., & DOWNS, R. M. (1991). The role of graphic representations in understanding the world. In R. M. Downs, L. S. Liben, & D. S. Palermo (Eds.), *Visions of aesthetics, the environment, and development: The legacy of Joachim F. Wohlwill* (pp. 139–180). Hillsdale, NJ: Erlbaum.

LIBEN, L. S., & DOWNS, R. M. (1993). Understanding person-space-map relations: Cartographic and developmental perspectives. *Developmental Psychology, 29*(4), 739–752.

LIBEN, L. S., KASTENS, K. A., & STEVENSON, L. M. (2002). Real-world knowledge through real-world maps: A developmental guide for navigating the educational terrain. *Developmental Review, 22*(2), 267–322.

LIBEN, L. S., MYERS, L. J., & CHRISTENSEN, A. E. (2010). Identifying locations and directions on field and representational mapping tasks: Predictors of success. *Spatial Cognition & Computation, 10*(2–3), 105–134.

LIBEN, L. S., MYERS, L. J., CHRISTENSEN, A. E., & BOWER, C. A. (2013). Environmental-scale map use in middle childhood: Links to spatial skills, strategies, and gender. *Child Development, 84*(6), 2047–2063.

LIBEN, L. S., MYERS, L. J., & KASTENS, K. A. (2008). Locating oneself on a map in relation to person qualities and map characteristics. In C. Freska, N. S. Newcombe, P. Gärdenfors,

& S. Wölfl (Eds.), *Spatial cognition VI: Learning, reasoning, and talking about space* (pp. 171–187). Berlin, Heidelberg: Springer.

LIBEN, L. S., & YEKEL, C. A. (1996). Preschoolers' understanding of plan and oblique maps: The role of geometric and representational correspondence. *Child Development, 67*(6), 2780–2796.

LINN, M. C., & PETERSEN, A. C. (1985) Emergence and characterization of sex differences in spatial ability: A meta-analysis. *Child Development, 56*(6), 1479–1498.

LIONETTI, F., KEIJSERS, L., DELLAGIULIA, A., & PASTORE, M. (2016). Evidence of factorial validity of parental knowledge, control and solicitation, and adolescent disclosure scales: When the ordered nature of likert scales matters. *Frontiers in Psychology, 7*, 1–10.

LITTLE, R. J. A. (1988). A test of missing completely at random for multivariate data with missing values. *Journal of the American Statistical Association, 83*(404), 1198–1202.

LITTLE, R. J. A., & RUBIN, D. B. (2002). *Statistical analysis with missing data* (2nd ed.). Hoboken, NJ: John Wiley & Sons.

LOBBEN, A. K. (2004). Tasks, strategies, and cognitive processes associated with navigational map reading: A review perspective. *The Professional Geographer, 56*(2), 270–281.

LOBBEN, A. K. (2007). Navigational map reading: Predicting performance and identifying relative influence of map-related abilities. *Annals of the Association of American Geographers, 97*(1), 64–85.

LOCKMAN, J. J., & PICK, H. L. (1984). Problems of scale in spatial development. In C. Sophian (Ed.), *Origins of cognitive skills* (pp. 3–26). Hillsdale, NJ: Erlbaum.

LOGAN, T. (2015). The influence of test mode and visuospatial ability on mathematics assessment performance. *Mathematics Education Research Journal, 27*(4), 423–441.

LOGAN, T., LOWRIE, T., & D iezmann, C. M. (2014). Co-thought gestures: Supporting students to successfully navigate map tasks. *Educational Studies in Mathematics, 87*(1), 87–102.

LOGAN, T., LOWRIE, T., & RAMFUL, A. (2017). Decoding map items through spatial orientation: Performance differences across grade and gender. In B. Kaur, W. K. Ho, T. L. Toh, & B. H. Choy (Eds.), *Proceedings of the PME 41* (Vol. 3, pp. 193–200). Singapore: PME.

LOHAUS A. (1999). *Räumliches Denken im Kindesalter*. Göttingen: Hogrefe.

LOHMAN, D. F. (1979). *Spatial ability: A review and reanalysis of the correlational literature*. Unpublished doctoral dissertation.

LOHMAN, D. F. (1982). Spatial abilities as traits, processes, and knowledge. In J. Sternberg (Ed.), *Advances in the psychology of human intelligence* (Vol. 4, pp. 181–248). Hillsdale, NJ: Erlbaum.

LOOMIS, J. M., KLATZKY, R. L., GOLLEDGE, R. G., & PHILBECK, J. W. (1999). Human navigation by path integration. In R. G. Golledge (Ed.), *Wayfinding behavior* (pp. 125–151). Baltimore, MD: John Hopkins University Press.

LORING-MEIER, S., & HALPERN, D. F. (1999). Sex differences in visuospatial working memory: Components of cognitive processing. *Psychonomic Bulletin & Review, 6*(3), 464–471.

LOWRIE, T., DIEZMANN CARMEL M., & LOGAN, T. (2011). Primary students' performance on map tasks: The role of context. *Proceedings of PME 36, 3*(3), 145–152.

LOWRIE, T., & LOGAN, T. (2007). Using spatial skills to interpret maps: Problem solving in realistic contexts. *Australian Primary Mathematics Classroom, 12*(4), 14–19.

LOWRIE, T., & LOGAN, T. (2015). The role of test-mode effect: Implications for assessment practices and item design. In C. Vistro-Yu (Ed.), *Proceedings of the 7th ICMI-East Asia regional conference on mathematics education* (pp. 649–655). Cebu, Philippines: Philippine Council of Mathematics Teacher Educators.

LOWRIE, T., & LOGAN, T. (2017). The influence of students' spatial reasoning on mathematics performance across different test mode formats. In B. Kaur, W. K. Ho, T. L. Toh, & B. H. Choy (Eds.), *Proceedings of the PME 41* (pp. 201–208). Singapore: PME.

LOWRIE, T., & LOGAN, T. (2018). The interaction between spatial reasoning constructs and mathematics understandings in elementary classrooms. In K. S. Mix & M. T. Battista (Eds.), *Visualizing mathematics. Research in mathematics education* (pp. 253–276). Cham: Springer.

LOWRIE, T., LOGAN, T., & RAMFUL, A. (2016). Spatial reasoning influences students' performance on mathematics tasks. *Annual Meeting of the Mathematics Education Research Group of Australasia*, 407–414.

LOWRIE, T., LOGAN, T., & RAMFUL, A. (2017). Visuospatial training improves elementary students' mathematics performance. *British Journal of Educational Psychology, 87*(2), 170–186.

LUMLEY, T. (2006). mitools: Tools for multiple imputation of missing data. *R Foundation for Statistical Computing*.

LÜTHJE, T. (2010). *Das räumliche Vorstellungsvermögen von Kindern im Vorschulalter: Ergebnisse einer Interviewstudie*. Hildesheim, BerlineDISSion Verlag im Verlag Franzbecker.

MACDONALD, J. D., & MACINTYRE, P. D. (1999). A rose is a rose: Effects of label change, education, and sex on attitudes toward mental disabilities. *Journal on Developmental Disabilities, 6*(2), 15–31.

MACEACHREN, A. M. (1992). Learning spatial information from maps: Can orientation-specificity be overcome? *The Professional Geographer, 44*(4), 431–443.

MACEACHREN, A. M. (1995). *How maps work: Representation, visualization, and design*. New York, NY: Guilford Press.

MAIER, P. H. (1999a). *Räumliches Vorstellungsvermögen: ein theoretischer Abriss des Phänomens räumliches Vorstellungsvermögen; mit didaktischen Hinweisen für den Unterricht*. Donauwörth: Auer.

MAIER, P. H. (1999b). Räumliches Vorstellungsvermögen – Raumgeometrie mit Raumvorstellung – Thesen, zur Neustrukturierung des Geometrieunterrichts. *Mathe-matikunterricht, 45*(3), 4–18.

MAIR, P., & HATZINGER, R. (2007). Extended Rasch modeling: The eRm package for the application of IRT models in R. *Journal of Statistical Software, 20*(9), 1–20.

MALINOWSKI, J. C. (2001). Mental rotation and real-world wayfinding. *Perceptual and Motor Skills, 92*(1), 19–30.

MALINOWSKI, J. C., & GILLESPIE, W. T. (2001). Individual differences in performance on a large-scale, real-world wayfinding. *Journal of Environmental Psychology, 21*(1), 73–82.

MALLOT, H. A. (1997). Behavior-oriented approaches to cognition: Theoretical perspectives. *Theory in Biosciences, 116*(115), 196–220.

MALLOT, H. A. (1999). Spatial cognition: Behavioral competences, neural mechanisms, and evolutionary scaling. *Kognitionswissenschaft, 8*(1), 40–48.

MANDLER, J. M. (1983). Representation. In C. U. Shantz, J. H. Flavell,& E. Markman (Eds.), *Cognitive development* (pp. 420–494). New York, NY: Wiley.

MANGER, T., & EIKELAND, O.-J. (1998). The effects of spatial visualization and students' sex on mathematical achievement. *British Journal of Psychology*, *89*(1), 17–25.

MARMOR, G. S. (1975). Development of kinetic images: When does the child first represent movement in mental images? *Cognitive Psychology*, *7*(4), 548–559.

MARR, D. (1982). *Vision: A computational investigation into the human representation and processing of visual information*. San Francisco, CA: Freeman.

MAY, M., & KLATZKY, R. L. (2000). Path integration while ignoring irrelevant movement. *Journal of Experimental Psychology: Learning, Memory, and Cognition*, *26*(1), 169–186.

MAY, M., & KLUWE, R. H. (2000). *Kognition im Umraum*. Wiesbaden: DUV: Kognitionswissenschaft.

McCLOSKEY, M. (2015). Spatial representation in mind and brain. In B. Rapp (Ed.), *The handbook of cognitive neuropsychology: What deficits reveal about the human mind* (pp. 101–132). Philadelphia: Psychology Press.

McDONALD, R. P. (1970). The theoretical foundations of principal factor analysis, canonical factor analysis, and alpha factor analysis. *British Journal of Mathematical and Statistical Psychology*, *23*(1), 1–21.

McDONALD, R. P. (1999). *Test theory: A unified treatment*. Mahwah, NJ: Erlbaum.

McGEE, M. G. (1979). Human spatial abilities: Psychometric studies and environmental, genetic, hormonal, and neurological influences. *Psychological Bulletin*, *86*(5), 889–918.

McNAMARA, T. P. (1986). Mental representations of spatial relations. *Cognitive Psychology*, *18*(1), 87–121.

MEILINGER, T. (2007). *Strategies of orientation in environmental spaces*. Doctoral dissertation. Retrieved from https://www.researchgate.net/publication/41781972_Strategies_of_Orientation_in_Environmental_Spaces (17.06.2019)

MEILINGER, T. (2008). The network of reference frames theory: A synthesis of graphs and cognitive maps. In C. Freska, N. S. Newcombe, P. Gärdenfors, & S. Wölfl (Eds.), *Spatial cognition VI: Learning, reasoning, and talking about space* (pp. 344–360). Berlin, Heidelberg: Springer.

MEILINGER, T., FRANKENSTEIN, J., WATANABE, K., BÜLTHOFF, H. H., & HÖLSCHER, C. (2015). Reference frames in learning from maps and navigation. *Psychological Research*, *79*(6), 1000–1008.

MEILINGER, T., & KNAUFF, M. (2008). Ask for directions or use a map: A field experiment on spatial orientation and wayfinding in an urban environment. *Journal of Spatial Science*, *53*(2), 13–23.

MEILINGER, T., KNAUFF, M., & BÜLTHOFF, H. H. (2008). Working memory in wayfinding – a dual task experiment in a virtual city. *Cognitive Science*, *32*(4), 755–770.

MEILINGER, T., RIECKE, B. E., & BÜLTHOFF, H. H. (2014). Local and global reference frames for environmental spaces. *The Quarterly Journal of Experimental Psychology*, *67*(3), 542–569.

MEISSNER, H. (2006). Projekt „DORF"- Raumvorstellungen verbessern. *Journal für Mathematik-Didaktik*, *27*(1), 28–51.

MICHAEL, W. B., GUILFORD, J. P., FRUCHTER, B., & ZIMMERMAN, W. S. (1957). The description of spatial-visualization abilities. *Educational and Psychological Measurement*, *17*(2), 185–199.

MICHAEL, W. B., ZIMMERMAN, W. S., & GUILFORD, J. P. (1950). An investigation of two hypotheses regarding the nature of the spatial-relations and visualization factors. *Educational and Psychological Measurement, 10*(2), 187–211.

MICHAEL, WILLIAM B., ZIMMERMAN, WAYNE S., & GUILFORD, JOY P. (1951). An investigation of the nature of the spatial-relations and visualization factors in two high school samples. *Educational and Psychological Measurement, 11*(4-1), 561–577

MICHELON, P., & ZACKS, J. M. (2006). Two kinds of visual perspective taking. *Perception & Psychophysics, 68*(2), 327–337.

MILLER, G. A. (1956). The magical number seven, plus or minus two: Some limits on our capacity for processing information. *Psychological Review, 63*(2), 81–97.

MILNE, A. A., & ROJAHN-DEYK, B. (Eds.). (1988). *Winnie-the-Pooh*. Stuttgart: Reclam.

MINISTÈRE DE L'ÉDUCATION ET DU DÉVELOPPEMENT DE LA PETITE ENFANCE. (2016). *Programme d'études mathématiques au primaire (4e année)*. Retrieved from http://www2.gnb. ca/content/dam/gnb/Departments/ed/pdf/K12/servped/Mathematiques/Mathematiques-4eAnnee.pdf (17.06.2019)

MIX, K. S., HAMBRICK, D. Z., SATYAM, V. R., BURGOYNE, A. P., & LEVINE, S. C. (2018). The latent structure of spatial skill: A test of the 2×2 typology. *Cognition, 180*, 268–278.

MIX, K. S., LEVINE, S. C., CHENG, Y. L., YOUNG, C., HAMBRICK, D. Z., PING, R., & KONSTANTOPOULOS, S. (2016). Separate but correlated: The latent structure of space and mathematics across development. *Journal of Experimental Psychology: General, 145*(9), 1206–1227.

MIYAKE, A., FRIEDMAN, N. P., RETTINGER, D. A., SHAH, P.,& HEGARTY, M. (2001). How are visuospatial working memory, executive functioning, and spatial abilities related? A latent-variable analysis. *Journal of Experimental Psychology: General, 130*(4), 621–640.

MIYAKE, A., & SHAH, P. (1999). *Models of working memory: Mechanisms of active maintenance and executive control*. Cambridge, UK: Cambridge University Press.

MIZZI, A. (2017). *The relationship between language and spatial ability: An analysis of spatial language for reconstructing the solving of spatial tasks*. Wiesbaden: Springer Spektrum.

MOFFAT, S. D., HAMPSON, E., & HATZIPANTELIS, M. (1998). Navigation in a "virtual" maze: Sex differences and correlation with psychometric measures of spatial ability in humans. *Evolution and Human Behavior, 19*(2), 73–87.

MONTELLO, D. R. (1992). The geometry of environmental knowledge. In A. U. Frank, I. Campari, & V. Formentini (Eds.), *Theories and methods of spatio-temporal reasoning in geographic space* (pp. 136–152). Berlin: Springer.

MONTELLO, D. R. (1993). Scale and multiple psychologies of space. In A. U. Frank& I. Campari (Eds.), *Spatial information theory: A theoretical basis for GIS. Proceedings of COSIT '93.* (pp. 312–321). Berlin: Springer.

MONTELLO, D. R. (2005). Navigation. In P. Shah & A. Miyake (Eds.), *The cambridge handbook of visuospatial thinking* (pp. 257–293). New York, NY: Cambridge University Press.

MONTELLO, D. R., & FREUNDSCHUH, S. M. (1995). Sources of spatial knowledge and their implications for GIS: An introduction. *Geographical Systems, 2*(1), 169–176.

MONTELLO, D. R., & PICK, H. L. (1993). Integrating knowledge of vertically aligned large-scale spaces. *Environment and Behavior, 25*(3), 457–484.

MONTELLO, D. R., WALLER, D. A., HEGARTY, M., & RICHARDSON, A. E. (2004). Spatial memory of real environments, virtual environments, and maps. In G. L. Allen & D. B. M.

Haun (Eds.), *Remembering where: Advances in understanding spatial memory* (pp. 251–285). Mahwah, NJ: Erlbaum.

MOOSBRUGGER, H. (2012). Item-Response-Theorie (IRT). In H. Moosbrugger & A. Kelava (Eds.), *Testtheorie und Fragebogenkonstruktion* (pp. 215–259). Berlin, Heidelberg: Springer.

MOOSBRUGGER, H., & KELAVA, A. (Eds.). (2012). *Testtheorie und Fragebogenkonstruktion.* Berlin, Heidelberg: Springer.

MOOSBRUGGER, H., & SCHERMELLEH-ENGEL, K. (2012). Exploratorische (EFA) und Konfirmatorische Faktorenanalyse (CFA). In H. Moosbrugger & A. Kelava (Eds.), *Testtheorie und Fragebogenkonstruktion* (pp. 325–362). Berlin, Heidelberg: Springer.

MUIR, S. P., & CHEEK, H. N. (1986). Mathematics and the map skill curriculum. *School Science and Mathematics, 86*(4), 284–291.

MÜLLER, M., & WEHNER, R. (1988). Path integration in desert ants, cataglyphis fortis. *Proceedings of the National Academy of Sciences, 85*(14), 5287–5290.

MULLIGAN, J. (2015). Looking within and beyond the geometry curriculum: Connecting spatial reasoning to mathematics learning. *ZDM, 47*(3), 511–517.

MUMAW, R. J., & PELLEGRINO, J. W. (1984). Individual differences in complex spatial processing. *Journal of Educational Psychology, 76*(5), 920–939.

MUMAW, R. J., PELLEGRINO, J. W., KAIL, R. V., & CARTER, P. (1984). Different slopes for different folks: Process analysis of spatial aptitude. *Memory & Cognition, 12*(5), 515–521.

MUTHÉN, B. O. (1984). A general structural equation model with dichotomous, ordered categorical, and continuous latent variable indicators. *Psychometrika, 49*(1), 115–132.

MUTHÉN, B. O. (2002). Beyond SEM: General latent variable modeling. *Behaviormetrika, 29*(1), 81–117.

MUTHÉN, L. K., & MUTHÉN, B. O. (2004). *Mplus user's guide: Statistical analysis with latent variables.* Los Angeles, CA: Muthén & Muthén.

NATHANS, L. L., OSWALD, F. L., & NIMON, K. (2012). Interpreting multiple linear regression: A guidebook of variable importance. *Practical Assessment, Research & Evaluation, 17*(9), 1–19.

NATIONAL RESEARCH COUNCIL. (2006). *Learning to think spatially: GIS as a support system in the K-12 curriculum.* Washington, DC: National Academy Press.

NCTM. (2000). *Commission on standards for school: Principles and standards for school mathematics.* Reston: National Council of Teachers of Mathematics.

NEIDHARDT, E., & POPP, M. (2010). Spatial tests, familiarity with the surroundings, and spatial activity experience. *Journal of Individual Differences, 31*(2), 59–63.

NEISSER, U. (1976). *Cognition and reality.* San Francisco, CA: Freeman.

NEISSER, U. (2014). *Cognitive psychology: Classic edition.* New York, NY: Psychology Press.

NERSESSIAN, N. J. (2002). The cognitive basis of model-based reasoning in science. In P. Carruthers, St. Stich, & M. Siegal (Eds.), *The cognitive basis of science* (pp. 133–153). Cambridge, UK: Cambridge University Press.

NEW SOUTH WALES BOARD OF STUDIES. (2012). *Mathematics K-10 syllabus : NSW syllabus for the australian curriculum.* Sydney: Sydney Board of Studies NSW.

NEWCOMBE, N. S. (1982). Development of spatial cognition and cognitive development. *New Directions for Child and Adolescent Development, 1982*(15), 65–81.

NEWCOMBE, N. S. (1989). The development of spatial perspective taking. *Advances in Child Development and Behavior, 22*, 203–247.

NEWCOMBE, N. S. (2002). Spatial cognition. In H. Pashler & D. Medin (Eds.), *Steven's handbook of experimental psychology* (Vol. 2, pp. 113–163). San Francisco, CA: Jossey-Bass.

NEWCOMBE, N. S. (2018). Three kinds of spatial cognition. In S. L. Thompson-Schill (Ed.), *Stevens' handbook of experimental psychology and cognitive neuroscience* (Vol. 4, pp. 521–552). Hoboken, NJ: John Wiley.

NEWCOMBE, N. S., & FRICK, A. (2010). Early education for spatial intelligence: Why, what, and how. *Mind, Brain, and Education, 4*(3), 102–111.

NEWCOMBE, N. S., & HUTTENLOCHER, J. (2003). *Making space: The development of spatial representation and reasoning*. Cambridge, MA: MIT Press.

NEWCOMBE, N. S., & SHIPLEY, T. F. (2015). Thinking about spatial thinking: New typology, new assessments. In J. S. Gero (Ed.), *Studying visual and spatial reasoning for design creativity* (pp. 179–192). Dordrecht: Springer.

NEWCOMBE, N. S., UTTAL, D. H., & SAUTER, M. (2013). Spatial development. In P. D. Zelazp (Ed.), *The Oxford handbook of developmental psychology* (Vol. 1, pp. 505–542). Oxford: Oxford University Press.

NIEDERMEYER, I. (2015). *Räumliche Perspektivübernahme am Schulanfang: Eine Interviewstudie zum Einfluss der Symmetrie*. Münster: Waxmann Verlag.

NORMAN, D. A. (2013). *Models of human memory*. New York, NY: Academic Press.

OECD. (2004). *The PISA 2003 assessment framework: Mathematics, reading, science and problem solving knowledge and skills*. Paris: OECD Publishing.

O'KEEFE, J.,& NADEL, L. (1978). *The hippocampus as a cognitive map*. Oxford, New York: Clarendon Press and Oxford University Press.

ORSINI, A., PASQUADIBISCEGLIE, M., PICONE, L., & TORTORA, R. (2001). Factors which influence the difficulty of the spatial path in Corsi's block-tapping test. *Perceptual and Motor Skills, 92*(3), 732–738.

OTTOSSON, T. (1986). Cognitive processes in orienteering: An outline of a theoretical frame of reference and some preliminary data. *Scientific Journal of Orienteering, 2*, 75–101.

OTTOSSON, T. (1987). *Map-reading and wayfinding*. Göteborg: Acta Universitatis Gothoburgensis.

PALMER, S. (1978). Fundamental aspects of cognitive representation. In E. Rosch & B. B. LLoyd (Eds.), *Cognition and categorization* (pp. 259–303). Hillsdale, NJ: Erlbaum.

PALMIERO, M., PICCARDI, L., GIANCOLA, M., NORI, R., D'AMICO, S., & OLIVETTI BELARDINELLI, M. (2019). The format of mental imagery: From a critical review to an integrated embodied representation approach. *Cognitive Processing*, 1–13.

PANORKOU, N., & PRATT, D. (2008). Mapping experience of dimension. *Proceedings of PME, 33*, 4281–288.

PARSONS, L. M. (1987a). Imagined spatial transformation of one's body. *Journal of Experimental Psychology: General, 116*(2), 172–191.

PARSONS, L. M. (1987b). Imagined spatial transformations of one's hands and feet. *Cognitive Psychology, 19*(2), 178–241.

PASHLER, H., MCDANIEL, M., ROHRER, D., & BJORK, R. (2008). Learning styles: Concepts and evidence. *Psychological Science In The Public Interest: A journal of the American Psychological Society, 9*(3), 105–119.

PAWLIK, K. (1966). Concepts in human cognition and aptitudes. In R. B. CATTELL (Ed.), *Handbook of multivariate experimental psychology*. Chicago: Rand McNally.

PAWLIK, K. (1968). *Dimensionen des Verhaltens*. Bern: Hans Huber.

PEARSON, J. L., & IALONGO, N. S. (1986). The relationship between spatial ability and environmental knowledge. *Journal of Environmental Psychology*, 6(4), 299–304.

PELLEGRINO, J. W., & KAIL, R. V. (1982). Process analyses of spatial aptitude. In J. Sternberg (Ed.), *Advances in the psychology of human intelligence* (Vol. 1, pp. 311–365). Hillsdale, NJ: Erlbaum.

PERRIN-GLORIAN, M. J., MATHÉ, A. C., & LECLERCQ, R. (2013). Comment peut-on penser la continuité de l'enseignement de la géométrie de 6 à 15 ans. *Le jeu sur les supports et les instruments. Repères-IREM*, 90, 5–41.

PIAGET, J., & INHELDER, B. (1956). *The child's conception of space*. New York, London: Norton. (Original work published 1948)

PIAGET, J., & INHELDER, B. (1969). *The psychology of the child*. New York, NY: Basic books.

PIAGET, J., & INHELDER, B. (1971). *Mental imagery in the child* (6th ed.). New York, NY: Basic books. (Original work published 1966)

PIAGET, J., INHELDER, B., & SZEMINSKA, A. (1981). *Child's conception of geometry*. New York, NY: Routledge.

PINKERNELL, G. (2003). *Räumliches Vorstellungsvermögen im Geometrieunterricht: Eine didaktische Analyse mit Fallstudien*. Hildesheim, Berlin: Franzbecker.

PITTALIS, M., & CHRISTOU, C. (2010). Types of reasoning in 3D geometry thinking and their relation with spatial ability. *Educational Studies in Mathematics*, 75(2), 191–212.

PITUCH, K. A., & STEVENS, J. P. (2015). *Applied multivariate statistics for the social sciences* (6th ed.). New York, NY: Routledge.

PLATH, M. (2014). *Räumliches Vorstellungsvermögen im vierten Schuljahr: Eine Interviewstudie zu Lösungsstrategien und möglichen Einflussbedingungen auf den Strategieeinsatz*. Hildesheim: Franzbecker.

PLORAN, E. J., ROVIRA, E., THOMPSON, J. C., & PARASURAMAN, R. (2015). Underlying spatial skills to support navigation through large, unconstrained environments. *Applied Cognitive Psychology*, 29(4), 608–613.

POUCET, B. (1993). Spatial cognitive maps in animals: New hypotheses on their structure and neural mechanisms. *Psychological Review*, 100(2), 163–182.

PRESSON, C. C. (1982a). The development of map-reading skills. *Child Development*, 53(1), 196–199.

PRESSON, C. C. (1982b). Strategies in spatial reasoning. *Journal of Experimental Psychology: Learning, Memory, and Cognition*, 8(3), 243–251.

PRESSON, C. C. (1987). The development of spatial cognition: Secondary uses of spatial information. In N. Eisenberg (Ed.), *Contemporary topics in developmental psychology* (pp. 77–112). New York, NY: Wiley.

PRESSON, C. C., DE LANGE, N., & HAZELRIGG, M. D. (1989). Orientation specificity in spatial memory: What makes a path different from a map of the path? *Journal of Experimental Psychology: Learning, Memory, and Cognition*, 15(5), 887–897.

PRESSON, C. C., & HAZELRIGG, M. D. (1984). Building spatial representations through primary and secondary learning. *Journal of Experimental Psychology: Learning, Memory, and Cognition*, 10(4), 716–722.

PRESSON, C. C., & MONTELLO, D. R. (1988). Points of reference in spatial cognition: Stalking the elusive landmark. *British Journal of Developmental Psychology*, 6(4), 378–381.

References

PRESSON, C. C., & SOMERVILLE, S. C. (1985). Beyond egocentrism: A new look at the beginnings of spatial representation. In H. Wellmann (Ed.), *The development of children's spatial search skill and spatial representation* (pp. 1–27). Hillsdale, NJ: Erlbaum.

PREVIC, F. H. (1998). The neuropsychology of 3D space. *Psychological Bulletin, 124*(2), 123–164.

PURSER, H. R. M., FARRAN, E. K., COURBOIS, Y., LEMAHIEU, A., MELLIER, D., SOCKEEL, P., & BLADES, M. (2012). Short-term memory, executive control, and children's route learning. *Journal of Experimental Child Psychology, 113*(2), 273–285.

QUAISER-POHL, C. (1998). *Die Fähigkeit zur räumlichen Vorstellung: zur Bedeutung von kognitiven und motivationalen Faktoren für geschlechtsspezifische Unterschiede.* Münster: Waxmann Verlag.

QUAISER-POHL, C., LEHMANN, W., & EID, M. (2004). The relationship between spatial abilities and representations of large-scale space in children–a structural equation modeling analysis. *Personality and Individual Differences, 36*(1), 95–107.

R CORE TEAM. (2018). *R: A language and environment for statistical computing* Doctoral dissertation Vienna, Austria. Retrieved from https://www.R-project.org/ (17.06.2019)

RAMMSTEDT, B. (2010). Reliabilität, Validität, Objektivität. In C. Wolf & H. Best (Eds.), *Handbuch der sozialwissenschaftlichen Datenanalyse* (pp. 239–258). Wiesbaden: VS Verlag für Sozialwissenschaften.

RASCH, G. (1960). *Probabilistic models for some intelligence and attainment tests.* Copenhagen: Nielsen EI Lydicke.

REINHOLD, S. (2007). *Mentale Rotation von Würfelkonfigurationen: theoretischer Abriss, mathematikdidaktische Perspektiven und Analysen zu Strategien von Grundschulkindern in einer konstruktiven Arbeitsumgebung.* Doctoral dissertation. Retrieved from https://www.tib.eu/de/suchen/id/TIBKAT%3A527630160/Mentale-Rotation-von-W/unhbox/voidbx/bgroup/accent127u/protect/penaltyM/hskip/zskip/egrouprfelkonfigurationen-theoretischer (17.06.2019)

RHEMTULLA, M., BROSSEAU-LIARD, P. É., & SAVALEI, V. (2012). When can categorical variables be treated as continuous? A comparison of robust continuous and categorical sem estimation methods under suboptimal conditions. *Psychological Methods, 17*(3), 354–373.

RICHARDSON, A. E., MONTELLO, D. R., & HEGARTY, M. (1999). Spatial knowledge acquisition from maps and from navigation in real and virtual environments. *Memory & Cognition, 27*(4), 741–750.

RIESER, J. J., GUTH, D. A., & HILL, E. W. (1986). Sensitivity to perspective structure while walking without vision. *Perception, 15*(2), 173–188.

ROBITZSCH, A., & PHAM, GIANG, YANAGIDA, TAKUYA. (2016) Fehlende Daten und Plausible Values. In S. Breit & C. Schreiner (Eds.), *Large-Scale Assessment mit R: Methodische Grundlagen der österreichischen Bildungsstandardüberprüfung* (pp. 259–293). Wien: facultas.

ROSS, A., & WILLSON, V. L. (2017). Hierarchical multiple regression analysis using at least two sets of variables (in two blocks). In A. Ross & V. L. Willson (Eds.), *Basic and advanced statistical tests* (pp. 61–74). Rotterdam: Sense Publishers.

ROSSANO, M. J., & MOAK, J. (1998). Spatial representations acquired from computer models: Cognitive load, orientation specificity and the acquisition of survey knowledge. *British Journal of Psychology, 89*(3), 481–497.

ROSSEEL, Y. (2012). lavaan: an R package for structural equation modeling. *Journal of Statistical Software*, *48*(2), 1–36.

ROST, D. H. (1977). *Raumvorstellung: psychologische und pädagogische Aspekte*. Weinheim: Beltz.

ROVINE, M. J., & WEISMAN, G. D. (1989). Sketch-map variables as predictors of way-finding performance. *Journal of Environmental Psychology*, *9*(3), 217–232.

RUBIN, D. B. (1976). Inference and missing data. *Biometrika*, *63*(3), 581–592.

RUPPERT, M., BAUER, A., HOHENWARTER, M.,& MEYER, K. (2013). Geogebra. In M. Ruppert & J. Wörler (Eds.), *Technologien im Mathematikunterricht* (pp. 1–37). Wiesbaden: Springer Fachmedien Wiesbaden.

SADALLA, E. K., BURROUGHS, W. J., & STAPLIN, L. J. (1980). Reference points in spatial cognition. *Journal of Experimental Psychology: Human Learning and Memory*, *6*(5), 516–528.

SAS, C., & MOHD NOOR, N. (2009). A meta-analysis on the correlation between measurements of spatial tasks and standardized tests of environmental spatial abilities. *Cognitive Processing*, *10*, 297–301.

SAYEKI, Y., COLE, M., ENGESTROM, Y., & VASQUEZ, O. (1997). Embodied spatial tranformations. "Body analogy" and the cognition of rotated figures. In M. Cole, Y. Engeström,& O. Vasquez (Eds.), *Mind, culture and activity* (pp. 90–99). Cambridge, UK: Cambridge University Press.

SCHULTZ, K. (1991). The contribution of solution strategy to spatial performance. *Canadian Journal of Psychology*, *45*(4), 474–491.

SEERY, N., BUCKLEY, J., & DELAHUNTY, T. (2015). Developing a spatial ability framework to support spatial ability research in engineering education. In *The 6th research in engineering education symposium*. Dublin, Ireland: Dublin Institute of Technology.

SEILER, R. (1996). Cognitive processes in orienteering-a review. *Scientific Journal of Orienteering*, *12*, 50–65.

SEMTOOLS CONTRIBUTORS. (2015). *semtools: Useful tools for structural equation modeling*.

SHELTON, A. L., & MCNAMARA, T. P. (2001). Systems of spatial reference in human memory. *Cognitive Psychology*, *43*(4), 274–310.

SHELTON, A. L., & MCNAMARA, T. P. (2004). Orientation and perspective dependence in route and survey learning. *Journal of Experimental Psychology: Learning, Memory, and Cognition*, *30*(1), 158–170.

SHELTON, A. L., & PIPPITT, H. A. (2007). Fixed versus dynamic orientations in environmental learning from ground-level and aerial perspectives. *Psychological Research*, *71*(3), 333–346.

SHELTON, A. L., & ZACKS, J. M. (2015). Spatial transformations of scene stimuli: It's an upright world. In J. S. Gero (Ed.), *Studying visual and spatial reasoning for design creativity* (pp. 245–266). Dordrecht: Springer.

SHEPARD, R. N., & COOPER, L. A. (1982). *Mental images and their transformations*. Cambridge, MA: MIT Press.

SHEPARD, R. N., & FENG, C. (1972). A chronometric study of mental paper folding. *Cognitive Psychology*, *3*(2), 228–243.

SHEPARD, R. N., & HURWITZ, S. (1984). Upward direction, mental rotation, and discrimination of left and right turns in maps. *Cognition*, *18*(1), 161–193.

SHEPARD, R. N., & METZLER, J. (1971). Mental rotation of three-dimensional objects. *Science*, *171*(3972), 701–703.

SHEPARD, R. N., & PODGORNY, P. (1978). Cognitive processes that resemble perceptual processes. In W. K. Estes (Ed.), *Handbook of learning and cognitive processes* (Vol. 5, pp. 189–237). Hillsdale, NJ: Erlbaum.

SHOLL, M. J. (1988). The relation between sense of direction and mental geographic updating. *Intelligence*, *12*(3), 299–314.

SHOLL, M. J. (1992). Landmarks, places, environments: Multiple mind-brain systems for spatial orientation. *Geoforum*, *23*(2), 151–164.

SHOLL, M. J., & FRAONE, S. K. (2004). Visuospatial working memory for different scales of space. In G. L. Allen (Ed.), *Human spatial memory: Remembering where* (pp. 67–100). Hillsdale, NJ: Erlbaum.

SI, Y., & REITER, J. P. (2013). Nonparametric bayesian multiple imputation for incomplete categorical variables in large-scale assessment surveys. *Journal of Educational and Behavioral Statistics*, *38*(5), 499–521.

SIEGEL, A. W. (1981). The externalisation of cognitive maps by children and adults: In search of better ways to ask questions. In L. S. Liben, A. H. Patterson,& N. S. Newcombe (Eds.), *Spatial representation and behavior across the life span* (pp. 167–194). New York, NY: Academic Press.

SIEGEL, A. W., HERMAN, J. F., ALLEN, G. L., & KIRASIC, K. C. (1979). The development of cognitive maps of large- and small-scale spaces. *Child Development*, *50*(2), 582–585.

SIEGEL, A. W., & WHITE, S. H. (1975). The development of spatial representations of large-scale environments. *Advances in Child Development and Behavior*, *10*, 9–55.

SIMON, H. A. (1954). Spurious correlation: A causal interpretation. *Journal of the American Statistical Association*, *49*(267), 467–479.

SIMONS, D. J., & WANG, R. F. (1998). Perceiving real-world viewpoint changes. *Psychological Science*, *9*(4), 315–320.

SINCLAIR, N., & BRUCE, C. D. (2015). New opportunities in geometry education at the primary school. *ZDM*, *47*(3), 319–329.

SMITH, E. R., & QUELLER, S. (2000). Mental representations. In A. Tesser & N. Schwarz (Eds.), *Blackwell handbook in social psychology* (Vol. 1, pp. 111–133). Blackwell: Oxford University Press.

SODIAN, B., THOERMER, C., & METZ, U. (2007). Now I see it but you don't: 14-month-olds can represent another person's visual perspective. *Developmental Science*, *10*(2), 199–204.

SOURY-LAVERGNE, S., & MASCHIETTO, M. (2015). Articulation of spatial and geometrical knowledge in problem solving with technology at primary school. *ZDM*, 47(3), 435–449.

SOUVIGNIER, E. (2000). *Förderung räumlicher Fähigkeiten*. Münster: Waxmann.

SPEARMAN, C. (1927). *The abilities of man*. New York, NW: Macmillan.

STEINKE, T., & LLOYD, R. (1983). Images of maps: A rotation experiment. *The Professional Geographer*, *35*(4)455–461.

TAHTA, D. (1980). About geometry. *For the Learning of Mathematics*, *1*(1), 2–9.

TARAMPI, M. R., HEYDARI, N., & HEGARTY, M. (2016). A tale of two types of perspective taking: sex differences in spatial ability. *Psychological Science*, *27*(11), 1507–1516.

TARTRE, L. A. (1990). Spatial orientation skill and mathematical problem solving. *Journal for Research in Mathematics Education*, *31*(3), 216–229.

TASCÓN, L., BOCCIA, M., PICCARDI, L., & CIMADEVILLA, J. M. (2017). Differences in spatial memory recognition due to cognitive style. *Frontiers in Pharmacology, 8*, 550.

TERGAN, S.-O. (1993). Psychologische Grundlagen der Erfassung individueller Wissensrepräsentationen: Teil I: Grundlagen der Wissensmodellierung. In J. Engelkamp & T. Pechmann (Eds.), *Mentale Repräsentation* (pp. 103–116). Bern, Göttingen, Toronto, Seattle: Huber.

THAGARD, P. (2005). *Mind: Introduction to cognitive science.* Cambridge, MA: MIT Press.

THORNDYKE, P. W., & HAYES-ROTH, B. (1982). Differences in spatial knowledge acquired from maps and navigation. *Cognitive Psychology, 14*(4), 560–589.

THURSTONE, L. L. (1938). *Primary mental abilities* Unpublished doctoral dissertation. Chicago, ILThe University og Chicago.

THURSTONE, L. L. (1947). *Multiple factor analysis.* Unpublished doctoral dissertation.

THURSTONE, L. L. (1949). *Mechanical aptitude III: Analysis of group tests.* Unpublished doctoral dissertation. Chicago University of Chicago, Psychometric Laboratory.

THURSTONE, L. L. (1950). Some primary abilities in visual thinking. *Proceedings of the American Philosophical Society, 94*(6), 517–521.

THURSTONE, L. L.,& THURSTONE, T. G. (1941). *Factorial studies of intelligence.* Chicago: The University of Chicago Press.

THURSTONE, L. L., & THURSTONE, T. G. (1949). *Mechanical aptitude II: Description of group tests.* Unpublished doctoral dissertation. University of Chicago, Psychometric Laboratory.

TOLMAN, E. C. (1948). Cognitive maps in rats and men. *Psychological Review, 55*(4), 189–208.

TVERSKY, B. (2005). Visuospatial reasoning. In K. Holyoak & R. G. Morrison (Eds.), *The cambridge handbook of thinking and reasoning* (pp. 209–240). Cambridge, UK: Cambridge University Press.

TYE, M. (2000). *The imagery debate.* Cambridge, MA: MIT Press.

UNGERLEIDER, L. G., & HAXBY, J. V. (1994). 'What' and 'where' in the human brain. *Current opinion in neurobiology, 4*(2), 157–165.

UTTAL, D. H. (1996). Angles and distances: Children's and adults' reconstruction and scaling of spatial configurations. *Child Development, 67*(6), 2763–2779.

UTTAL, D. H. (2000a) Maps and spatial thinking: A two-way street. *Developmental Science, 3*(3), 283–286.

UTTAL, D. H. (2000b) Seeing the big picture: Map use and the development of spatial cognition. *Developmental Science, 3*(3), 247–264.

UTTAL, D. H., MEADOW, N. G., TIPTON, E., H and L. L., ALDEN, A. R., WARREN, C., & NEWCOMBE, N. S. (2013). The malleability of spatial skills: A meta-analysis of training studies. *Psychological Bulletin, 139*(2), 352–402.

VAN DEN HEUVEL-PANHUIZEN, M., ELIA, I., & ROBITZSCH, A. (2015). Kindergartners' performance in two types of imaginary perspective-taking. *ZDM, 47*(3), 345–362.

VANDENBERG, S. G., & KUSE, A. R. (1978). Mental rotations, a group test of three-dimensional spatial visualization. *Perceptual and Motor Skills, 47*(2), 599–604.

VANDER HEYDEN, K. M., HUIZINGA, M., KAN, K.-J., & JOLLES, J. (2016). A developmental perspective on spatial reasoning: Dissociating object transformation from viewer transformation ability. *Cognitive Development, 38*(1), 63–74.

VANDER HEYDEN, K. M., HUIZINGA, M., RAIJMAKERS, M. E. J. , & JOLLES, J. (2017). Children's representations of another person's spatial perspective: Different strategies for different viewpoints? *Journal of Experimental Child Psychology*, 15357–73.

VEDERHUS, L., & KREKLING, S. (1996). Sex differences in visual spatial ability in 9-year-old children. *Intelligence, 23*(1), 33–43.

VÖLKL, K., & KORB, C. (2018). Drittvariablenkontrolle. In K. Völkl & C. Korb (Eds.), *Deskriptive statistik* (pp. 255–290). Wiesbaden: Springer.

VOYER, D., VOYER, S., & BRYDEN, M. P. (1995). Magnitude of sex differences in spatial abilities: A meta-analysis and consideration of critical variables. *Psychological Bulletin, 117*(2), 250–270.

WALLER, D. A. (1999). *An assessment of individual differences in spatial knowledge of real and virtual environments.* Unpublished doctoral dissertation. University of Washington.

WALLER, D. A., LOOMIS, J. M., & HAUN, D. B. M. (2004). Body-based senses enhance knowledge of directions in large-scale environments. *Psychonomic Bulletin & Review, 11*(1)157–163.

WALLER, D. A., LOOMIS, J. M., & STECK, S. D. (2003). Inertial cues do not enhance knowledge of environmental layout. *Psychonomic Bulletin & Review, 10*(4), 987–993.

WALLER, D. A., MONTELLO, D. R., RICHARDSON, A. E., & HEGARTY, M. (2002). Orientation specificity and spatial updating of memories for layouts. *Journal of Experimental Psychology: Learning, Memory, and Cognition, 28*(6)1051–1063.

WALSH, D. A., KRAUSS, I. K., & REGNIER, V. A. (1981). Spatial ability, environmental knowledge, and environmental use: The elderly. In L. S. Liben, A. H. Patterson, & N. S. Newcombe (Eds.), *Spatial representation and behavior across the life span* (pp. 321–357). New York, NY: Academic Press.

WANG, L., COHEN, A. S., & CARR, M. (2014). Spatial ability at two scales of representation: A meta-analysis. *Learning and Individual Differences, 36*(1), 140–144.

WANG, Q., MANRIQUE-VALLIER, D., REITER, J. P., HU, J., & WANG, M. Q. (2014). Package npbayesimpute: Non-parametric bayesian multiple imputation for categorical data.

WANG, R. F., & SIMONS, D. J. (1999). Active and passive scene recognition across views. *Cognition, 70*(2), 191–210.

WEATHERFORD, D. L. (1982). Spatial cognition as a function of size and scale of the environment. In R. Cohen (Ed.), *New directions for child development: Vol. 15. Children's conceptions of spatial relationships* (Vol. 1982, pp. 5–18). San Francisco: Jossey-Bass.

WEISMAN, J. (1981). Evaluating architectural legibility way-finding in the built environment. *Environment and Behavior, 13*(2), 189–204.

WELWOOD, J. (1977). On psychological space. *The Journal of Transpersonal Psychology, 9*(2), 97–118.

WERNER, S., & SCHMIDT, K. (1999). Environmental reference systems for large-scale spaces. *Spatial Cognition and Computation, 1*(4), 447–473.

WEST, S. G., FINCH, J. F., & CURRAN, P. J. (1995). Structural equation models with nonnormal variables: Problems and remedies. In R. H. Hoyle (Ed.), *Structural equation modeling: Issues, concepts, and applications* (pp. 56–75). Newbury Park, CA: Sage.

WIEGAND, P. (2006). *Learning and teaching with maps.* New York, NY: Routledge.

WIENER, J. M., BÜCHNER, S. J., & HÖLSCHER, C. (2009). Taxonomy of human wayfinding tasks: A knowledge-based approach. *Spatial Cognition & Computation, 9*(2), 152–165.

WILSON, M. (2005). *Constructing measures: An item response modeling approach.* Mawah: Lawrence Erlbaum Associates.

WILSON, M.,& GOCHYYEV, P. (2013). Psychometrics. In T. Teo (Ed.), *Handbook of quantitative methods for educational research* (pp. 3–30). Rotterdam, Boston, Taipei: Sense Publishers.

WIMMER, M. C., MARAS, K. L., ROBINSON, E. J., & THOMAS, C. (2016). The format of children's mental images: Evidence from mental scanning. *Cognition, 154*, 49–54.

WITTMANN, E. C. (1999). Konstruktion eines Geometriecurriculums ausgehend von Grundideen der Elementargeometrie. In H. Henning (Ed.), *Mathematik lernen durch Handeln und Erfahrung. Festschrift zum 75. Geburtstag von Heinrich Besuden.* (pp. 205–223). Oldenburg: Bültmann & Gerriets.

WOLBERS, T., & HEGARTY, M. (2010). What determines our navigational abilities? *Trends in Cognitive Sciences, 14*(3), 138–146.

WOLBERS, T., & WIENER, J. M. (2014). Challenges for identifying the neural mechanisms that support spatial navigation: The impact of spatial scale. *Frontiers in Human Neuroscience, 8*, 1–12.

WOLF, E. J., HARRINGTON, K. M., CLARK, S. L., & MILLER, M. W. (2013) Sample size requirements for structural equation models: An evaluation of power, bias, and solution propriety. *Educational and Psychological Measurement, 73*(6), 913–934.

WRAGA, M., CREEM, S. H., & PROFFITT, D. R. (2000). Updating displays after imagined object and viewer rotations. *Journal of Experimental Psychology: Learning, Memory, and Cognition, 26*(1), 151–168.

WRAGA, M., CREEM-REGEHR, S. H., & PROFFITT, D. R. (2004). Spatial updating of virtual displays. *Memory & Cognition, 32*(3), 399–415.

WRAGA, M., SHEPHARD, J. M., CHURCH, J. A., INATI, S., & KOSSLYN, S. M. (2005). Imagined rotations of self versus objects: an fMRI study. *Neuropsychologia, 43*(9), 1351–1361.

WRENGER, K. (2015). *Kartengestützte Orientierung im Realraum unter besonderer Berücksichtigung der Einflussgröße Raum: Eine empirische Studie mit Schülerinnen und Schülern zu Beginn der Sekundarstufe I.* Münster: Monsenstein und Vannerdat.

WU, M. (2013). Using item response theory as a tool in educational measurement. In M. M. C. Mok (Ed.), *Self-directed learning oriented assessments in the asia-pacific* (pp. 157–185). Dordrecht: Springer.

WU, M. L., ADAMS, R. J., & WILSON, M. (2015). *ACER ConQuest: Generalised item reponse modelling software.* Melbourne: Acer Press.

XISTOURI, X., & PITTA-PANTAZI, D. (2011). Elementary students' transformational geometry abilities and cognitive style. *CERME 7, Rzeszów, Poland.*

YOUNG, C., LEVINE, S. C., & MIX, K. S. (2018). What processes underlie the relation between spatial skill and mathematics? In K. S. Mix & M. T. Battista (Eds.), *Visualizing mathematics. Research in mathematics education* (pp. 117–148). Cham: Springer.

YU, A. B., & ZACKS, J. M. (2017). Transformations and representations supporting spatial perspective taking. *Spatial Cognition & Computation, 17*(4), 304–337.

ZACKS, J. M. (2008). Neuroimaging studies of mental rotation: A meta-analysis and review. *Journal of Cognitive Neuroscience, 20*(1), 1–19.

ZACKS, J. M., & MICHELON, P. (2005). Transformations of visuospatial images. *Behavioral and Cognitive Neuroscience Reviews, 4*(2), 96–118.

ZACKS, J. M., MIRES, J., TVERSKY, B., & HAZELTINE, E. (2000). Mental spatial transformations of objects and perspective. *Spatial Cognition and Computation*, 2(4), 315–332.

ZACKS, J. M., RYPMA, B., GABRIELI, J. D., TVERSKY, B., & GLOVER, G. H. (1999). Imagined transformations of bodies: An fMRI investigation. *Neuropsychologia*, 37(9), 1029–1040.

ZACKS, J. M., & TVERSKY, B. (2005). Multiple systems for spatial imagery: Transformations of objects and bodies. *Spatial Cognition and Computation*, 5(4), 271–306.

ZACKS, J. M., VETTEL, J. M. & MICHELON, P. (2003). Imagined viewer and object rotations dissociated with event-related fMRI. *Journal of Cognitive Neuroscience*, 15(7), 1002–1018.

ZIMMER, H. D. (1993). Von Repräsentationen, Modalitäten und Modulen. In J. Engelkamp & T. Pechmann (Eds.), *Mentale Repräsentation* (pp. 93–102). Bern, Göttingen, Toronto, Seattle: Huber.

ZIMMER, H. D., MÜNZER, S., & UMLA-RNGE, K. (2010). Visuo-spatial working memory as a limited resource of cognitive processing. In M. W. Corcker & J. H. Siekmann (Eds.), *Cognitive technologies, resource-adaptive cognitive processes* (pp. 13–34). Berlin, London: Springer.

Printed in the United States
By Bookmasters